Automatic Supervision in Manufacturing

Advanced Manufacturing Series
Series Editor: D.T. Pham

AUTOMATIC SUPERVISION IN MANUFACTURING

Edited by

Maciej Szafarczyk

With 172 Figures

Springer-Verlag
London   Berlin   Heidelberg   New York
Paris   Tokyo   Hong Kong
Barcelona   Budapest

Professor Maciej Szafarczyk, PhD
Warsaw Univeristy of Technology, Narbutta 86, 02–524 Warsaw,
Poland

Series Editor

Professor D. T. Pham, PhD
University of Wales College of Cardiff
School of Electrical, Electronic and Systems Engineering
P.O. Box 917, Cardiff CF2 1XH

ISBN-13: 978-1-4471-3460-2   e-ISBN-13: 978-1-4471-3458-9
DOI: 10.1007/978-1-4471-3458-9

British Library Cataloguing in Publication Data
Automatic Supervision in Manufacturing.–
(Advanced Manufacturing Series)
  I. Szafarczyk, Maciej II. Series
  629.8
ISBN-13: 978-1-4471-3460-2

Library of Congress Cataloging-in-Publication Data
Automatic supervision in manufacturing/edited by Maciej Szafarczyk.
      p.   cm. – (Advanced manufacturing series)
  Includes bibliographical references and index.
  ISBN-13: 978-1-4471-3460-2
  1. Production management—Automation. 2. Supervisors, Industrial.
I. Szafarczyk, Maciej, 1931– . II. Series.
TS155.4.A98 1994   93–46279
670.42'7—dc20   CIP

Typeset by Concept Typesetting Ltd, Salisbury, Wiltshire

69/3830–543210 Printed on acid-free paper

Contents

Preface

Automation has certainly become the main trend in the development of modern and advanced manufacturing production. Automatic supervision in manufacturing (ASM), dealing with unavoidable disturbances during production, results in the highest level of automation in which comprehensive and reliable supervision makes possible the untended (unmanned) functioning of manufacturing systems. That is a good reason for writing a book on ASM and why this book is included in Springer's Advanced Manufacturing Series.

Several parts may be distinguished in the book. The first chapter and the short Glossary at the end of the book, by Szafarczyk, deal with the concept of automatic supervision, the classification of supervisory systems and their functions, and terms and definitions in the new field of ASM.

The main part of the book consists of six chapters that describe the latest achievements of research work and industrial experience on automatic supervision in different manufacturing techniques. Lindström presents the state of the art on supervision in turning and boring. Tlusty discusses automatic supervision in milling and Westkämper in surface grinding. Chodnikiewicz and Olejnik provide information about automatic supervision in metal forming. Kruth deals with supervision in physical and chemical machining, a group of machining techniques still frequently viewed as unconventional. Arnström and Onori describe problems of supervision, and their solutions, in the important field of assembly automation.

The two following chapters are devoted to the specific problems of supervising the proper functioning of the main parts of manufacturing systems. Spiewak presents the state of the art in the supervision of machine tools and other mechanical modules of the system, while Pritschow does the same for the supervision of the ever more complicated controllers of manufacturing systems.

The main and most sophisticated part of automatic supervision – signal processing for monitoring purposes – is described, in a separate chapter, by Dornfeld. He includes in it the new techniques of artificial intelligence (AI).

Another separate chapter, by Peklenik, deals with the principles of so-called Geometrical Adaptive Control – automatic supervision aimed at the quality of workpieces.

Acknowledgements

My thanks are extended to authors of all the chapters for their contributions; to the Series editor, Professor D.T. Pham, for his encouragement; and to Mr Nicholas Pinfield and Mrs Imke Mowbray (Springer-Verlag) for their help with the production of this book.

M. Szafarczyk

List of Contributors

Professor A. Arnström
Department of Manufacturing Systems/Assembly Systems
KTH Royal Institute of Technology
100 44 Stockholm 70, Sweden

Dr K. Chodnikiewicz
Warsaw University of Technology
Narbutta 85, 02-524 Warsaw, Poland

Professor D.A. Dornfeld
Department of Mechanical Engineering
University of California
Berkeley, CA 94720, USA

Professor J.P. Kruth
Katholieke Universiteit Leuven
Mechanical Engineering Department, Production Engineering Division
Celestijnenlaan 300B, B–3001 Heverlee, Belgium

Professor B. Lindström
Department of Materials Processing, Production Engineering
Faculty of Materials and Mechanical Engineering
KTH Royal Institute of Technology
100 44 Stockholm 70, Sweden

Dr L. Olejnik
Warsaw University of Technology
Narbutta 85, 02-524 Warsaw, Poland

M.A. Onori
Department of Manufacturing Systems/Assembly Systems
KTH Royal Institute of Technology
100 44 Stockholm 70, Sweden

Professor J. Peklenik
Department of Control & Manufacturing Systems
University, Faculty of Mechanical Engineering
Aškerčeva 6, 61000 Ljubljana, Slovenia

Professor G. Pritschow
Inst. für Steuerungstech. der Werkzeugmaschinen
Universität Stuttgart
Seidenstrasse 36, D-70174 Stuttgart, Germany

Assistant Professor S.A. Spiewak
University of Wisconsin-Madison
Department of Mechanical Engineering
1513 University Ave, Madison, WI 53706, USA

Professor M. Szafarczyk
Warsaw University of Technology
Narbutta 86, 02-524 Warsaw, Poland

Professor J. Tlusty
Mechanical Engineering, Department
University of Florida
Gainesville, FLO 32611, USA

Prof. Dr.-Ing. E. Westkämper
Institut für Werkzeugmaschinen und Fertigungstechnik
der Technischen Universität Braunschweig
Langer Kamp 19B, D-38106 Braunschweig, Germany

1 Principles of Automatic Supervision in Manufacturing and Classification of Supervisory Systems

M. Szafarczyk

1.1 Introduction

Automatic Supervision in Manufacturing (ASM) is a relatively new term, but the need for such supervision has been recognized since the very beginning of manufacturing production. Supervision of production involves dealing with all the stochastic events (that is, disturbances) which might influence the production run and its end products. Those involved in production have always had to supervise such things as the state of machine tools, the performance of manufacturing processes, the behaviour of other people etc., to ensure that the end products were obtained as quickly and as cheaply as possible.

However, with the ever increasing automation of production processes, more and more supervisory tasks of necessity demanded automatic monitoring. Complete automation of a production system requires a comprehensive automatic supervisory system that monitors all the vital parameters which might conceivably vary during the operation. The historical influence of various developments on the on-line involvement of people in manufacturing production is illustrated in Fig. 1.1.

After being expelled from the Garden of Eden, mankind was forced to invent tools and recognize that much hard work would be needed if the dream of manufacturing production was to be fulfilled. People also started to look for means to make their tasks easier. The invention of the machine tool has helped this process, for it automatically guides a tool along a pre-defined path such as a straight line or a circle. The machine tool with a driving system has relieved people from the physical effort, but not from the mental activity and stress, connected with the manual control of a manufacturing process. Although the mental involvement has been reduced by developments in automatic control, even with an automatic control system covering the whole manufacturing cycle it is necessary to have some personnel in attendance to monitor unforeseen disturbances that could disrupt the production process. Only a fully comprehensive automatic supervisory system would enable the manufacturing equipment to operate unattended without on-line human involvement. According to a survey conducted in Germany by Eversheim *et al.* [1], unattended machining during second and third shifts, as well as on Saturdays, Sundays and holidays, can increase the effective use of production equipment by a significant factor.

The need for full automation of production is a very important impetus for the

Fig. 1.1. Reduction of on-line involvement of people in manufacturing production.

research work into the development and application of ASM systems, but not the only one. With the growing complexity of manufacturing equipment, as well as the ever increasing parameters of production processes (speeds, forces etc.), human intervention might not be able to react as quickly and reliably as is necessary. In many cases automatic supervisory systems are essential. They are even more vital for processes such as electro-discharge or electro-chemical machining which are difficult or impossible for humans to monitor.

The level of supervision is strongly related to quality in manufacturing – both quality of products and quality of the production processes.

1.2 Quality of Products and Quality of Production

With regard to quality and supervision in manufacturing, the ISO definition of quality [2] may be used:

the totality of features and characteristics of a product or service that bear on its ability to satisfy stated or implied needs.

Such a general concept of quality covers both quality of products and quality of manufacturing processes. It also includes quality assessed from such different points of view as precision of the product, aesthetic values, useful life of the product, low production costs, quick response to market demands etc. The document further specifies:

> 3. Needs are usually translated into features and characteristics with specific criteria. Needs may include aspects of usability, safety, availability, reliability, maintainability, economics and environment. . .

But ISO standard also notes:

> 4. The term "quality" is not used to express a degree of excellence in a comparative sense nor is it used in a quantitative sense for technical evaluations. In these cases a qualifying adjective shall be used. For example, use can be made of the following terms: . . .

In industry, demands for a high level of quality must be translated into specific requirements based on features that can be measured. The results of the measurements of physical quantities are used to calculate a measure of quality, as allowed by ISO. Szafarczyk and Chisholm [3] proposed this measure be called "*quality index*" and designated by q. The bigger the value of quality index, the higher the quality level. The term "quality index" is similar to "performance index" commonly used in optimization of processes, but has a broader meaning because it may be applied not only to processes but also in relation to objects such as workpieces, products and manufacturing equipment.

In manufacturing departments the quality of a product is still frequently reduced to conformance of the properties of the end product to technical specifications. In specific cases of dimensional accuracy this means that all dimensions of the workpiece after machining should be kept within prescribed tolerance ranges. This then leads to the simplest, binary assessment of workpiece quality, as "good" or "bad" (Fig. 1.2a). The quality index q may have one of two values: $q = 1$ ("good") when the diameter is inside the permitted range and $q = 0$ ("bad") when it is outside. The modern trend, set first by Japanese industry, is a steady attempt to increase the uniformity of workpieces in a batch and to narrow the dimensional tolerances required (Fig. 1.2b). It is a much more ambitious task for a manufacturing team to express quality in terms of a multi-value quality index, even for the simplest case of say a single workpiece diameter.

When considering the quality level of manufacturing processes, at least two aspects may be distinguished: (1) quality assessment from the safety or break-down point of view; and (2) quality assessment from the efficiency or cost-of-production point of view. In the first case, quality level is associated with a two-value quality index; in the second, with a multi-value quality index. An example of the relationships between the parameters of machining and the production cost is shown in Fig. 1.3. This enables optimization of the process to be specified. The cost of the turning process may be minimized by the proper selection of cutting speed and feed. The quality index q is then inversely proportional to E – the cost of turning per unit of workpiece surface.

A multi-value quality index may be assessed *a priori* (i.e. before machining), but the actual value of the index may differ considerably from the assessed one if disturbances have occurred during the process.

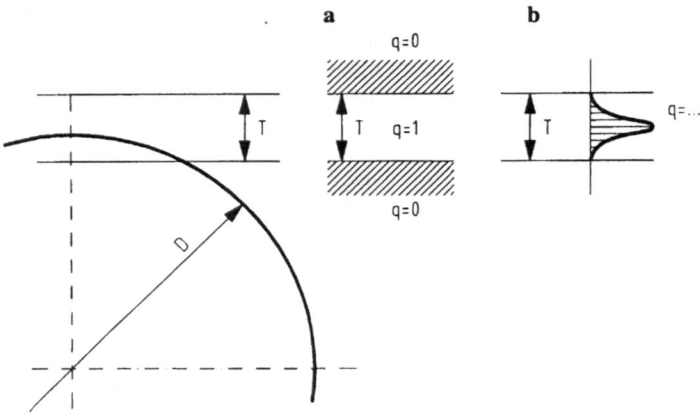

Fig. 1.2. Assessment of workpiece quality from the point of view of its dimension: **a** binary assessment; **b** multi-value assessment.

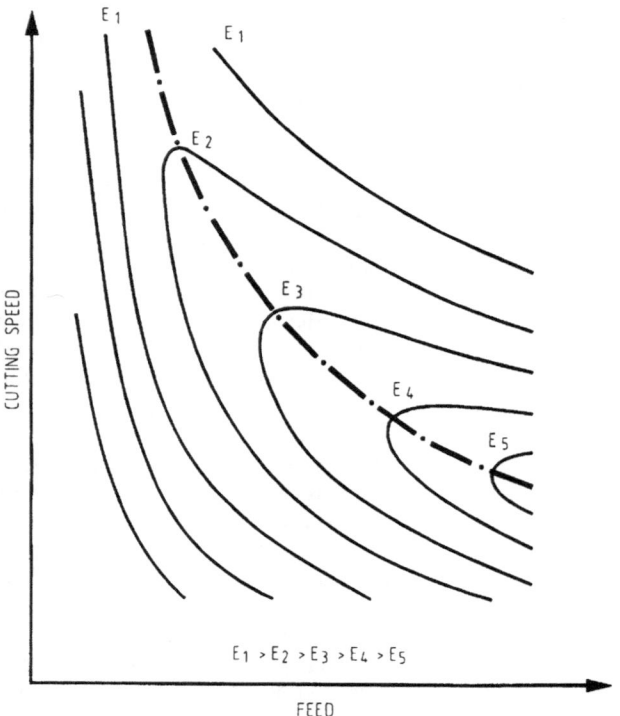

Fig. 1.3. Influence of feed and cutting speed on the cost of the turning process. E = cost of turning per unit of workpiece surface.

1.3   Disturbances, Adaptation and Supervision

Any unplanned influence on the quality index should be treated as a disturbance. There are many kinds of disturbance. Some arise inside a manufacturing system (wear of the tool, chatter, insufficient lubrication of a bearing, improper operation of a motor etc.). Others enter a manufacturing system via one of the inputs: material input (hardness of the workpiece different from that planned, an incorrect tool etc.), energy input (lack of power supply, pressure of compressed air different from that specified etc.) or information input (error in the numerical control (NC) part program etc.). They may even be the result of an environmental influence, such as sunlight (warming up a machine tool).

The influence of certain types of disturbance may be reduced or even eliminated by an appropriate design of the manufacturing system. This is achieved either by lowering its sensitivity to the particular types of disturbance involved or by building adaptability into the manufacturing system. Use of materials with a low coefficient of thermal expansion for machine tool elements demonstrates a good example where sensitivity to thermal deformations is lowered. If the influence is so small that it can be neglected, the disturbance ceases to be significant.

Another way of dealing with disturbances is to use an adaptable manufacturing system. Adaptability is the characteristic feature of a manufacturing system that enables it to reduce the influence of disturbances on the quality index by appropriate changes in the functioning of the system. Adaptability may be a natural feature of the manufacturing system, or it can be built into it through the adaptive functions of the control system. Two examples of systems designed to control the diameter of cylindrical components, in grinding and turning, are shown in Fig. 1.4a and Fig. 1.4b respectively. Figure 1.4a illustrates a basic control system for grinding cylindrical workpieces using feedback information on the actual workpiece diameter. The grinding cycle is adapting to any disturbances (except those that influence the measurement of the diameter) in such a way that, after grinding, the diameter of each workpiece has the desired value.

Figure 1.4b illustrates the use of an additional control system, a supervisory system, for a more difficult case of controlling the diameter of workpieces. Because of the different characteristics of the turning process there is no practical way of measuring and correcting the diameter during the turning operation itself. The diameter must be measured after the operation has been completed, and the result of this measurement is then used by the supervisory system to calculate any unwanted trends and to adjust accordingly the diameter of the workpieces which are to be machined subsequently.

Because of the different characteristics of various types of possible disturbance and their diversified influences on manufacturing processes, there is a need for many different supervisory functions and systems.

1.4   Classification of Disturbances and Supervisory Functions

There are many possible ways of classifying disturbances in manufacturing. From the point of view of supervisory functions the most important one seems to be that based on the outcomes of the of disturbances.

Disturbances may lead (Fig. 1.5) to: break-downs, "production" of rejects or

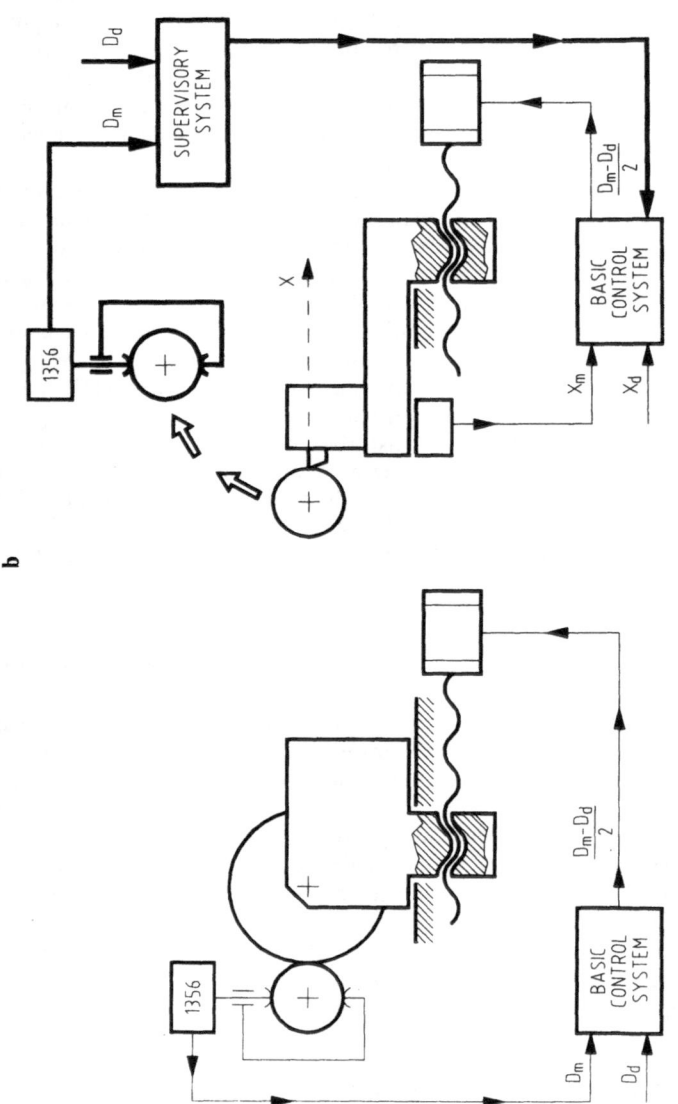

Fig. 1.4. a Adaptability built into the basic control system or **b** introduced by a supervisory control system [3].

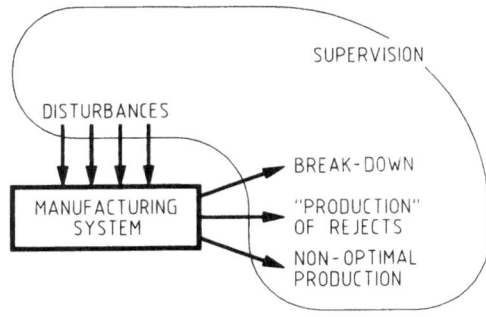

Fig. 1.5. Three classes of consequences of disturbance during manufacturing.

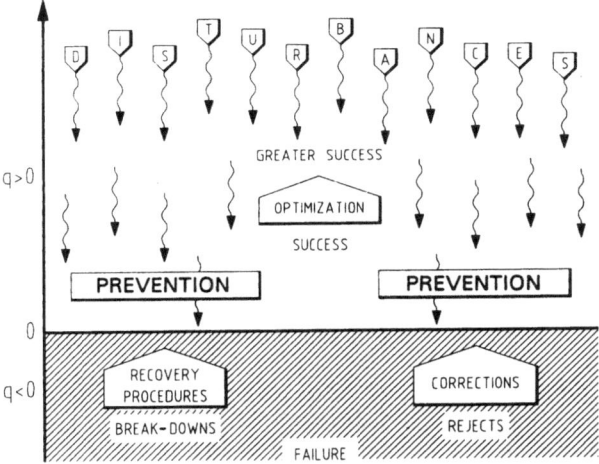

Fig. 1.6. Supervisory functions in manufacturing.

non-optimal production. Break-downs are caused by the class of disturbance that does not permit further operation of the manufacturing system. For example, it may be a catastrophic failure of the tool or a break in power supply to the machine tool. Another class includes disturbances that lead to the production of rejects (scrap) – the manufacturing system may still work but its operation is futile because the results are unacceptable. For example, the diameter of the workpiece after turning is outside tolerance because of tool wear or thermal deformation of the machine tool. Disturbances belonging to the third class lead to the least serious consequences: products are acceptable but production is suboptimal, which means that all possibilities of improvement have not been exploited. For example, the workpiece before turning has a smaller diameter than that assumed during process planning, when the cutting parameters were established, and the chosen parameters are therefore lower than they should be. The cost of the operation may be lowered by turning at higher cutting parameters than those specified in process planning. A comprehensive supervisory system would set the cutting parameters at values that would optimize the turning process.

The different supervisory functions needed in manufacturing are illustrated in Fig. 1.6. The vertical axis represents the quality index q. It is assumed that positive values of q mean that the manufacturing process is successful; the bigger the value of q , the better the results of manufacturing. If q is negative or equal to zero, the manufacturing is a failure. The bigger the negative value of q, the higher the losses caused by the failure. In the case of disturbances leading to break-downs the main supervisory function is to *prevent* failure. However, if this is not possible, the supervisory system should attempt to reduce losses, probably by switching off the machine tool as quickly as possible (emergency stop), and then to perform a *recovery procedure* which brings manufacturing back to the successful state.

In cases of disturbance that lead to manufacturing products which cannot be accepted, the supervisory function is to *prevent* production of rejects. However, if it can only make *corrections*, it must change the manufacturing process in such a way that the parameters of the products are within the required tolerances. In cases of successful production, the role of supervision is to *optimize*, that is to make success even greater by increasing the quality index.

Supervisory functions may be performed by humans, built into the automatic control system of the working cycle (see, for example, Fig. 1.4a) or need a special supervisory system (see, for example, Fig. 1.4b).

1.5 Monitoring, Diagnosing, Supervising

The terms "monitoring" and "diagnosing" are frequently used in automatic supervision, sometimes interchangeably, and hence may cause confusion. Monitoring may be considered as a special kind of diagnosing, but the terms are not synonymous.

In everyday terms, to "monitor" means "to observe or record (the activity or performance) of (an engine or other device)" [4]. In professional technical literature terms, "monitoring" has been defined as "the art of measuring indicative change or condition of function signals as warnings that possible corrections are required" [5]. *Monitoring*, in the context of automatic supervision in manufacturing, can be defined, with the use of the term "quality index", as:

> watching over or inspecting the chosen features of the process, the product or the production equipment with the aim of gathering information as an aid to ensuring the required value of the quality index or maximizing value of the quality index.

Machine tool operators can themselves monitor the manufacturing process using their sensing organs and knowledge based on experience. They may be helped by a monitoring system that measures the chosen features of the machining process and/or the production equipment and that sends signals in cases of malfunction.

Sometimes the term "monitoring system" is wrongly used for the term "supervisory system", and vice versa. In automatic supervision a monitoring system (or rather a subsystem in this case) is part of an ASM system, but *only* part of it. In addition to monitoring, automatic supervision must have the capability to carry out a control action which influences the production process according to a chosen supervising strategy.

"Diagnosis" is a term commonly used in medicine; it is the process of

AUTOMATIC SUPERVISION

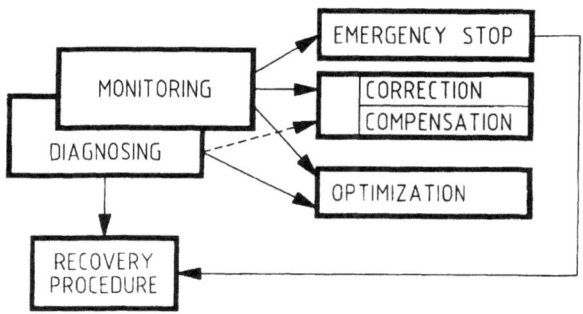

Fig. 1.7. Relationship between monitoring and diagnosing in supervision.

determining whether a patient is ill or not, and if so what disease is responsible. *"Diagnostic"* has been defined more widely as "pertaining to the detection and isolation of a malfunctioning" [4]. Diagnosing has a broader meaning than monitoring. Monitoring is a kind of on-line diagnosing with stress on the quick determination of what has happened – of what is the result of the disturbances. The detailed cause of a failure is frequently established later by a more sophisticated diagnostic process.

Symptom is another word adapted from medicine, where it means the change in the body's condition that indicates the illness. In ASM, "symptom" means a characteristic change in the manufacturing process or equipment that indicates a failure, significant probability of a failure or non-optimal condition of the process. Symptom is recognized by the monitoring system and signalled by its output. In many cases a symptom is just signalling that "something is wrong" because conditions are significantly different from normal. This change of conditions may be characteristic of a whole class of failure. Further diagnosing is needed to determine which specific kind of failure has occurred. Also the determination of a reliable recovery procedure after an emergency stop may need more detailed diagnosing (Fig. 1.7).

Identification of a failure can be made more accurately on the basis of a set of symptoms. A set of symptoms characteristic for the particular failure is called a *syndrome*, which is also a word of medical origin. The concurrent measurement and evaluation of several features of the monitored system make possible a more specific determination of the failure. In many cases, however, it is more important that it increases the credibility of monitoring. In the field of ASM this technique is called *sensor fusion* and is very important because of the high level of measurement noise in the workshop and the complexity of the system to be monitored.

1.6 Automatic Supervisory Systems

According to the classification of supervisory functions presented in Section 1.4, three types of ASM systems may be distinguished: *safety control, Geometrical*

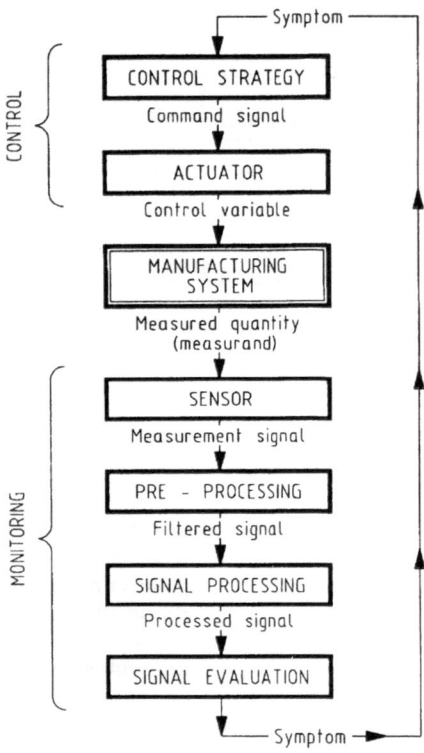

Fig. 1.8. Typical structure of an automatic supervisory system [3].

Adaptive Control (GAC) and *Technological Adaptive Control* (TAC). Safety systems deal with disturbances that lead to break-downs. GAC systems look after the quality indices of the products (in the case of machining, this means the quality indices of the workpieces) while TAC systems are concerned with the quality indices of the technological processes.

Adaptive control is a term that has been used for many years in control theory, but with a meaning not identical to the one it has in ASM. In ASM, adaptive control simply means a control system which makes changes so that actual conditions are adapted in such a way that the quality index is kept as high as possible. To accentuate the difference, the adjectives "Geometrical" and "Technological" are now placed as the first words in the terms. In the literature on ASM these adjectives used to be placed as the final words – authors used the terms ACG (Adaptive Control Geometrical) and ACT (Adaptive Control Technological).

The above classification of ASM systems based on their supervisory functions is very useful, but it is important to remember that, in practical cases, the supervisory systems may fulfil different functions. For example, a TAC system for turning based on the measurement of cutting force with the aim of increasing productivity may at the same time be called upon to safeguard the manufacturing system against overload.

The typical structure of an elementary system of automatic supervision is presented in Fig. 1.8. The quantity being measured (sometimes called the "measurand" [4]) is transformed by a sensor into a measurement signal; this is nearly always an electrical quantity. This signal is then pre-processed, generally by amplification, tare, filtering out needless frequency bands and other methods used to improve signal-to-noise ratio. Pre-processing is also called *signal conditioning*. The pre-processed signal undergoes the main processing (e.g. Fourier analysis, regression analysis, time series analysis) in order to present it in a form best suited for the extraction of the signal parameter which best represents the feature of the manufacturing system chosen for investigation. The processed signal is then evaluated from the point of view of the symptom of malfunctioning or non-optimal functioning of the manufacturing system. This process is termed *feature extraction.*

The symptom is signalled by the output of the monitoring part which is sent to the control part of the supervisory system. When the signal indicates the appearance of a symptom, an appropriate command signal is formed according to the control strategy. This command signal is sent to the actuator and changes the *controlled variable* of the manufacturing system.

The system presented in Fig. 1.8 is called "elementary", because only one parameter of the manufacturing system is measured, and as a result of supervision only one control variable may be changed. In practice, more complicated systems are in use, which employ several sensors for the measurement and several control variables to influence the operation of the manufacturing system.

Automatic supervisory systems, especially safety systems, may differ considerably from the typical structure presented in Fig. 1.8. In one extreme case, it may be a very simple hardware system, such as an end-switch pressed by a guard and connected to the electric circuit of the machine tool in such a way that it prevents the spindle from rotation when the guard is open. In another extreme case, it may be pure software, such as a special program residing in the computer numerically controlled (CNC) unit, which checks the NC part programs from the point of view of the formal requirements of programming.

1.7   Monitoring and Measurements

In an automatic supervisory system, monitoring is associated with measurements of the chosen features of the process and/or the production equipment. The general rule is to use a measuring installation that is as simple and reliable as possible. The installation should not interfere with the manufacturing functions and be immune to rough workshop conditions. Whenever possible, the signals already existing in the manufacturing system should be used for monitoring purposes. Nevertheless, in most cases, special sensors have to be employed. These are usually resistive (e.g. wire strain gauges), piezoelectric, inductive or optical. The sensors are used for the measurements of dimensions, forces, vibrations, material waves (so called acoustic emission (AE)) etc.

The choice of a quantity to be measured depends in most cases on the aim of the supervisory system (formulation of the quality index) and the characteristics of the manufacturing process to be supervised. The general rule is that the measured quantity should represent the quality index as closely as possible. In some specific cases the measured quantity should represent the value of the main disturbance and not the value of the quality index.

The choice of the sensor for measurement of the selected quantity depends on the character of the manufacturing process and the configuration of the manufacturing

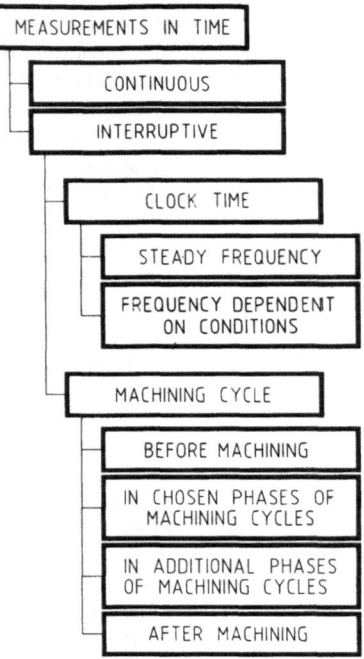

Fig. 1.9. Classification according to the timing of the measurements.

system, as well as the static and dynamic requirements relating to the results of the measurement. The cost of the sensor and its associated measuring installation must of course be considered, but reliability and robustness are usually more important than cost where monitoring is concerned. Various examples of sensors and measuring systems are presented in subsequent chapters of this book.

The rapidly developing integration techniques for electronic circuits make it feasible to assemble, as one element, a sensor together with a large part of the measuring system that will perform the signal conditioning, the self-calibration and the self-diagnosing – or even feature extraction. The term *intelligent sensor* is frequently used in such cases, but *integrated sensor* seems to be more appropriate.

From the general point of view, measurements made for monitoring purposes can be classified on the basis of the timing of the measurement, as presented in Fig. 1.9. Firstly, measurement may be continuous or interruptive. The discrete character of manufacturing production calls for interruptive measurements and a classification which accords to timing relationships of the machining cycle rather than to those of "clock time". The choice of timing for the monitoring measurements depends on the character of the manufacturing process, the nature of the expected disturbances and, of course, the practicality of measurements. For example, turning makes the measurement of the workpiece diameter during turning impracticable, as already discussed in Section 1.3. It is possible to *measure* the diameter *after machining* and to make, if needed, an appropriate correction before machining the subsequent workpieces of the same production batch (Fig. 1.4b). As a matter of fact, such a GAC system supervises the turning process of the whole batch and not just of individual workpieces. It may be used when disturbances change slowly, with recognizable trends, and when the production batch is big enough.

When the diameter before turning may vary from one workpiece to another (e.g. the blank is made by free-form forging) this may be the principal disturbance because of the changing depth of the first cut. In such cases a supervisory system may be used in which the diameter of the blank is measured (the *measurement before machining*), then the expected disturbance can be established and appropriate action taken.

In specific cases, *measurements* taken *in additional phases of the machining cycle* may be useful. An example has been described by Szafarczyk *et al.* [6]. In order to obtain a very high accuracy of diameter after boring, the final cut is divided into two passes. During the first pass, half of the machining allowance is removed, then the resulting diameter is automatically measured and its value is compared with the expected value. Any difference is used for automatic correction of the cutting edge position in the boring bar before the final pass, using the same cutting parameters as during the first pass. All conditions during the second pass are expected to be the same as during the first one and hence, because of the correction, the resulting value of diameter is expected to be close to that desired.

1.8   Supervisory Actions and Control Strategy

Apart from monitoring, a supervisory system should be capable of taking appropriate action, when necessary, to influence operation of the manufacturing system. Some of the possible actions are shown in the Fig. 1.10.

Changing the machining process parameters is the most common action taken in TAC systems. For example, a TAC system can be used to change the feed value during rough turning in order to increase efficiency without exceeding the permissible value of the measured cutting force. In some cases a change of tool condition may be needed, for example, automatic redressing of the grinding wheel to improve its condition. When automatic changing of the tool condition in the machine tool is not possible, the worn out tool may be automatically replaced by a

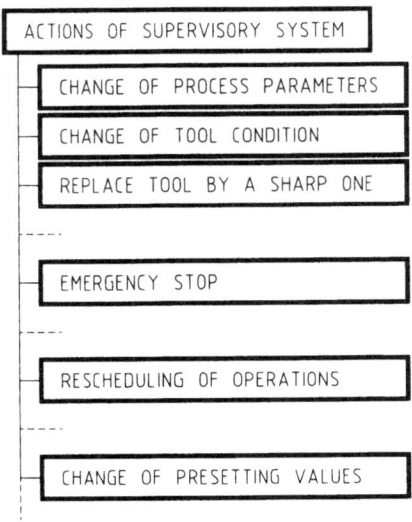

Fig. 1.10. Actions of supervisory systems.

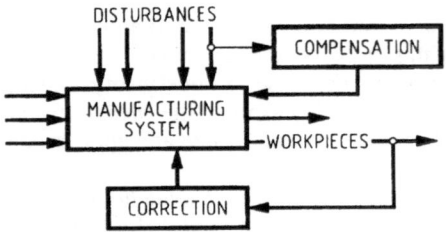

Fig. 1.11. Correcting and compensating.

new one. This is an easy operation for machining centres, since they are equipped with magazines for the automatic changing of tools.

In some cases of disturbance, such as a catastrophic failure of the cutting tool or a collision between the tool and the production equipment, the ASM system must interrupt the machining process as quickly as possible. An automatic recovery procedure may be applied later. This may involve changes to the process plans by omitting certain types of machining operation, or rescheduling and omitting the machining of certain workpieces.

In GAC systems, corrections of workpiece dimensions are frequently made by changing the pre-set values of the tool or the workpiece in NC machine tools. Existing facilities of NC control systems are available for this purpose.

In a typical system of automatic supervision, the results of the manufacturing process (e.g. the dimension of the workpiece or the value of the cutting force) are measured and appropriate corrections are made to the process (Fig. 1.11). In such a closed-loop system the influences of all disturbances on the measured result of manufacturing may be corrected, but *a posteriori* (that is, after they have already influenced the result). Another solution is to measure the value of the main disturbance and, knowing how it will influence the result, compensate this influence by appropriate changes in the work of the manufacturing system. In such a supervisory system, only the influence of a measured disturbance may be eliminated or reduced.

The ASM systems may have different structures and work according to the different control strategies involved. This is also especially true for TAC systems optimizing on-line manufacturing processes. Because of complicated, not-fully-identified models of the processes and difficulties in process monitoring, a variety of techniques is frequently used: expert systems, artificial intelligence and learning procedures. Two main types of TAC systems are specified in the literature on automatic supervision: ACC (Adaptive Control Constraints) and ACO (Adaptive Control Optimization). These names may be misleading, because both types are intended to optimize the manufacturing process. Which of them should be used depends on the form of the relationship between the process parameter chosen as the controlled variable (the input variable to the manufacturing system) and the quality index.

Two different relationships between the quality index q and the process parameter C, which is to be changed in order to influence the performance of the system, are illustrated in Fig. 1.12. In the case of the monotonic function of q within the permissible range of controlled variable C (between C_{min} and C_{max}), as shown in Fig. 1.12a, the biggest value of q is on the edge of the range, and an ACC system may be used for optimization. The ACC system will always increase C up to the constraint, thus increasing the value of q. In cases where there is an extremum of q

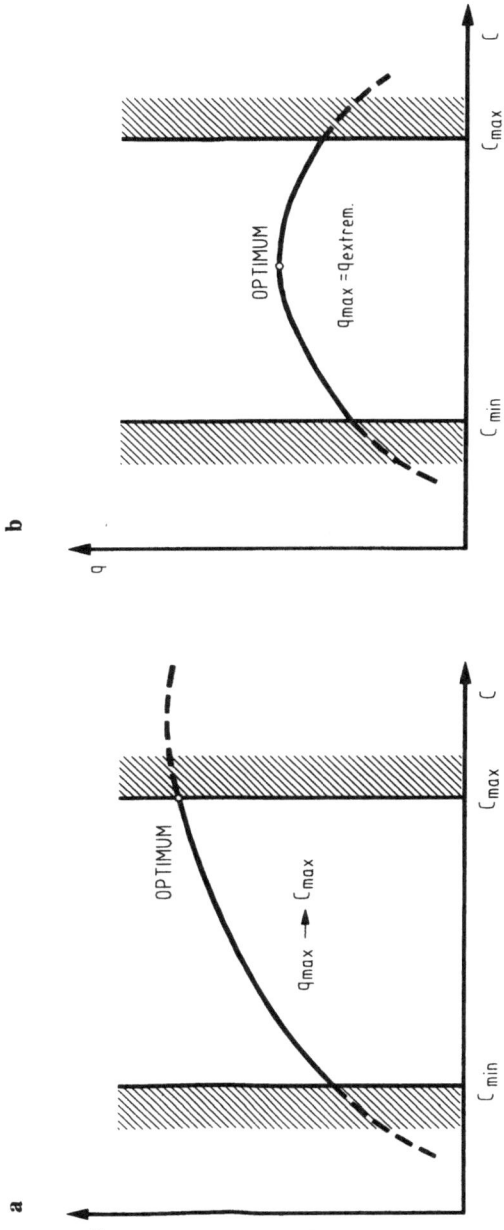

Fig. 1.12. Quality index as a function of controlled variable: **a** a monotonic function and **b** a function with extremum [3].

within the permissible range of C, as shown in Fig. 1.12b, an ACC system cannot work properly, and an ACO system should be used.

References

1. Eversheim W, König W, Weck M, Pfeifer T. Tagungsband des AWK'84. Aachener Werkzeug-maschinen-Kolloquium, 1984
2. ISO 8402 – 1986; BS 4778 Part 1 – 1987: quality – vocabulary. British Standards Institution, London
3. Szafarczyk M, Chisholm AWJ. Automatic supervision in manufacturing systems. In: Proc 3rd Intl Conf Automatic Supervision, Monitoring and Adaptive Control in Manufacturing, AC'90, CIRP, Rydzyna, Poland, 1990
4. Glossary of terms and definitions. International Measurement Confederation, IMEKO. Committee TC 10 – Technical Diagnostics 2nd ed, 1983
5. Collins dictionary of the English language 2nd ed. Collins, London, 1986
6. Szafarczyk M, Szala W, Holnicki A, Klimkowski J. Precise boring on unmanned machining center. In: MATADOR Conf, Manchester, 1988

2 Automatic Supervision in Turning and Boring

B. Lindström

2.1 Introduction

Of all machining operations, turning and boring are perhaps the most used in industry. These processes have also been studied by researchers for almost a century, and a great deal of knowledge has been formalized in such a way that systematic control of these processes is now within reach. Industrial applications of automatic supervision systems for machining are also perhaps the most frequent for these processes. Nonetheless, the systems that are commercially available cannot be regarded as sophisticated.

In this chapter, some of the parameters that can be monitored are discussed. Some of the systems available on the market are also briefly described. Since it is not the intention of this book to provide a complete list of available systems, only the principles behind some of the systems are mentioned.

2.2 Control of the Machining Process

2.2.1 The Process and the Machine Tool

2.2.1.1 Machining Data Defined

All machining operations are performed using a set of operation variables called cutting or machining data. Cutting speed, feed and depth of cut are examples of such basic machining data that are used in turning and boring. Also, a set of additional specifications pertaining to the tool and tool geometry must be specified for each operation. Theoretically, basic machining data can easily be changed during the process, whereas additional machining data must usually be kept constant.

Machining data can be optimized with different optimization goals in mind. Technical limitations in the operation may be the lack of power or torque, excessive force, poor surface finish or feed/speed range limitations. The aim of an optimizing machining control (ACO) system should be to adjust the basic machining data continuously in such a way that the operation will work with optimal data regardless of changes in the process during the operation. With an ACO system, optimal data will be achieved during machining. This obviously implies the necessity of updating dynamically the machining data bank in order to provide more accurate initial data.

2.2.1.2   Machining Process Quantities that Can Be Measured

In designing a well-developed supervision system, the desire to measure certain quantities is usually greater than the actual ability to do so. For instance, information about tool life in the form of tool wear is an important issue in rough turning steel. Unfortunately, tool wear cannot be measured directly in-process. Because of the severe conditions in the cutting zone, there is no way of inspecting the tool wear optically during cutting. Such inspections have to be performed between individual cuts, which of course makes this method less reliable for industrial applications. Tool wear must therefore be evaluated from indirect measurements using other principles. In most cases, tool wear is evaluated from the behaviour of the cutting forces, i.e. an increase of one force component or the changes in relationships between two or three force components, or even the increase in cutting power. Since the measured forces also depend on several factors other than tool wear, estimation of tool wear may be hazardous.

In the supervision systems used in industrial applications at present, most factors are evaluated by indirect measurement methods. Among the quantities and events that could be measured and detected directly or indirectly in-process in a machining system during turning in order to make performance of the system smooth are:

Tool wear
Tool failure
Tool chipping
Tool collisions
Missing tool
Cutting forces
Vibrations
Chip breaking
Diameter and tolerances
Surface roughness

Tool wear in turning and boring operations usually comprises estimations of flank wear on the relief side or crater wear on the rake side of the tool. The mechanisms deciding what wear type will occur in a specific operation depend on the tool and workpiece material combination, and the cutting data used. In addition to these important wear types, plastic deformation of the tool tip and chipping of the cutting edge play important roles in tool wear.

In an automatic supervision system, tool wear is an important parameter to monitor. Unfortunately, only flank wear can be measured using optical methods and an image processor system. Crater wear is, as such, more complicated to measure, and the common occurrence of sintered chip-breaking grooves on modern tools complicates crater-wear measurements even more. Almost all attempts at tool-wear measurements have therefore been as flank wear. As a consequence, tool-wear measurements using systems that measure flank wear will be unreliable if crater wear or plastic wear is the dominating wear type.

Tool failure means that the tool has been completely worn out or broken. This state is usually easier to record than the slower tool-wear progress. If the tool fails during a cutting operation, it is possible to detect breakage using the signal analysis of cutting forces. Usually, when a tool breaks, the cutting force will temporarily vanish or be drastically reduced. The reason for this behaviour is that when the tool collapses, no actual cutting takes place for a short period of time. If the feed is not turned off, the remaining part of the tool will start to cut with very bad performance after an interval of usually some hundred milliseconds. This conduct is used in some systems as an in-process indication of a broken tool. Needless to say, the control system has to react fast enough to prevent further cutting after the force drop. If it does not do so, severe damages to the tool holder, the workpiece and the machine tool may occur. Combined with the missing tool function, touch probes can be used between cuts as a tool-failure sensor.

Tool chipping means that small parts of the active cutting edge are removed. This is usually a sign of unfavourable cutting conditions or that the tool grade used was not the best choice. Tool chipping may sometimes be detected by signal analysis of the dynamic component of the cutting force. The tool will usually continue to function for a while, but larger and larger chipping will ultimately lead to catastrophic failure. Some systems used in turning and boring operations are able to analyse this phenomenon.

Tool collision is a state in which the numerical control programmer has omitted to take the geometrical limitations of the workpiece or its clamping devices into consideration. Clamping devices may also have been changed. This state may typically be reached when the first part in a batch is machined or during program testing prior to production. A tool collision means that forces in the tool will increase drastically, very fast and permanently. The action taken is usually to shut the whole system off, since a fast reaction is needed and the strategy to retract the tool is usually difficult to analyse.

Missing tool literally means that the tool is missing in its holder. This state is more common in drilling than in turning. Usually a simple touch probe is used to feel if the tool is there. This type of sensor is sometimes also used as a broken tool detector between passes.

Cutting forces are the most common parameters measured in metal-cutting science. As a measure of metal cutting process quality, it has both advantages and disadvantages. With modern technology, it is relatively easy to measure cutting forces with reasonable precision, even in production. From the cutting force, metal-cutting power can be calculated knowing the spindle speed, the feed and the active diameter. Metal-cutting power in itself is of interest only when calculations of the spindle motor power are at hand.

The cutting force is also frequently used as an indicator of tool wear, assuming that cutting forces increase with tool wear. This is very often the case but there are exceptions. Crater wear of the tool will very often diminish the cutting force, since the effective rake angle is increased. Extensive crater wear will also increase chip breaking, thus also creating greater cutting forces. On the other hand, flank wear usually increases the main cutting force. The result may sometimes be that balanced tool wear will only generate a slight increase in cutting force. Tool wear consisting mainly of flank wear, as for instance in the cutting of cast iron, will produce a marked increase in the cutting force.

The wear behaviour of cutting tools will also affect the relationships between the different cutting-force components (main, feed and radial). This fact has been used in some systems in which relationships between these factors are used. The cutting-force component that is usually the most dependent on tool wear is the feed force, which, in most materials, shows a significant increase with tool wear. Also, the increase of the feed force in relation to the main force or the resultant cutting force has been frequently used as a tool-wear criterion.

Vibrations sometimes occur during cutting and are then usually called chatter vibrations. This type of vibration is a closed-loop phenomenon that, to a certain extent, can be affected by the cutting data chosen. This is a complicated task to analyse, and vibrations are therefore usually treated as a catastrophic state that has to be handled in some way by the operator. Coupled with modern signal analysis methods, the onset of vibrations can be an important future parameter to study in-process.

Chip breaking, or rather the absence of chip breaking, is one of the most common causes of disturbances in turning and even more so in boring. Usually, the chip form can be checked at the start of the machining of a component, but changes in tool shape or workpiece machinability variations may drastically change chip form and chip breakability. Some devices have been proposed to detect unbroken chips, but these systems will usually not work properly in an industrial environment. In boring operations, the problem of detecting bad chip form is even more critical since the space for chips is usually much smaller, and the ability to visualize the chip is poor. So far, chip-breaking monitoring is one of the unsolved problems of automatic supervision – but it is one of the most important problems.

Diameter and *tolerance* of the machined workpiece are important parameters of all machining operations. However, there are only a few primitive devices to measure those important properties available at present, mainly of the touch trigger probe type. As for tolerances, the ability to measure the diameter is the limiting factor. Very often, the positioning tolerances of NC systems are as good as the measuring device itself, resulting in conflict in interpreting the results.

Surface roughness is of interest especially in finishing operations. Since the classification of a surface according to standards is a rather complex procedure, industrial applications of this method have been somewhat meagre. No measuring device for the in-process measurement of surface roughness in turning and boring can be found on the market at present.

2.2.1.3   Machine Tool Quantities that Can Be Measured

Generally speaking, quantities to be used for automatic supervision in manufacturing that stem from the machine tool are easier to measure than quantities that are related to the cutting process. Machine tool power, temperature and deflections are all factors that may more or less readily be measured. Nevertheless, the question of what to do with the measured values is quite another matter. Since the information most frequently needed is related to the cutting process, most factors concerning machining supervision have to be related to the cutting process itself.

Important exceptions to this are control systems dealing with the geometric control of the machined surface, especially diameters and tolerances. In these cases, deflections of the machine tool can be directly measured or estimated from forces, and corrections of the tool position can then be made. This is a method more frequently used in grinding, and no commercial systems for geometric adaptive control in turning or boring are used in industry. In turning operations, the gripping force of the chuck has also been the subject of in-process control.

2.2.1.4 Sensors for Adaptive Control

In adaptive control systems for turning and boring, conventional sensors able to measure one or more of the following are commonly used:

Force, direct measurement
— static and/or dynamic

Force, indirect
— power or torque

Vibrations
— acceleration, speed, displacement

Acoustic emission

Tool wear, optical measurement

Tool presence

Diameter and tolerance

Surface roughness

Direct Force Measurements. The direct measurement of cutting forces in turning and boring is relatively easy to perform with acceptable reliability. The existing measuring platforms for force measurement are usually employed in laboratory work only. For applications in productive machine tools, special sensors are usually used. Cutting forces can be evaluated by applying a force transducer in a device that transmits the force. The most common types of sensors for this purpose are piezo-electric transducers and strain gauges.

A piezo-electric transducer is built around a material that is electrically polarized with a proportional potential when a mechanical strain is applied. By measuring the potential, a measurement of the applied strain and also the force can be made. Piezo-electric materials are usually ceramic, but crystals applied to plastic foil are also available.

The advantage of the piezo-electric type of transducer is its high resolution, which makes the choice of placement less critical, and its great stiffness, which makes its dynamic response very good. The main disadvantage of piezo-electric transducers is their sensitivity at high temperatures, meaning that the piezo effect will be lost at excessively high temperatures. Maximum temperatures of 250–400°C are the usual range for ceramic materials. Also, the combination of thermal sensitivity coupled with high stiffness will, for most applications, make a piezo-electric transducer more a good thermometer than a force transducer.

A piezo-electric force transducer has to be pre-loaded, and several transducers may be combined in such a way that three-component force vectors can be measured. Piezo-electric transducers are also best suited for dynamic measurements, since the static force is represented by an electric potential that dissipates in the amplifier. Static measurements can thus only be performed using amplifiers having extremely high impedance, but today a realistic time of 15 minutes with acceptable static signal level is possible. Static measurements are therefore always quasi-static. Because of this and the temperature drift problem, piezo-electric transducers have to be reset prior to measurement, thus guaranteeing a correct measuring value.

In many applications of piezo-electric transducers, the whole force is transmitted through the transducer. This also makes the transducer sensitive to an overload, the transducer is then often destroyed. In some recent applications, only part of the force is transmitted through the transducer, while the main force is passed through the machine structure surrounding the transducer. This "by-pass" can be achieved by careful installation and by the selection of load cell pre-load. Although this diminishes the accuracy of the measurement, it also considerably increases protection against transducer failure. Despite all their drawbacks, piezo-electric transducers may prove very useful provided that their disadvantages are properly taken into account.

Strain-gauge-based transducers are still used in many industrial applications. The measurement of strain in a machine structure is performed by measuring the resistance changes in a very thin wire. Different electrical configurations capable of doing this can be found, but the main principle is to construct a bridge consisting of various parts, with one part being affected by strain. This technique also makes it possible to include temperature compensation, making these transducers almost independent of temperature variations during measurement. Strain gauges thus permit measurements to be truly static over a long period of time.

A major disadvantage is that strain gauge transducers are less sensitive than piezo-electric transducers. One common way to increase sensitivity is to make the structure where the transducer is applied more flexible. This is of course a major disadvantage, since the whole machine tool structure will be more liable to vibrate. Another way is to use modern calculating tools, i.e. finite element analysis, to find the optimal placement of the transducer.

Strain gauge transducers are generally rugged and robust, and hence still popular as measuring instruments.

Indirect Force Measurements. Cutting forces can also be evaluated by indirect measurements. This involves using the torque of rotating parts or the power in a motor. In turning, several safety systems are built to operate on information from the feed motor power. Since this information is gathered much further away from the cutting process, it will also contain data on friction in the mechanical drive chain. Nevertheless, such information is very simple to acquire and no special transducer installations are needed. This type of measurement has therefore become widespread. Such a system is cheap to install when manufacturing the machine tool, and still cheap to fit into an already existing machine on the shopfloor. Recent research work also shows substantial potential for such measurement techniques using modern signal analysis.

Other indirect measuring methods employ the use of the torque in the feed drive or the torque of the spindle motor. These measurements are usually much more difficult to carry out in practice since a special torque transducer is needed. In most cases, difficulties with signal transmission from the rotating part are encountered. Some tests with radio transmitted signals have been performed. This problem usually is solved using rotating rings and mercury baths.

In general, all indirect force measurement methods have less accuracy than direct measurements. In contrast, the measuring problems are often easier to solve for indirect measurements.

Vibrations. Transducers for vibration are most often piezo-electric accelerometers. They are generally easy to handle and to place, but their environmental sensitivity sometimes hampers the possibilities of using them in machine tools. Accelerometers do, as the name implies, measure acceleration. Both velocity and displacement can be acquired by electric signal integration. Care should always be taken in interpreting signals from accelerometers, since their dynamic characteristics may sometimes influence measuring results in an undesirable way. The integration of signals may also lead to misinterpretations if the real movement is not in accordance with the assumed vibration waveform. In some systems, vibrations as such are not measured, but instead the dynamic cutting-force variations are used as an indication of vibrations.

Also, sound variations recorded by a microphone have been tested as an indirect vibration transducer, with very good results.

Acoustic Emission. Acoustic emission (AE) is a relatively new measuring technique that stems from non-destructive materials testing. In that field, crack signals pertaining to crystal structure failures are observed. The usual frequency range is 100 kHz to 1 MHz. Two different types of acoustic emission signals can be identified. Continuous emissions have low amplitude and high frequency and are caused by the cracking of the crystal structure in the material. Discontinuous or burst type emissions have lower frequency and high amplitude, and appear as randomized outbursts caused by the microcracking of the surface as well as by friction phenomena.

In adaptive control, the use of acoustic emission is a way of analysing high-frequency patterns originating from the tool or the cutting process. In that sense, there is a discrepancy between acoustic emission in machining and in materials testing. One factor making this difference is that, in machining, the acoustic emission signal usually has to pass through several different surfaces, with the resulting signal reflections and damping. The actual placement of the transducer is therefore very important and should be as close to the cutting zone as possible. The AE technique has been applied to detect tool wear, chip breaking and chip form. The acoustic emission transducers employed in machining processes are usually high-frequency accelerometers.

Optical Measurement of Tool Wear. As earlier indicated, the tool wear usually encountered in industrial applications is flank wear. Flank wear can be evaluated using a charge coupled device (CCD) camera and image processing software. The recording of flank wear is not so complicated or difficult provided there is sufficient illumination. In Fig. 2.1, flank wear is shown as recorded by an image system of this kind. Tool life can be estimated from the information on tool wear at different machining times. In order to obtain complete information on tool wear, this type of transducer should present the possibility of looking at the tool from different angles, i.e. have the possibility of moving along the tool edge and around the nose radius.

The drawback of this method in turning and boring operations is the inability to record such images in-process. At present, only laboratory prototype systems have the possibility of recording tool wear by optical inspection.

Tool Presence. The transducer for tool presence is usually a touch trigger probe, also called a broken tool detector (BTD). This device can be used in two different ways. In the first, only information about the actual presence of the tool is used, i.e. the tool is there or not. In the second, information about tool-tip damages or excessive

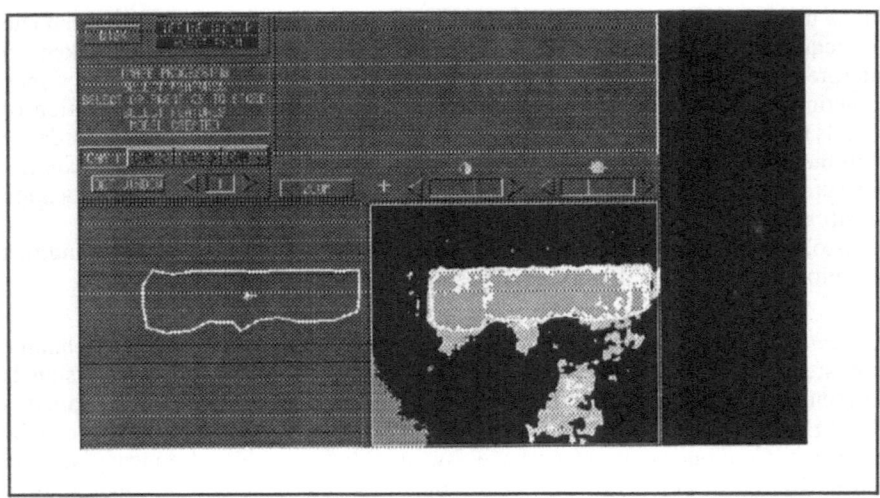

Fig. 2.1 Flank wear estimation from a CCD camera. Original image with wear surface marked (right) and extracted wear borders (left) [Lundholm, KTH].

tool wear can be obtained. Since the probe uses the numerically controlled (NC) system for movement, the accuracy of the probe measuring system is not better than the accuracy of the NC system. The relatively small amount of dimensional changes created by normal tool flank wear is thus often beyond the reach of this type of sensor.

Diameter and Tolerance. The most common transducer for dimensional measurement in machining is the touch trigger probe, especially the Renishaw probe. This transducer can easily be connected to the computer numerically controlled (CNC) system, and information on workpiece dimension and tool position can readily be obtained. In external turning, systems with a mechanical device similar to large vernier callipers are often used. Some experimental devices utilizing lasers for dimensional measurements have also been tested. Because of the techniques employed, the diameter information can only be used as a correction for the next workpiece. For internal boring, the problem is much more difficult since the surface is concave and the diameter is often rather small. No solution to this problem is known as yet.

Surface Roughness. Surface roughness is the quality measure of the surface generated in machining. The standards for surface roughness are all founded on measurements of the surface profile using special devices, usually in a laboratory environment. In actual machining, no standardized method of performing surface roughness measurement has been established. Some efforts to use lasers have been made. The reflected light from a laser beam carries some information on roughness, but this information cannot unequivocally be related to the existing measures for surface roughness.

2.2.2 Safety Systems

2.2.2.1 Safety Systems Defined

Safety systems, sometimes also called monitoring systems in industry, are on-line diagnosing systems used to prevent break-downs or to minimize the damage caused by catastrophic tool failure. These systems usually monitor the process, and the control action that is taken is a simple on/off signal. As indicated by the name, a safety system may be regarded as a "safety net" for the machine tool, the tool or the workpiece. A safety system is a system that monitors the process and halts execution at dangerous levels. For example, in a turning system, such a system may monitor the cutting force and as soon as the force exceeds a pre-set limit, machining will be stopped. The control action taken is thus to stop the process, and the system could therefore also be described as an electronic "shear pin".

Of all the supervision systems used in industry today, safety systems are the most common, especially in turning and boring. The existing systems all belong to the emergency type of safety systems.

2.2.2.2 Different Generations of Safety Systems

As stated above, safety systems are the most common of all supervision systems used in industry for turning and boring operations. The first systems emerged on the market in the mid-1970s. Since then, the systems have developed to such an extent that one could even speak of different generations.

Safety systems of the first generation usually consist of a monitoring device made up of a transducer, an amplifier and electronic devices that analyse the measured signal. They also work with teaching techniques, meaning that information about the measured process quantity is recorded and memorized together with NC information. This in turn means that the system records the process parameter, for example the cutting force, for each NC-block used for machining a component.

After the information related to all the machining involved in making a completed component is stored, the actual monitoring phase can take place for the next workpieces. As soon as the instantaneously measured process parameter exceeds the unit calculated on the recorded value, the process is stopped and an alarm signal is activated. Since the cutting process is a more-or-less stochastic process, the measured values will always differ from the recorded values. In order to handle this, an appropriate tolerance band has to be defined, i.e. the process is stopped only when the measured value is outside this tolerance of, for instance ±20%. An example of such a system using cutting forces with tolerance bands is shown in Fig. 2.2.

One big disadvantage of this kind of system is that the operator must calibrate it by using the first workpiece as a calibrating device. The time and memory used for this can be enormous, especially for more complex workpieces using long NC-programs. Furthermore, in modern small-batch production, the number of workpieces can be so small that even just one workpiece can be a considerable percentage of the whole batch.

In the second generation of safety systems, signal processing and signal evaluation of the system have become more advanced. Also, the use of more advanced transducers is typical. Most systems now used in industrial applications belong to this generation. These systems may still have the teaching strategy, meaning that at least one initial workpiece has to be used as a calibrating instrument.

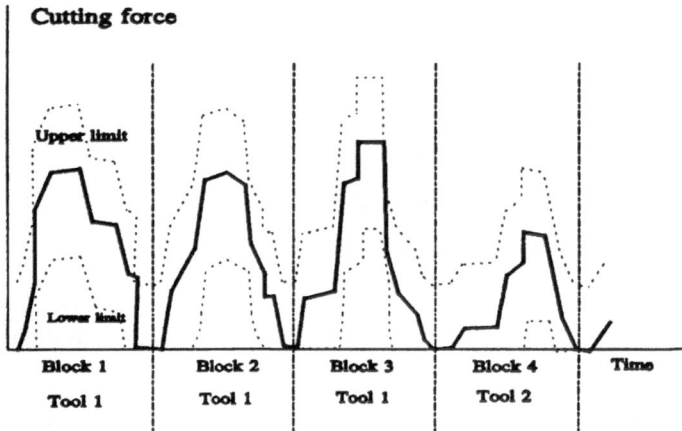

Fig. 2.2 Teaching system with tolerance bands. Blocks indicate different NC-program blocks. Different cutting force levels are indicated.

The third generation of safety systems has eliminated the necessity for the teaching mode. This means that such systems can work adequately from the first workpiece, and they are thus more suited for small-batch production. These systems also have a higher level of intelligence, meaning that they can distinguish between different situations such as tool breakage, tool chipping and tool wear. At the same time, they all retain the safety features of the earlier generations of systems. Yet another advance is their computational ability, which makes it possible to store historic information that can be of considerable help to the operator. Their communication abilities with the operator, in the form of screens and printouts, have also been improved.

Examples of Existing Safety Systems. Typical cutting-force monitoring systems using current measurements in the spindle and the feed force motors are available on the market. Other similar systems are also available, sometimes with active force measurements through special force transducers. Most of the systems available at this time belong to this category. Examples of manufacturers of power monitoring systems are Artis, Valenite Power Monitoring and Niigata. Somewhat more sophisticated systems are manufactured by Promess and Sandvik.

The safety system usually has two different operational modes. In the first, maximum cutting forces are memorized for each NC-block during the machining of the first component. This learning process is sometimes called teach-in. From these values, minimum and maximum cutting-force limits are created for each NC-block. Three different limits are established: tool wear limit, tool breakage limit and minimum limit. The minimum limit is used for the checking of a missing workpiece or tool, or completely broken tools. When components are machined after teach-in, actual cutting forces are monitored and checked to ensure they are within the established limits. Another global maximum force limit for the machine tool, acting as an electronic shear pin, is determined by the operator according to the size of the machine tool and the spindle motor power available.

The tool-wear function assumes that the cutting force will increase with tool wear. Since this is not always the case, one should be aware of the misinterpretations that may occur. The force limit is decided, for instance, as 130% of the recorded force value. As soon as the tool-wear force limit is exceeded after a steady increase of the force, a signal from the system will tell the NC system that a new tool is needed. If a new tool should be available in the turret or in the tool magazine, a change can be activated. Otherwise an alarm signal will alert the operator to a malfunction.

The breakage force limit is decided, for instance, as 150% of the recorded force value. When the breakage force limit is exceeded for a certain pre-determined time after a fast increase of the force, the spindle and the feed are immediately stopped and an alarm will alert the operator to investigate the machine.

The minimum force limit with the typical value of 0 N will usually not stop the machine tool immediately, but only after the on-going NC-block has come to an end.

One common problem with these systems is that machine tool friction in slides, drives and spindles will vary with time, both during the day and over a longer time span. The system therefore must be calibrated during idle running of the NC program. The stored values must then be deducted from the measured values in order to obtain the true machining values. While this may be a nuisance, the recorded values will also yield information about the general machine tool condition.

The main advantage of this type of system is that integration with the NC system is very easy and, especially when current-measuring systems are used, the installation can easily be carried out. The disadvantage, as many industrial installations will show, is the frequent number of false alarms, often forcing the operator to disconnect the system.

Because of the nature of the measured values, safety systems of this type are usually only of use for medium and heavy cuts. Cutting forces for finishing operations will usually be too close to the friction forces in the system.

2.2.2.3   Prometec

A series of safety systems of different degrees of sophistication are manufactured by the German firm Prometec. Two different families of these are the Process Monitor system series and the Tool Monitor system series utilizing static thresholds and dynamic thresholds, respectively. In the former family of systems, four static thresholds are created during teach-in, enabling the system to distinguish between collisions, breakage, tool wear and empty cut – see Fig. 2.3. These systems are of use for pre-machined parts, extruded parts and bar material where hardness fluctuations are small.

In the more advanced Tool Monitor system series, tool wear is monitored using dynamic thresholds – see Fig. 2.4. These are formed instantaneously with the force signal, forming both an upper limit (UL) and a lower limit (LL) which are not permanently stored. Upon tool breakage or tool wear, the cutting force will rapidly exceed these slightly delayed thresholds. For tool breakage, the upper limit is crossed in both directions. If the lower threshold is then crossed within a very short time, a breakage signal will be initiated. In cases where the upper limit is not crossed, a breakage signal occurs only if the lower threshold is crossed for a certain time span. The dynamic thresholds tolerate force fluctuations due to variations in the

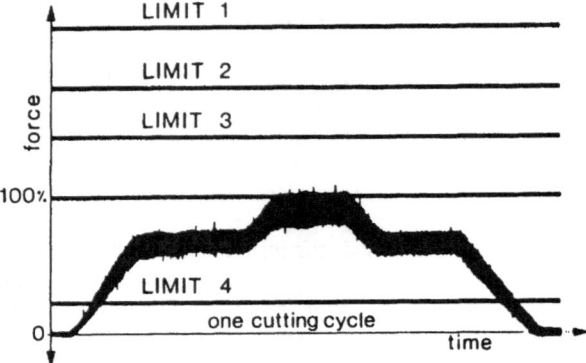

Fig. 2.3 Prometec monitoring system using several force limits for different safety events [Prometec].

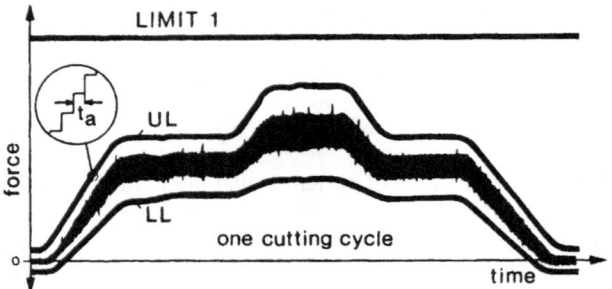

Fig. 2.4 Dynamic threshold (UL) is the upper limit of the cutting force. LL denotes the lower limit [Prometec].

depth of cut or workpiece hardness up to a specified ratio without misinterpretation.

For normal tool wear detection, the force F_n is evaluated by means of a sliding mean value F_{Gn} (see Fig. 2.5). The final force tool wear detection value F_E is calculated from the initial force F_A. The tool is considered worn out as soon as the cutting force exceeds F_E or when cutting forces decrease. The main application area of the Tool Monitor system series is in the machining of blanks or cast and forged parts with extremely rapid detection of tool breakage.

Cutting forces for both system types are measured with piezo-electric force transducers placed in a measuring plate between the turret housing and the cross slide, or between the turret and the turret housing. The plate is 10 to 15 μm thinner than the transducers, and, by pre-loading the arrangement, the transducers are compressed until the measuring plate takes up the load. This prevents the transducers from overload, since only part of the full load passes through the transducer. At the same time, the measuring accuracy is diminished. Nevertheless, cutting-force variations in the order of 10 N are reported to have been measured. Other transducer solutions such as piezo-electric strain transducers bolted to the exterior of the turret housing or headstock may also be used.

This supervision system has the ability of distinguishing between tool breakage

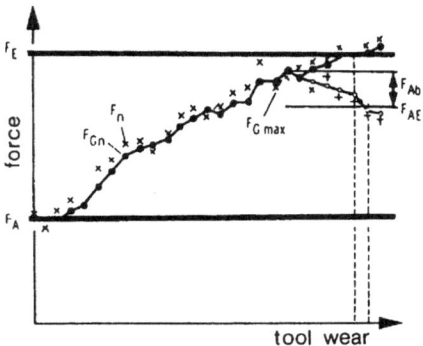

Fig. 2.5 Cutting force increase due to tool wear [Prometec]. F_A = initial cutting force value; F_n = individual cutting force value; F_{Gn} = sliding mean value; F_E = maximal cutting force value, after increase due to wear; $F_{Ab} - F_{AE}$ = strategic interval for detection of wear.

and tool wear. Tool breakage is detected in the special behaviour of cutting forces when the tool breaks. The typical indication of a tool breakage is a fast increase of the force from the steady-state force level followed by a temporary drop to zero force.

2.2.2.4 Montronix

Perhaps the most advanced safety system on the market is the US Montronix system, originally developed and marketed by Kennametal. This system can be considered as belonging to the third generation of safety systems, mainly because no teach-in is necessary. This means that the first component can be monitored and the system is thus also suitable even for batch production down to a single component. This is also a valuable feature for extremely expensive parts.

The transducer installation is somewhat similar to that of the Prometec transducer installation, and three-component piezo-electric transducers are used. A by-pass of forces similar to that of Prometec is used.

The system is capable of discriminating between tool wear, tool breakage, collision and missing tool. Tool wear, in this case meaning flank wear, is sensed by monitoring the relative changes between the three cutting-force components. A new tool signifies a relative wear index of 100%, and a worn out tool signifies an index of 0%. The relative change in tool wear can be pre-set by the operator, meaning that various wear magnitudes of the tool can be used. The exact way in which calculations of this wear index are done has been kept secret, but force-component ratios are probably employed.

Tool breakage is sensed by using advanced pattern recognition of changes in the cutting force. This is done by simultaneously comparing the cutting force to stored cutting-force patterns. Several different patterns are stored in the system, each signifying a different tool breakage event, or typical patterns for different tool materials or work materials. Standard installations of 16 different patterns are made, but customer design of new patterns is possible. As soon as a pattern is recognized, an alarm indicating tool breakage is activated.

The collision detection will activate the alarm when the cutting force exceeds a pre-set value. In order not to generate false alarms arising from sudden high and very short cutting-force levels, a collision time delay may be pre-set. As soon as the force level has exceeded the force limit for a time period longer than this collision delay, an alarm signal will be sent to the emergency shutdown of the CNC system.

The response times of the monitoring system are fast enough to allow for this check without the occurrence of any serious damage.

Cutting tests performed at KTH, Sweden, have shown that the Montronix system is able to detect tool wear and tool breakage with extremely good reliability.

2.2.3   Adaptive Control with Constraints

The family of adaptive control systems termed adaptive control with constraints (ACC) is under development at several research institutes. The aim of an ACC system is to regulate a variable in such a way that excessive forces are avoided. In a turning operation, for instance, the cutting force may be the monitored variable and the feed may be the controlled variable. The system will control the feed in such a way that the cutting force just reaches the limiting value. The control action taken by these systems is thus the adjustment of an important variable to reach the pre-set limit of the monitored variable. ACC systems seldom involve the control of more than one machining variable.

The purpose of most ACC systems is to increase productivity, usually the metal removal rate. Increases ranging from 20% to 75% have been reported, the highest values being for machining operations where heavy changes in cutting depths occur or where CNC programming has been somewhat careless. One important task of an ACC system in turning and boring is to help reduce machining times simply by increasing the feed when the tool is "cutting air". The ACC system thus corrects for bad programming of the CNC machine tool. This advantage has, of course, nothing to do with the real purpose of such a system.

It is understandable that ACC systems are mostly intended for rough cutting operations.

2.2.4   Adaptive Control with Optimization

As has been stated earlier in this chapter, workable adaptive control systems that optimize turning and boring operations are still not to be found in industry. Nevertheless, much research and development effort is being spent worldwide in order to realize such systems.

One important problem still to be solved in ACO systems is the measurement of tool wear. Tool wear is the most important parameter for the calculation of economic cutting data in roughing operations. Different approaches to measure tool wear in-process have been tried in various laboratories, but no robust measuring technique able to withstand the severe environmental conditions of industrial production has yet been put forward. Recent research findings involving advanced signal analysis present some hope that this problem will eventually be solved.

One important question still unanswered is the economic benefits of optimizing systems compared with ACC systems. One investigation showed that the biggest gain was reached in going from uncontrolled machining to the implementation of a safety system. The implementation of an optimizing system then only increased the economic benefit slightly because of the much heavier investment needed. Certainly, the justification for more complex adaptive control systems has to be found in factors that are difficult to quantify.

In the complex world of manufacturing today, many problems have to do with humans. Labour costs are usually very high for a skilled workforce. On top of that, a skilled workforce, in itself, is difficult to find. The general environment of workshops does not help the situation, and no significant change can as yet be seen. Another important factor is that lack of experience on the workshop floor will lead to a serious shortage of experienced planners. This also means that reliable machining data, as such, will be more important in the future. All this leads to the conclusion that the machine tool of tomorrow must be able to monitor the process and diagnose all occurring events in an almost intelligent way. Experience with today's safety systems show that they only work properly when they are carefully installed and attended. These are actions that have to be taken by humans. This will become impossible in the future if the number of skilled people surrounding the individual machine decreases drastically. The solution to this is Adaptive Control with Optimization.

The optimization in an ACO system involves the calculation of some performance index, i.e. machining cost or average metal removal rate. Real global optimization of the complete machining of a component will need the measurement of all important variables such as tool wear, cutting forces, vibrations and surface quality.

2.2.5   Geometrical Adaptive Control

It is no exaggeration to say that the need for Geometric Adaptive Control (GAC) systems is much more obvious than the number of available systems. In reality, no systems worth the name are available at all for turning and boring. As has been stated earlier, surface roughness and workpiece dimensions are the parameters to be measured. Only a few rather primitive systems for workpiece diameter control can be found.

Usually, GAC systems are intended for finishing operations. The aim of such systems is to achieve dimensional accuracy or consistency in surface finish. Most systems for GAC are in reality off-line systems in which dimension corrections from measurements on one workpiece are used by the CNC system while machining the next workpiece. In this way, compensation for tool wear, i.e. flank wear, can be achieved. Important obstacles to GAC systems are the heavy cost and the difficulty in using optical systems due to the adverse environmental conditions caused by metal chips and the cutting fluid.

2.2.6   Existing Experimental Systems

There are several advanced automatic supervision systems at present under development around the world. Some of these control only the machine tool according to variations in one parameter and can be termed ACC systems, while others have a more ambitious system approach whereby many parameters are monitored. These latter systems can be divided into systems that perform suboptimization, i.e. optimization of one parameter, or optimization of the process, where the economics of the machining operation are also considered.

Tool wear is monitored using CCD cameras in several systems for turning. Since the tool can only be inspected between passes, only semi-in-process tool wear can be recorded. Tool wear therefore has to be monitored in-process using other measuring means. The analysis of dynamic cutting-force components is one approach being tried out at several different laboratories. Information on tool wear is needed for optimizing systems.

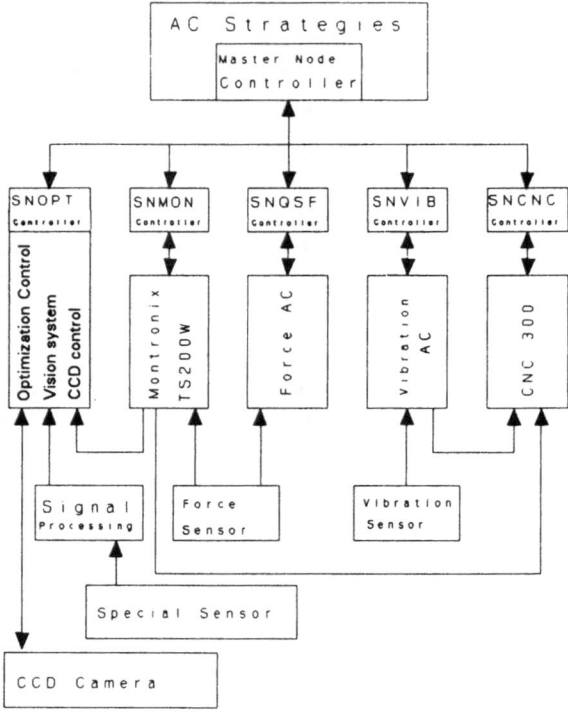

Fig. 2.6 The KTH Machining Control System. SNOPT = slave node optimization; SNMON = slave node monitoring; SNQSF = slave node quasi-static forces; SNVIB = slave node vibration; SNCNC = slave node computer numerical control.

Vibrations in turning and boring operations represent another severe limitation to the actual process. The fast analysis of the growth of vibrations is therefore one important issue in future machining systems. At present, prediction algorithms for vibrations are under development.

In Sweden, extensive research and development programmes in the field of automatic supervision for turning have been carried out. A very substantial National Programme on Adaptive Control has been running for almost a decade. The almost completed prototype system developed at the Royal Institute of Technology can be seen as an investigation into the possibilities and needs of future supervision systems (Fig. 2.6). Information from the different sensor systems, vibration, force and tool wear are treated locally in different modules. Information from various slave nodes is then analysed in the master node where AC strategies are formulated. Information from the force transducers is used both for ACC handling and in the Montronix safety system module which acts as a "safety net". The "special sensor" marked in Fig. 2.6 carries out an analysis of the dynamic forces in order to obtain tool wear information in-process. Information from the vibration module is treated in the master node and can also directly influence the CNC system. This system will work as an ACC system with added ACO features whenever possible. All events will be permanently guarded by the safety system.

2.3 Future Needs and Developments

The automatic supervision of machining operations is still a technology that can be considered to be in the first stage of development. Experience from practical applications shows that the first systems were over-simple and unreliable. Confidence in such installations has sometimes become very low and many installations have been decommissioned. For future developments, simple installation and use coupled with better reliability are necessary. The need for supervisory systems will nevertheless grow, especially since less-manned production will become ever more important. Systems with adaptive functions will also be more in demand owing to this evolution.

Computer developments during the previous decade have also enabled the use of multiple-sensor systems, also called sensor fusion, to become established. Such systems will need much better models for effective use during analysing and decision making. More sophisticated methods of analysis, such as Fast Fourier Transform (FFT) and acoustic emission, will also be incorporated. The present systems where only an alarm is activated or the machine tool is switched off will then become obsolete. Systems that incorporate advanced monitoring, analysing, decision making and control will be common in the great majority of cases.

The lack of suitable sensors for tool-wear estimation is still one of the most serious obstacles to be overcome. Co-operative work between machine tool designers will perhaps solve the present problem of costly sensor installations. The cost of actually installing sensors will then be very low, and at least one current problem of automatic supervision will have diminished.

Further Reading

1. Bejhem M, Rosendahl B. Evaluation of the monitoring system TS200W. Department of Production Engineering, KTH Stockholm, 1990
2. Chryssolouris G, Domroese M, Zsoldos L. A decision-making strategy for machining control. Annals of the CIRP 1990; 39(1): 501–504
3. Dornfeld DA, Lan MS. Experimental studies of tool wear via acoustic emission analysis. In: Proc 10th North American Manufacturing Research Conf, 1982
4. Harder L. An adaptive control constraint system for cutting force control in turning. Licentiate thesis, Department of Production Engineering, KTH, Stockholm, 1992
5. Harder L, Lundholm T, Lindström B. An adaptive control subsystem for quasi-static cutting force control in turning. In: Computers in engineering, Book No G0639B. Springer, New York, 1991
6. Jiang CY, Zhang YZ, Xu HJ. In-process monitoring of tool wear stage by the frequency band energy method. Annals of the CIRP 1987; 36(1): 45–48
7. Kals HJ, van Houten FJAM. On flexible manufacture on a production information management system. In: Proc 14th CIRP Int Seminar Manufacturing Systems, Trondheim, Norway, 1982
8. Kinnander A, Svenningsson I. Strategies to control variations in machinability in mixed manufacturing. Annals of the CIRP 1984; 33(1): 11–14
9. Kluft W. Werkzeugüberwachungssysteme für die Drehbearbeitung. Dissertation D82, TH Aachen, 1983
10. Koren Y, Ulsoy AG, Danai K. Tool wear and breakage detection using a process model. Annals of the CIRP 1986; 35(1): 283–288
11. Lee M, Wildes DG, Hayashl SR, Keramati B. Effects of tool geometry on acoustic emission intensity. Annals of the CIRP 1988; 37(1): 57–60
12. Lundholm T, Yngen M, Lindström B. Advanced process monitoring – a major step towards adaptive control. Robotics and Computer-Integrated Manufacturing 1988; 4(3/4): 413–421
13. Lundholm T. A flexible real-time solution to modular design of an adaptive control system for turning. Dissertation, Department of Production Engineering, KTH, Stockholm, 1990
14. Lundholm T. A flexible real-time system for turning. Annals of the CIRP 1991; 40(1): 441–444

15. Lundholm T, Bergström E, Enarsson D, Harder L, Lindström B, Nicolescu M, Nilsson B. New techniques applied to adaptive controlled machining. Robotics and Computer-Integrated Manufacturing 1992; 4(5): 383–389
16. Mannan MA, Broms S. Monitoring and adaptive control of cutting process by means of motor power and current measurements. Annals of the CIRP 1989; 38(1); 347–350
17. Montronix Inc. TS200W tool monitor user's guide. Raleigh, North Carolina
18. Moriwaki I. Application of acoustic emission measurement to sensing of wear and breakage of cutting tools. Bull Japan Society of Precision Engineering 1982; 17(3): 154–160
19. Prometec GmbH Aachen. Schnittstellenbeschreibung zum TOOL MONITOR SYSTEM
20. Varma AH, Kline WA. Force transducer applications on CNC lathes. In: SME Advanced Machining Technology III Conf, 1990
21. Sandvik Coromant Automation GmbH. TM20000 process monitors. Viernheim, Germany, 1990
22. Sandvik Coromant. Tool monitoring system – tool monitor. Viernheim, Germany, 1986
23. Tlusty J, Andrews GC. A critical review of sensors for unmanned machining. Keynote paper, Annals of the CIRP 1983; 32(2): 563–572
24. Tönshoff HK, Wolfsberg JP, Kals HJJ, König W, van Luttervelt CA. Developments and trends in monitoring and control of machining processes. Annals of the CIRP 1988; 37(2): 611–622
25. Weck M. Machine diagnostics in automated production. J Manufacturing Systems 1983; 2(2): 101–106
26. Yngen M, Varma A, Valerius E. Spürnasen in der Maschine. Maschinenmarkt, Würzburg 1988; 19: 39
27. Yngen M, Varma A, Valerius E. Erkennen und Reagieren. Maschinenmarkt, Würzburg 1988; 19: 42
28. Yngen M, Lundholm T, Lindström B. Adaptive controlled machining for high productivity CIM. Manufacturing International '88, Atlanta, Georgia, USA, 1988, vol 1, pp 271–276

3 Automatic Supervision in Milling

J. Tlusty

3.1 Introduction: Basic Formulations

The problems arising in milling which need automatic detection and control may be briefly listed as follows:

1. Sudden *force overload*, especially that arising from rapid traverse approach and entry of tool into the workpiece – this may be a component of constant force adaptive control.
2. *Torque overload* of the spindle drive causing spindle stall and a catastrophic build-up of cutting load – this may cause cutter breakage and/or spindle damage.
3. Milling *cutter tooth* chipping and *breakage*, which can occur for several reasons. Although not usually associated with other damage, this can nevertheless develop if the fault is not quickly detected and corrected.
4. *Tool wear* develops gradually and should be detectable before it reaches a critical value.
5. *Chatter vibrations* may cause damage to the tool edge and affect bearing life during roughing operations. They are liable to spoil the quality of machined surfaces during finishing passes, by generating deep chatter marks.
6. Severe *resonant forced vibrations* may develop in fine-finish end milling cuts on thin-walled parts, leading to bad surface finish and chipping of the cutting edge.

It is necessary to sense and detect each of the problems listed, and then to formulate *corrective actions*.

In order to define the basic terms of a milling operation, the parameters of a milling pass are illustrated in Fig. 3.1. The cutter diameter d (mm), number of teeth m, spindle rotational speed n (rev/min), radial depth (width) a (mm), axial depth (depth) b (mm), chip load (feed per tooth) c and feedrate f (mm/min) are all specified. The radial immersion (or simply immersion) $i = a/d$, the ratio of radial depth to cutter diameter is also used. Cuts with $a = d, i = 1$, are called "slotting" cuts. The cutting speed v (m/min) is obtained from:

$$v = \pi dn/1000 \tag{3.1}$$

and the metal removal rate M (mm^3/min) from:

$$M = abcmn \tag{3.2}$$

The cutting force resulting from the simultaneous action of potentially several

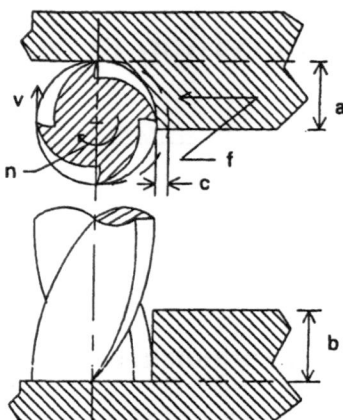

Fig. 3.1. Basic parameters of the milling process.

teeth may be defined as the cumulative effect of the tangential components on the torque T (N m) and of the radial components on the resulting radial force F_r (N), or alternatively as the cumulative components F_x (N) and F_y (N) along the X and Y axes. These axes are defined to be in the plane perpendicular to the spindle axis: generally X is the table motion and Y is perpendicular to X and is executed by the saddle or by the headstock. The individual forces and the torque usually consist of a static (DC) and a periodically variable (AC) component. The fundamental frequency of the variable component is called the tooth frequency f_t (Hz):

$$f_t = mn/60 \qquad (3.3)$$

The variable force component also contains harmonics of f_t and, as a result of the run-out of the cutter teeth which occurs once per revolution, there is also a run-out frequency f_{ro}:

$$f_{ro} = n/60 \qquad (3.4)$$

Later, it will be shown that ideally, for a cutter with zero run-out and m an integer multiple of 4, the variable force and torque components in slotting cuts are zero; all of them are purely DC. However, we may define the average values of the torque and the radial force, and also the power P_{av} (W) used in the operation:

$$P_{av} \text{ (W)} = MK_s/60\ 000 \qquad (3.5)$$

where M is in mm^3/min and K_s, the specific force, is in N/mm^2. For illustration, the usual values of K_s are 2000 for steels, 1400 for cast irons and 700 for aluminium alloys. Then:

$$T_{av} \text{ (N m)} = 60P_{av}/(2\pi n) = 9.55\ P_{av}/n = 1.59 \times 10^{-4}\ K_s\ abcm \qquad (3.6)$$

and

$$F_{t,av} \text{ (N)} = 2000T_{av}/d = 0.318\ K_s(a/d)\ bcm \qquad (3.7)$$

3.2   General Discussion

The first three problems listed above (force overload, torque overload and cutter breakage) are related yet different. For force overload and torque overload we try to prevent damage. For cutter breakage, once damage has occurred but is still limited to the cutter teeth, it should be detectable before it spreads and leads to force or torque overload, or causes damage to the spindle or the workpiece. Spindle stall is initiated by torque overload, which may be the result of an unexpected (unprogrammed) or badly programmed increase in the axial or radial depth of the cut. Thus, for instance, the blank of the workpiece may be higher than nominal, or during end milling the axial pull due to the helical edges may pull the cutter out of the collet so increasing the depth of the cut. The spindle then starts to slow down, but the feedrate continues unchanged, so the actual chip load starts to increase as does the torque. The spindle consequently slows down still more, and this process continues until the sensor in the electric drive stops the feed; the latter operation may take a long time to occur. Breakage of the cutter may save the spindle from damage, but this may not happen. Obviously spindle stall is a most serious event; early *detection of torque overload* is needed, followed by a fast feed stop, if it is to be prevented. We shall show a *fast feed stop* is necessary to overcome all the listed problems, except for tool wear.

The problem of force overload differs from the problem of torque overload just discussed. Force overload occurs, especially on small-diameter cutters, when the *force limit* is exceeded but not the torque. Spindle stall is not involved, but the excessive force may cause breakage of the stem of a slender, long end mill, or of the cutting edge on stronger cutters. This happens generally when the cutter is approaching or entering the workpiece in rapid traverse. Damage to the cutter and to other parts of the system may be avoided if any rapid increase of cutting force is quickly detected and the feed slowed down with minimal delay. This is best done by stopping very quickly and then slowly increasing the feedrate to the nominal value corresponding to the milling operation. Any flexibility inherent in the system such as in the case of a slender end mill, will limit the build-up of the force if the operation is stopped quickly enough. This kind of action may also be useful in collision situations where collision occurs at a point which is included in the force-sensing area, and where some flexibility is also involved. For cutting forces the force sensor may be in the spindle system, but for collisions it is better located in the feed drive. It can be difficult to determine the triggering level of the force except in the case of a long, slender end mill. It is then preferable to base the triggering level on rate of force increase instead.

The problem of cutter tooth chipping and breakage differs from force and torque overload in that the tooth of a cutter can break off without either the torque or the total limits being exceeded. This can happen because of a hard spot in the work material or because of a weak spot in the cutter tooth, or simply because the limit of the total force overload system had to be set higher than the force needed to break the tooth. It is, therefore, necessary to distinguish a tooth breakage in its own right, and to stop the feed very quickly before the overload on the tooth following the broken one causes the damage to spread.

The case of tool wear is rather different from the three preceding ones. Tool wear develops gradually and a certain amount must be tolerated until it starts to harm the milling operation. There are two criteria to define the maximum permissible amount of wear. One is that, from that point on, the wear rate normally accelerates and starts to cause edge chipping. Either a limit value can be set for each planned operation or, instead of the wear itself, its rate of increase is measured. This approach is also useful for schemes designed to optimize the economy of the process. The other criterion applies to finishing operations where tool wear may affect the location of the machined surface and the dimensions of the workpiece. Here, of course, it is better to use a dimensional gauge to determine when the tool should be changed. In the former case, where it is necessary to detect the presence and type of wear, the best approach is to use visual inspection of each tool between cuts. Although some research has been carried out in this area, so far as is known, no system is as yet commercially available. Currently, the usual approach is to register the actual working time of each tool, and initiate tool change when a pre-set total cutting time has been reached. We will not discuss the details of such systems here.

3.3   Description of Individual Systems

The solutions discussed in this section have resulted from research work carried out at the Machine Tool Research Laboratory at the University of Florida. They have been tested in the laboratory and to some extent also in industrial applications. Some of the test results will be included. It should be understood that we have chosen solutions that were considered best when taking into account the published research results of other workers in this field. Other and better solutions may arise following further work by the machine tool research community.

3.3.1   Force Overload: Fast Adaptive Control

Firstly, let us briefly deal with the problem of *sensing the cutting force*. In laboratory work it is conventional to use table type dynamometers with three-dimensional piezo-electric cells. These are indeed excellent elements with very high stiffness-to-sensitivity ratios. Even so, considering the mass of the table and the workpiece, the bandwidth of this type of dynamometer is limited. The frequency response is shown in Fig. 3.2 The graph is plotted in co-ordinates of the ratio $R = F_{meas}/F$ of the dynamometer output over the force acting on the workpiece against the frequency f of the force. It can be seen that there is a peak response of $\times 22.7$ at 365 Hz followed by a dip at 500 Hz, and after 1000 Hz the response drops to about 0.2. There is, of course, also a phase shift which varies strongly with frequency. The calibration varies to some extent with the motion of the force F over the workpiece and it changes with the mass of the workpiece. Strictly speaking, the response is satisfactory only from 0 to 100 Hz.

Apart from its rather limited frequency response, the table dynamometer is also bulky and occupies too much precious workspace. Bishoff *et al.* [1] have described work where the piezo-electric layer was applied directly on to the cutting insert, or under the insert and the shim in the cutter head, in applications for face mills. This solution would certainly yield a very high frequency bandwidth, but it is not applicable to solid end mills and it is rather expensive because the dynamometer needs to be connected to each insert on each face mill, and the connection of the wiring through the toolholder and spindle to the transmitter between the rotating spindle and the housing is a further complication.

We chose to use displacement sensors between the spindle housing and the spindle surface in front of the bearing (see Fig. 3.3). This is a diagram of a complete supervisory system which overcomes all six of the problems listed in Section 3.1. It uses a directional microphone M for sensing tool and workpiece vibrations and sensors S on the spindle. A variety of sensors may be used; we tried inductance probes fed by a 50 kHz frequency and eddy current probes with a 500 kHz carrier. The latter sensors need a highly conductive ring of aluminium or silver coating on

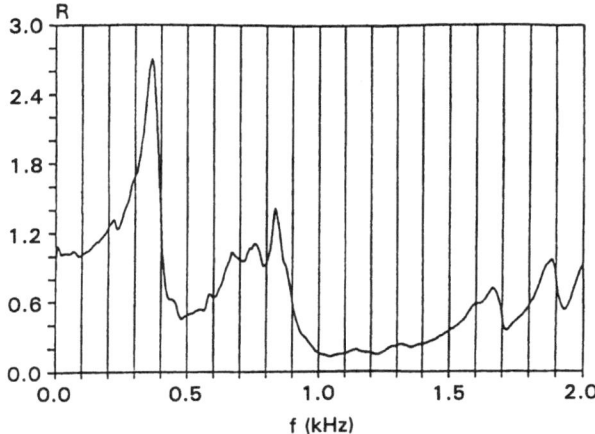

Fig. 3.2. Frequency response of a table dynamometer.

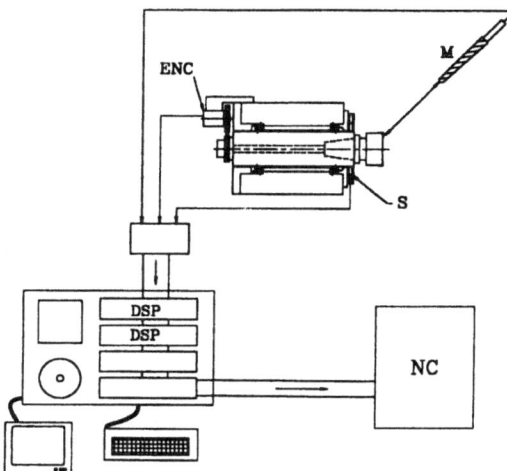

Fig. 3.3. The supervision system for milling.

the spindle. In both instances two pairs of probes were used along the X and Y directions, in a differential connection; this eliminates the effects of thermal spindle growth on the DC part of the signal. A resolution of 0.5 μm is achievable in practice. The frequency response is again rather limited – it is determined by the transfer function between the force acting on the cutter and the relative displacement between the spindle and the housing. An example of the response is shown in Fig. 3.4 as it applies to a machine tool with a spindle having 100 mm bore bearings, and an M50 taper and face mill diameter of 100 mm attached to a 150 mm extension holder.

The graph is analogous to that shown in Fig. 3.2, except that a factor of 0.5 was used. The response is affected by the headstock and spindle modes, the lowest being

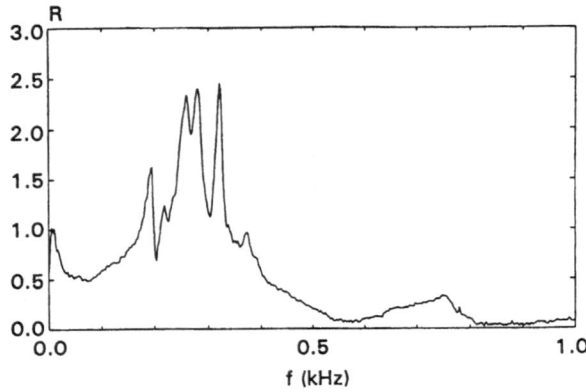

Fig. 3.4. Frequency response of force sensors on the spindle.

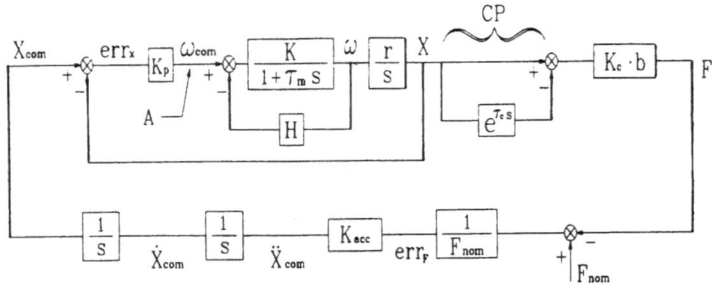

Fig. 3.5. Block diagram of the constant force adaptive control.

at about 18 Hz. This peak is unusual, and would not be found on most machine tools. In this particular case it is caused by the heavy headstock vibrating vertically on the spring of the leadscrew. The response may again be acceptable within the range of 0–100 Hz. Outside this range there is a ×2.5 peak at 280–330 Hz and after 500 Hz the response is strongly attenuated. This result was obtained for a rather heavy tool; the bandwidth would be wider for a lighter tool. It will be shown later that we are not dealing so much with actual forces as with features of force variations, and that these tend to be reasonably preserved through the practical frequency ranges. In any case, the response obtained from the displacement sensors on the spindle is just as bad as that from the table dynamometer, although these sensors are much less invasive of the workspace of the machine.

It will be explained in Section 3.3.3 that for the cutter breakage algorithm the *force must be sampled in synchronization* with the spindle rotation. However, it is very useful for the force overload problem to establish an average force per tooth period AV, the moving sum of p samples where $p = P/m$ and P is the number of samples per spindle revolution. Generally we take $P = 120$; so for a cutter with 6 teeth, $p = 20$. The AV signal is re-established for every single sample. The sampling is triggered from an encoder ENC (Fig. 3.3) attached to the spindle. In high-speed spindles where it might be difficult to drive the encoder, it is usual to machine serrations in a band of the spindle shaft and use a proximity sensor to supply the sampling signals. There is also a separate once-per-spindle-revolution synchronizing pulse.

We shall now describe a *force overload elimination* scheme as part of a "*constant force adaptive control*". This may be useful in die sinking, in the finishing operation, using a long, small-diameter end mill [2]. The tool has to remove material that has been left after a roughing operation where a larger-diameter tool was used that could not enter into some of the corners of the die shape. Consequently, the finishing tool is idling most of the time but occasionally runs into chunks of material. To save time the idling motion is performed in rapid traverse. As the tool enters a cut it is necessary to slow down quickly to such a feedrate that the resulting force on the cutter is kept just below the maximum force which would snap the end mill shank at its root: this is termed the "nominal force F_{nom}". A block diagram of the system is shown in Fig. 3.5.*

The NC positional command X_{com} is input to the positional discriminator where the encoder feedback X is also obtained. Their difference is the tracking error err_x. It is amplified by gain K_p and used as the speed command ω_{com} for the servomotor. The speed feedback loop includes the tachogenerator with gain H. The motor is modelled as first-order system with gain K and motor time constant τ_m. The servomotor with the tacho feedback responds quickly with a time constant typically of about 8–10 ms. Between the motor speed ω and the table displacement X is a leadscrew and nut transmission ratio r and integration. The positional feedback loop X/X_{com} is approximately second order and, still more approximately, of dominant time constant that is much longer as a result of the integration – it may typically be 100 ms. The cutting process generates the force F which can be measured and compared with the desired force F_{nom}. The difference $(F_{nom}-F)/F_{nom}$ is the relative force error err_F. If the force F is smaller than F_{nom}, the error is positive and we want to go faster, i.e. we want to accelerate. Most simply, we establish a commanded acceleration $\ddot{X}_{com} = K_{acc} \times err_F$. If $F > F_{nom}$, the error is negative and we decelerate. The variable $\ddot{X}_{com}$ is integrated twice to close the force loop and establish X_{com}. This loop is rather slow because of the two additional integrations. The gain K_{acc} must be kept low to avoid instability, and the response may typically be expressed by a dominant time constant of 1 s.

As described, the system would fail upon the sudden entry of the tool into the cut since the transient force would rise to about a value which would correspond to milling with rapid traverse, and therefore the tool would break. To prevent this happening a special, non-linear algorithm is used. As soon as a pre-set rate of increase of the AV force signal is detected, the connection between K_p and ω_{com} is interrupted at point A. This amounts to a step command down to zero speed applied to the tacho feedback servomotor loop. It responds quickly. At the same time the generation of the command X_{com} is stopped. After a pre-set period when ω is close to zero, the whole system should be reconnected. However, a large error err_x has accumulated in rapid traverse and this has not dissipated during the brief slow-down of the motor. Hence the system would jump on reconnection. The number err_x must be taken out of the positional discriminator and returned to the command generator, or some other technique must be used to empty the discriminator. For example, a

---

* The diagram in Fig. 3.5 should have an input from the CNC controller. The measured value of the cutting force should only influence the speed of movement – the value of the feed. [Editor]

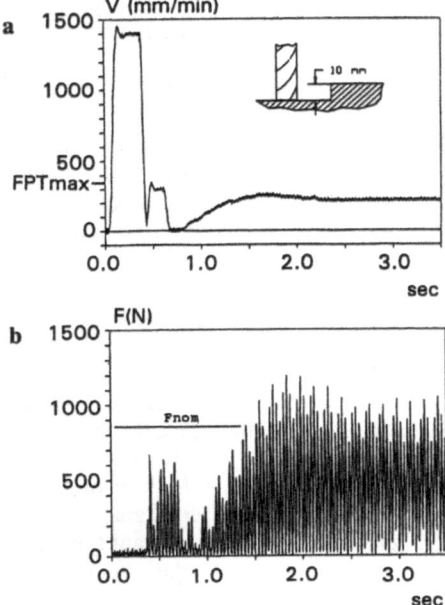

Fig. 3.6. Experimental cut in aluminium with force constraint: **a** feed; **b** force.

low-voltage command is brought from outside to the servomotor, which moves slowly while X_{com} is held constant. After this emptying period the system is reconnected and X_{com} restarted.

The functioning of the system is illustrated in the records of Fig. 3.6. The upper graph is the tachogenerator signal representing the feedrate. Away from the workpiece the tool starts in idling motion. It then accelerates to the rapid traverse rate. The lower graph is the force signal. The force is zero during the idling tool motion. As soon as the force rises, on contact of the tool with the workpiece, the stop routine is triggered. It is followed by a period of slow feed and low cutting force during which the value err_x in the discriminator decreases. When it has decreased to zero the system is reconnected. On reconnection, the system slowly accelerates into the cut and the force is regulated to the level of the pre-set $F_{nom} = 850\,N$.

Consider how the system might be affected by the distorted frequency response of measuring the force as illustrated in Fig. 3.4. Obviously, at the first tool contact we are not much interested in the actual value of the force. A rate of increase of the sensor signal may be determined experimentally for various tool sizes. When it comes to regulating to F_{nom}, as we are already dealing with a slow response servo we may deliberately low pass the signal to deal with the quasi-static (average) force for which the calibration is known.

More often, larger-diameter end mills are used which would not snap except at very high forces. The problem is then breakage of the cutting edge. This does not depend on the resulting cutting force but on the force per unit length of the cutting edge. To maintain its desired value means maintaining a pre-determined safe value of the chop load. The system is, therefore, most often used to regulate the transition

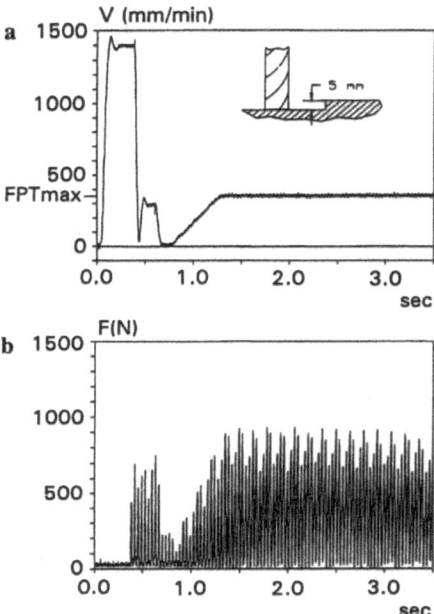

Fig. 3.7. Experimental cut in aluminium with feed constraint: a feed; b force.

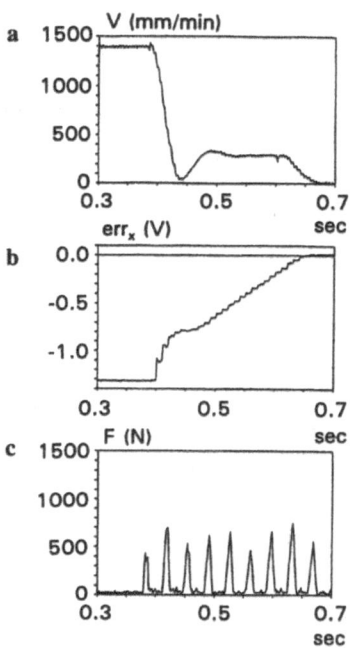

Fig. 3.8. Details of the record of Fig. 3.7: a feed; b positional error; c force.

from rapid traverse to a pre-set value of working feedrate. The force feedback loop of Fig. 3.5 is not used and the block diagram reduces that of a positional servo with an internal velocity loop. However, the same special fast stopping routine is utilized as was described previously. An example of this function is shown in Fig. 3.7. The graph is similar to the one for constant force. A detail of the first part of the record is shown in Fig. 3.8 on an expanded time scale. The record in the middle is the value err_x, initially -1.4 V. Gradually it increases up to zero, upon which the system is reconnected.

At any given time the algorithm operates in one of two possible states, the fast-feed or the nominal feed state. The system performs simple processing on spindle displacement data to produce two features that tell it whether or not actual milling is occurring; based on this result, the algorithm knows which feed state should be current. It uses the displacement probes with synchronous sampling of the D signal. This signal is used as the fast-acting indicator to detect tool collision with the workpiece. When in the fast-feed mode, it is continuously compared with a threshold value to ensure that only small vibrations are sensed. When this threshold is exceeded, the algorithm declares the "cut-in-progress" state and the transition from fast-feed to the nominal feed rate occurs. The algorithm remains in the nominal feed state until a second criterion is met. The second state-determining feature is computed from the average resultant displacement of the spindle over the previous revolution; when this DC feature falls below a minimum threshold value, the adaptive feed algorithm re-enters the fast-feed state.

When the system is in the fast-feed state and initially detects a tool crash, a fast-stop is immediately issued to relieve the rapid build-up of force; following this, the feeds are resumed at the nominal feedrate. When the algorithm re-enters the fast-feed state, the new feed is commanded directly, without stopping. The feedrate changes are performed by commanding a percentage feed override value to the machine tool controller. In this way the system does not actually need to know the feedrate that is being commanded by the part program.

The Fast-Stop routine is also attached to the Torque Overload and Cutter Tooth Breakage functions. It is however not followed by an automatic resumption of the feed motion.

3.3.2   Torque Overload: Spindle Stall

It has already been explained in Section 3.2 that this is an especially dangerous event. As the torque is exceeded and the spindle starts to slow down it may become an unstable, force-runaway case unless feed is stopped more quickly than the spindle. The chip load is obtained as $c = f/(nm)$. Obviously, if the spindle speed n decreases more quickly than the feedrate f, the chip load increases, the overload increases, the spindle slows down faster, etc. It is important first to detect overload and then command Fast Stop without stopping the spindle. This can be stopped later.

The overload is preferably detected by the motor current exceeding a set threshold. This will take different forms for a DC drive than for an AC drive or for a variable-frequency AC drive. Alternatively, the trigger may come from the detection of an incipient slow-down of spindle speed as compared with the commanded speed. The trigger starts the Fast-Stop routine described earlier in Section 3.3.1.

3.3.3   Milling Cutter Tooth Breakage Detection

The cutter breakage detection (CBD) function does not deal with gradual tool wear. It concentrates on the sudden events of tooth breakage that are detrimental to the tool, the workpiece and the machine tool. It should be especially useful for unmanned, unsupervised work on machining centres. Experience has shown that many Flexible Manufacturing Systems operate in practice without an effective CBD function. However, most work on easy-to-machine materials, such as aluminium and cast iron, and use very conservative chip loads (feed per tooth values). At low chip loads (0.05–0.15 mm), tool breakage is a very rare occurrence if correct tool materials are used. The probability of breakage increases with chip load and becomes rather high above 0.3 mm. A robust CBD function will permit an increase of feedrate to about 0.2–0.4 mm chip load on cast iron, and still practically eliminate damage from tooth breakage. For materials difficult to machine (such as titanium and nickel alloys) where tooth chipping and breakage occur even at low depths and widths of cut, any increase of these parameters under protection of the breakage detection systems will be extremely useful.

The CBD function is available in two modes, each having different advantages and characteristics. The operator can select which one to use. Both algorithms use displacement probes to sense the deflection of the spindle at the point closest to the milling cutter (see Fig. 3.3). The sensors are oriented to sense spindle deflections in two orthogonal directions perpendicular to the spindle axis. The displacement

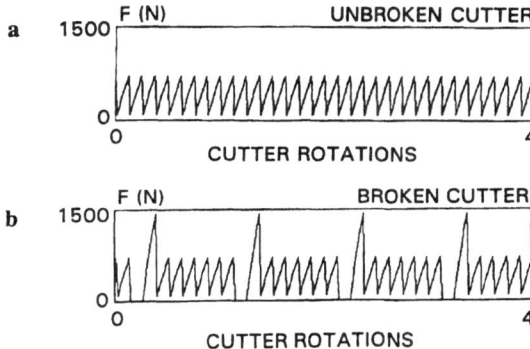

Fig. 3.9. Simulated cutting force; $i = 0.20$, $m = 8$, up-milling: **a** unbroken cutter; **b** broken cutter.

sensors produce signals that are related to the instantaneous cutting force through the transfer function between the tool and spindle at the sensor location. A spindle-mounted encoder provides 120 pulse per revolution clocking which is used to sample the displacement sensors. In each sampling period, the two displacement signals are vector summed to produce a resultant displacement D that is interpreted as the force on the cutter at that instant. The basic feature derived from this signal is the first difference FD obtained as $D_n - D_{n-i}$, where i may be the integer number of samples per tooth period or per revolution. The former approach (per tooth), termed Mode T, is used because it gives an immediate and permanent breakage identification and is robustly insensitive to transients such as entry into cut or milling over a slot. It does not distinguish between the effects of cutter run-out and tool damage on the force. This, however, may be taken as a positive characteristic assuming that a large run-out is as bad as breakage. Mode R (differencing per revolution) is used if it is intended to exclude the run-out. In this mode, however, the breakage detection feature is available only once, during the first revolution after breakage, and then it disappears. Correspondingly, while Mode T in which the tool breakage detection feature remains preserved will trigger an alarm also at the start of a cut, with a cutter that has been damaged previously, or which has been assembled with a large run-out of teeth, this will not happen in Mode R which acts only at the actual instant of breakage. The user of the machine will decide in which circumstances to engage each of these different modes.

Mode T. The first difference in the cutting force between successive tooth periods produces a characteristic signature when one of the cutters is broken or damaged, [3]. Figure 3.9a shows a simulation of the resultant force seen by an unbroken eight-tooth cutter with zero run-out in a quarter immersion up-milling cut. The figure shows the periodic and uniform force pulses produced by successive tooth engagements. Figure 3.9b shows the same cut, but with one broken tooth. The series of force pulses is irregular because the broken tooth does not engage the workpiece, producing no force (for that tooth), and the following tooth engagement must take twice the chip load, producing a large force.

It is obvious that the same signal as the one in Fig. 3.9b would also be obtained if one tooth of the cutter was set with a negative *run-out* equal to the chip load; it would be hidden behind the preceding tooth and the following tooth would see

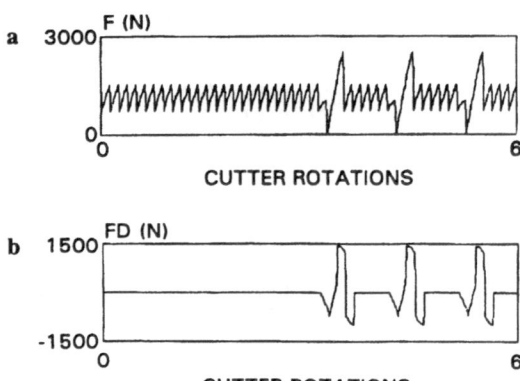

Fig. 3.10. Cutting force and the FD feature for simulated cut; $i = 0.50$, $m = 8$, up-milling: **a** cutting force; **b** FD.

double the chip load. The graph shown in Fig. 3.9a has been generated assuming that all the teeth of the cutter are perfectly set on exactly equal radii. This is never the case in practice and, depending on the amount of run-out, the system in Mode T will trigger the alarm equally on large run-out and breakage. This could be seen as an advantage because a large run-out causes as much overload on the highest tooth as a broken and missing tooth and, justifiably, the cut with such a large run-out should be prevented. If, however, for any reason the user felt otherwise and wanted to distinguish clearly between run-out and breakage, mode R should be used, as described later.

Figure 3.10a shows the simulated force for a 0.2 radial immersion cut with a breakage occurring in the middle of the plot. Figure 3.10b shows the first-difference FD signal obtained by subtracting from every sample signal the sample recorded at an instant one tooth period back. It is seen that over the region without breakage, FD is zero but the FD signal exhibits a negative–positive–negative variation in the region after the tooth breakage. The positive FD values are summed over a period, as well as the negative peaks, to produce the first-difference sums FDS that are used for further processing. It is important to notice that the characteristic negative–positive–negative FD sequence remains for the rest of the cut, and it would still be there if we started a new cut with the damaged cutter.

Because of the equivalence of the effect of the radial tooth run-out and cutter breakage, the FD signal is almost never zero or even close to zero. It is, therefore, necessary carefully to derive the threshold value for the FDS signal at which the alarm will be triggered [4]. The magnitudes of the FS and FDS signals depend on the axial depth of cut b and on the chip load c; it is actually, in a first approximation, proportional to both of them. To eliminate these effects, signal DC is produced as the average of AV over one spindle revolution. It is also proportional to b and c. Both FD and FDS are then divided by DC. The result of these actions is illustrated in Fig. 3.11 for depth of cut b, and in Fig. 3.12 for the chip load c. Furthermore, the FD and FDS signals are also affected by the radial immersion i and by the number of teeth on the cutter m. This is illustrated in Fig. 3.13. The number m is known beforehand, but radial immersion i is not known and it may vary during the cut. It

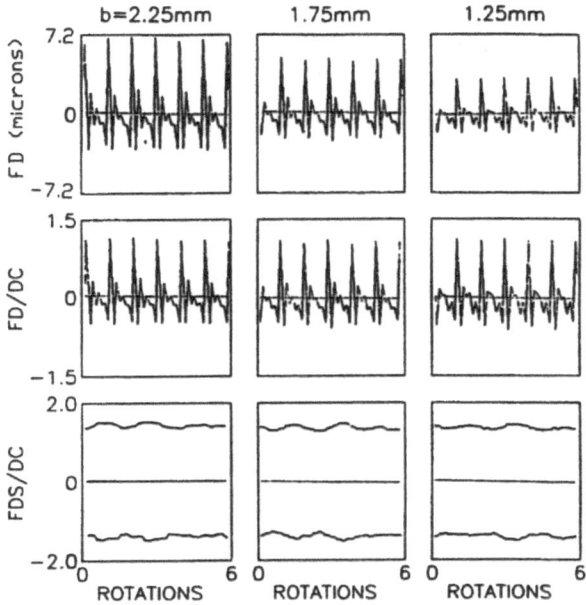

Fig. 3.11. Dividing FD and FDS by DC eliminates effects of changing axial immersion b. Conditions: $i = 0.25$, $c = 0.25$ mm, $n = 600$ rpm, milling cast iron, $m = 8$.

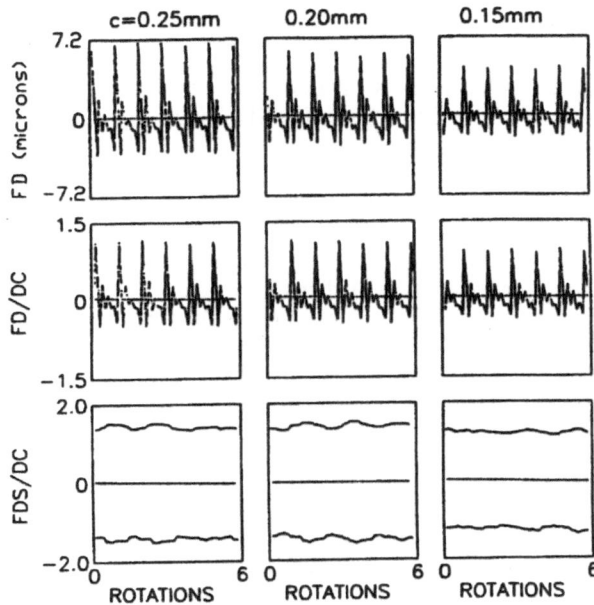

Fig. 3.12. Dividing FD and FDS by DC eliminates effects of changing chip load c. Conditions: $i = 0.25$, $b = 2.25$ mm, $n = 600$ rpm, milling cast iron, $m = 8$.

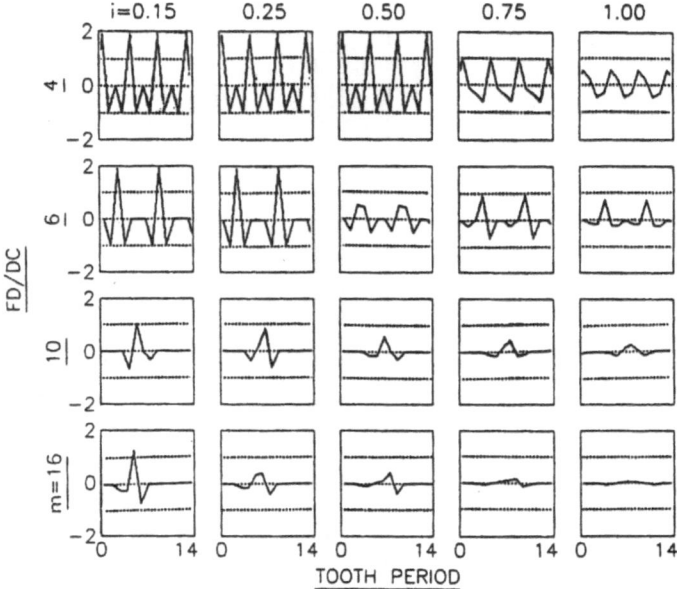

Fig. 3.13. The cutter breakage signal level in FD is related to the radial immersion i and the number of teeth m on the cutter, $b = 2.25$ mm, $n = 600$ rpm, up-milling.

can be derived from the ratio of the average max.–min. displacement range per revolution AC versus the value DC. In our system the signals AV, FD, FDS, DC and AC are combined to form the features FDS/DC and DC/AC in order to determine the threshold value which remains invariant through changes of b, c and i. We speak about an "adaptive threshold".

Now let us look at examples of records that illustrate the functioning of Mode T. Firstly, records will be shown for cutters with very small run-outs. In these examples the signal will be processed only through the FD parameter in order to show better the distinction of cutter damage.

In Fig. 3.14 the AV and FD signals are shown for a complete cycle of entry, steady state and exit of a cut in cast iron with a 4 inch face mill with 8 teeth, 0.1 inch depth, 2.0 inch width of cut, 0.010 inch chip load, at 600 rpm. On the left side are the records for an undamaged cutter with very low run-out and on the right side those for a cutter with one grossly damaged tooth. It is seen that the FD signal distinguishes clearly between the two cutters.

The algorithm is robust enough and not disturbed by transients such as entry into a cut or exit from it, or even milling over a slot – see the records in Fig. 3.15. On the left is the per tooth average AV and below it the first difference FD, for an undamaged cutter. On the right are AV and FD for a cutter with a missing tooth. The damage was already there before the cut occurred.

In Fig. 3.16 tooth breakage and its spread over other teeth has been captured. At the top is AV and below it the first difference FD. Breakage 1 occurred at the end of the entry into the cut, and then spread to teeth 2 and 3.

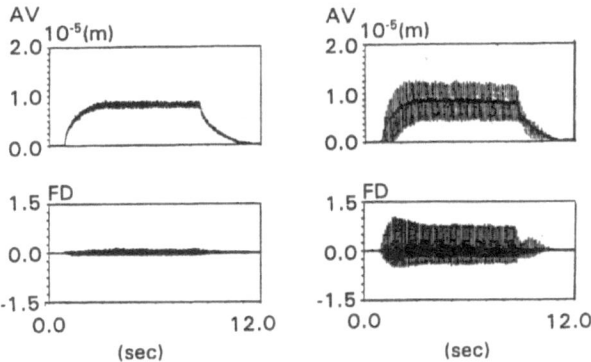

Fig. 3.14. Average displacement AV and first difference FD of the signals in the complete cutting cycle.

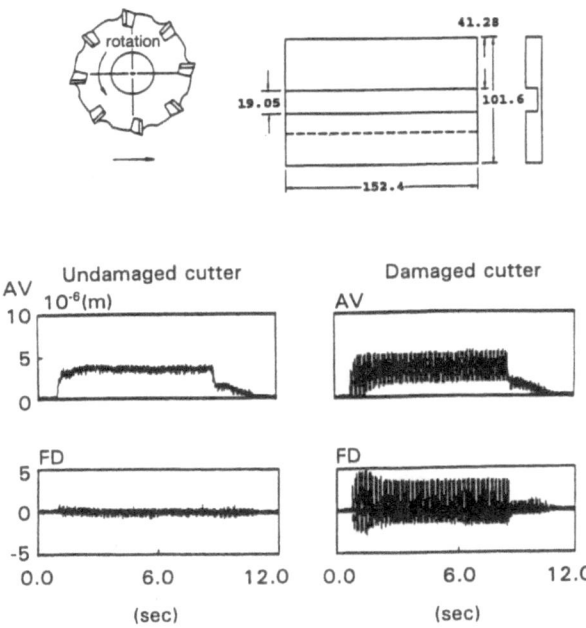

Fig. 3.15. Average displacement AV and first difference FD signals during milling over a slot.

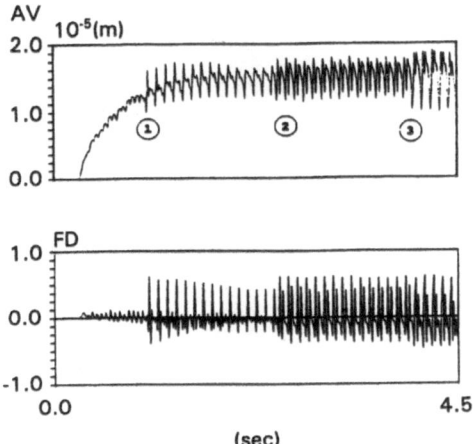

Fig. 3.16. Average displacement AV and first difference FD signals with three consecutive tool breakages.

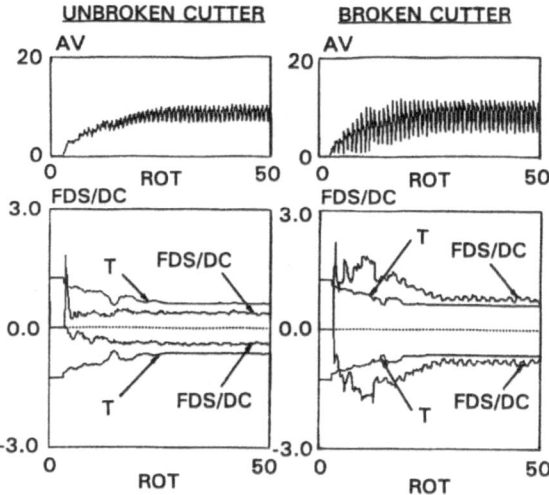

Fig. 3.17. Typical cutter breakage tests with an unbroken and a broken cutter; demonstrates adaptive threshold with entry transient. Mode T.

Signals obtained in tests with cutters with a common amount of run-out and processed through the automatic thresholding routine are shown in Fig. 3.17 and Fig. 3.18. These tests were carried out with a face mill of diameter 105 mm with 8 teeth, milling cast iron at 600 rpm in depth of cut 2.5 mm and width of cut 42 mm. Both figures are arranged so that the left plots are of an unbroken cutter and the right plots are of a cutter with one tooth missing from the beginning. The AV feature is shown in the top plots, and the bottom plots contain the FDS/DC feature with

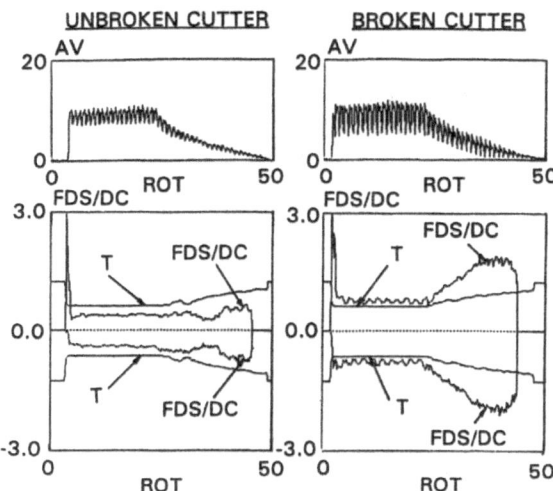

Fig. 3.18. Typical cutter breakage tests with an unbroken and a broken cutter; demonstrates sudden entry and exit transients. Mode T.

thresholds labelled T. Results in Fig. 3.17 are for a cut with a regular entry transient. The system automatically assumes a non-cutting condition when the DC level is below 1 micrometre and the FDS/DC feature is then set to zero. This must be done to keep the FDS/DC from becoming unusually large as the DC feature becomes very small. Therefore, the record begins with a zero value. Upon entry, the cutter violates the positive threshold but not the negative one, and this is then not interpreted as breakage. The FDS/DC signal remains within the threshold limits throughout the rest of the record. With the damaged cutter, however, both thresholds are violated at entry and remain so throughout the cut. The automatic alarm and stop function was deliberately disconnected to make this record possible. Results of the test in Fig. 3.18 are very similar; this is a cut with sudden entry and regular exit from the cut. As in the previous case, breakage is not signalled for the unbroken cutter but it is correctly identified for the damaged cutter.

Mode R. In this mode, after signal conditioning through a once per tooth averaging, the signals are differenced with signals one cutter revolution back [5]. Reproducing in Fig. 3.19 the graph of Fig. 3.10a, we can easily see that before breakage the difference indicated by arrow 1 will yield zero; at the instant of breakage the difference indicated by arrow 2 will give a negative value followed, arrow 3, by a positive value. After another revolution, see arrows 3 and 4, the signal goes down to zero again: breakage is now forgotten. However, it is obvious that any run-out is eliminated because every signal subtracts from the signal on the same tooth one revolution back.

The actual technique is illustrated using simulation runs for a face mill with 8 teeth. Figure 3.20a shows breakage during entry into a 20% radial immersion cut. Run-out is 5% of chip load and noise is 5% of cutting force. The left graph is the filtered signal AV and the right graph is first difference FD. The breakage event is

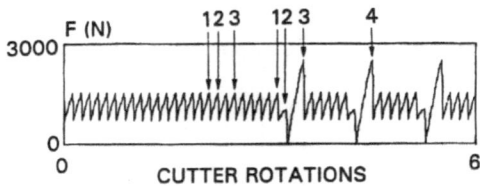

Fig. 3.19. Cutting force for simulated cut; $i = 0.50$, $m = 8$, up-milling. Mode R explained.

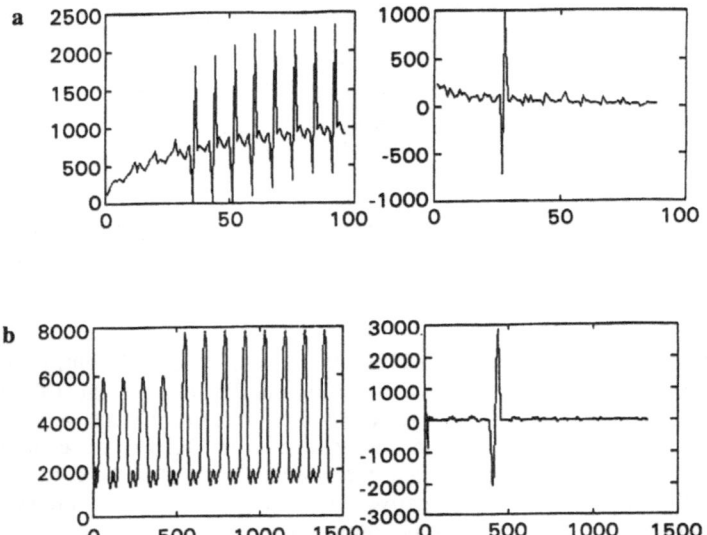

Fig. 3.20. Simulated breakage for 8 tooth face mill: **a** during entry into 20% radial immersion cut; **b** entry into 90% radial immersion cut. Mode R.

clearly seen in the minus–plus sequence. It is also seen that this feature disappears after one revolution. Figure 3.20b is an entry into a 90% radial immersion cut with a cutter with a huge run-out of three times the chip load. On the left is the signal AV and on the right is FD. This example illustrates how well this mode disregards run-out. The breakage feature is again well recognized.

It should be repeated here that, unlike in Mode T, if the cutter were broken before the cut was made, the system would not recognize it in Mode R.

In Mode R, the problem of thresholding is minimized. Because of the run-out effect, elimination of the signal is normally very small and clearly stands out upon breakage. The problem to deal with is that of transients. However, even that can be overcome very easily.

An example of an experimental record is shown in Fig. 3.21. This applies to a cut in steel with a 100-mm diameter face mill with 8 inserts, axial depth of cut 2.5 mm, 25% radial immersion, 400 rpm and 0.5 mm in chip load. The record shows the

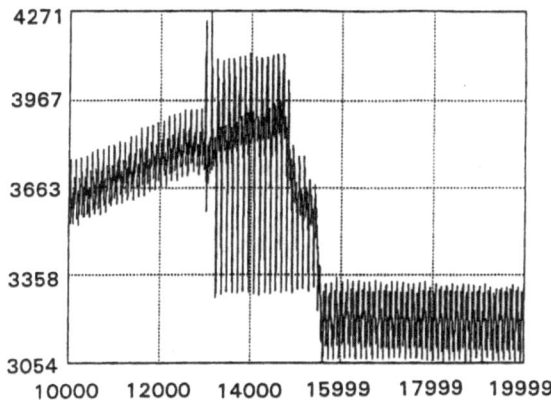

Fig. 3.21. Experimental record of cutter breakage for cut in steel with a 100 mm diameter face mill with 8 inserts. Axial depth of cut 2.5 mm, 25% radial immersion, 400 rpm and 0.5 mm chip load. Mode R.

averaged signal AV. At about sample 13 000 a tooth broke. At 15 000 it starts exiting the cut. The FD signal is shown in Fig. 3.22a and in an expanded scale again in Fig. 3.22b. The latter graph shows unusual behaviour. At the instant of breakage B1 we see a plus–minus sequence, and after another two revolutions at B2 a large minus–plus sequence. We can only surmise that the broken piece first jammed and, instead of giving a missing signal, it produced an increased signal. Full breakage did not happen until B2. In the record each vertical dotted line marks one revolution. The feed stopping routine was disconnected so as to make the record possible. In the regular function the system, which watches for a violation of both the negative and positive threshold within one revolution, would already have triggered the alarm at B1.

3.3.4 Chatter Recognition and Control (CRAC)

The CRAC system *detects chatter*, as distinguished from regular forced vibrations, and stops the feed rate in a fast action [6, 7]. In this way it protects the spindle and tool against the detrimental effects of the high amplitudes of vibration and of force associated with chatter. It is much more clearly suited for this function than other probes which do not distinguish chatter and react to a certain vibration velocity, the threshold of which cannot be set by any logical criterion. These other kinds of probes may not detect chatter but at the same time shut down on a regular machining operation.

Furthermore, the system *helps to eliminate chatter* in high-speed milling operations. It will not suppress chatter at all depths of cut, but it will increase the chatter-free metal removal rate (MRR) by regulating the spindle speed into zones of high stability against chatter. These zones are encountered only in high-speed milling operations such as the milling of aluminium or brass, or using Si_3N_4 tools on cast iron.

The basic significance of stiffness for chatter limits is generally valid and therefore CRAC is especially valuable as a help in situations where stiffness of the

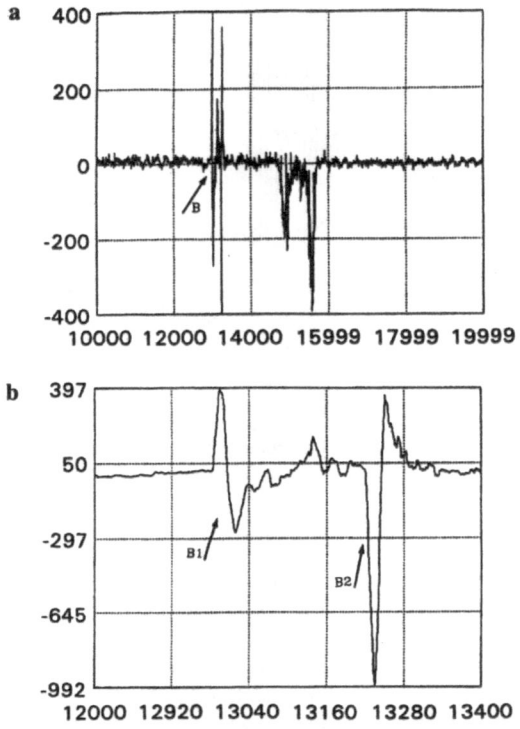

Fig. 3.22. The FD signal: **a** unexpanded; **b** expanded. Mode R.

tool is low, such as end milling (long, slender cutter bodies) and milling with a cutter attached to the end of a long extended spindle or a long, slender spindle attachment.

In summary, the CRAC system is recommended when all the following conditions apply:

1. High-speed milling of aluminium or copper alloys, or milling of cast iron with Si_3N_4 inserts.
2. Limited stiffness: using long end mills, milling with cutters at the end of long extension spindles.
3. Operations with a high proportion of milling time.
4. The machine is intended for universal work.

When applied under the above conditions, the system is an excellent means for improving the quality and quantity of work. It is not a substitute for other more fundamental measures such as ensuring the maximum possible dynamic stiffness between the tool and the workpiece by using the best spindle design and the shortest possible tool holder and tool. However, the system goes beyond that, and by selecting the most stable speed, it produces an improvement of the cut area by 1.5 to 4 times, depending on the complexity of the dynamics of the structure of each particular case.

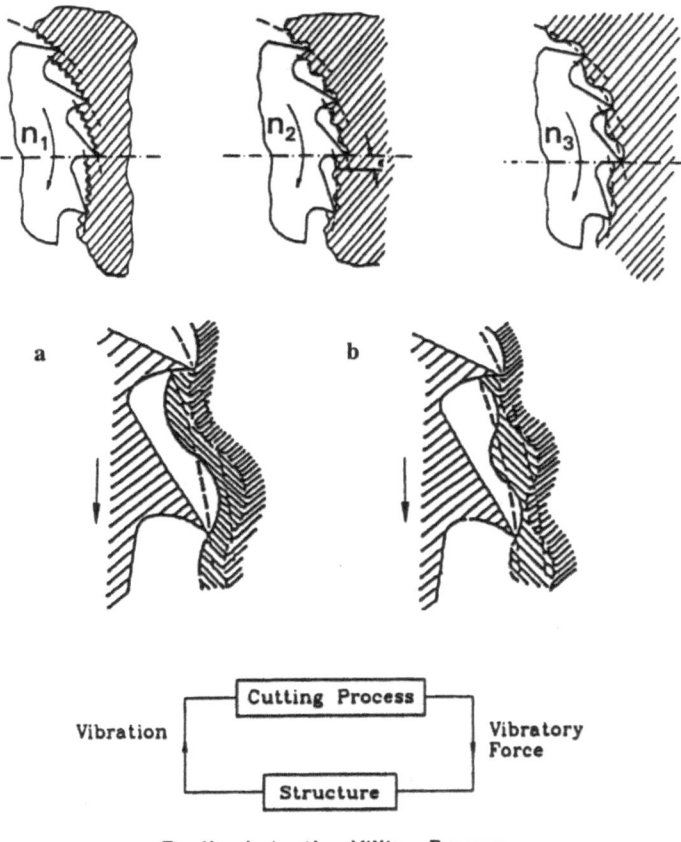

Feedback in the Milling Process

Fig. 3.23. Regeneration of surface waves in high-speed milling.

Chatter is a self-excited kind of vibration which is detrimental to tool life and spoils the machined surface by producing "chatter marks". The self-excitation comes from the regeneration of waviness on the cut surface – see Fig. 3.23. If the tool vibrates, each tooth leaves behind a wavy surface and the following tooth cuts into it and "regenerates" the undulations. The structure and the cutting process are involved in a feedback loop: vibration causes a modulation of the cutting force and the variable cutting force, in turn, excites vibration.

Whether or not this self-excitation effect occurs, i.e. whether this vibration occurs, depends mainly on the dynamic stiffness of the machine–spindle–holder–tool structural system on one hand, and on the "gain" in the cutting process, on the other hand. This gain is obviously directly proportional to the depth of cut b – see Fig. 3.1; by doubling it, the same vibration produces twice the variable force. For a sufficiently low depth of cut, chatter does not occur; by increasing b, a limit is reached above which chatter develops. A similar effect is produced by increasing the width of cut a and by increasing the number of teeth m on the cutter; both these changes increase the number of teeth simultaneously participating in the regenerative process. All these parameters are included in the equation (3.2) for metal removal rate:

$$M = abcmn$$

where c is the feed per tooth (chip load) and n is the spindle speed, which will be discussed separately in the next section. Here we see that trying to increase M by increasing either a or b or m causes chatter to occur: chatter is a strong limiting factor on M. Moreover, the machined surface is spoiled by "chatter marks". The actual M limit is of course dependent on the dynamic stiffness of the structure.

The only cutting process parameter which may give both an increase of M and an upgrade of the chatter limit, with regard to the depth and width of cut, is the spindle speed n.

Let us now return to Fig. 3.23 which depicts the regeneration of waviness on the cut surface. The frequency of the vibration does not essentially change with speed or any other one of the cutting conditions; it is dictated by the dominant natural frequency, f_n, of the structure and it is very close to it. Most often, this frequency depends on the design of the spindle and on the mass and flexibility of the tool and holder. Typically, for a 100 mm diameter face mill on an M50 taper holder, the frequency is between 550 and 700 Hz, and for a 25 mm diameter end mill 100 mm long, it is between 1600 and 2500 Hz. When the spindle speed is changed, the wavelength of the undulations changes in proportion. At low speed, n_1 in the top row of Fig. 3.23, the waves are short but frequent between successive teeth. At higher speed, n_2, they are longer, and at yet higher speed n_3, longer still. In case n_1, the flank of the tool may be rubbing on the steep down-slopes of the wavy path, and this produces "process damping" which prevents chatter, but this effect is lost at higher speeds.

At high speeds, the waves may become so long that there is only one full wave between teeth – see Fig. 3.23a; in this case, the "tooth frequency" $f_1 = nm/60$ equals the natural frequency f_n (the "resonant" case) and, as is seen, despite vibration there is no chip thickness variation, no force variation and no self-excitation. In Fig. 3.23b, one and one-half waves fall between successive teeth and the same magnitude of vibration causes a strong chip-thickness variation. These illustrations show that at high speeds, in the range around resonance ($nm/60 = f_n$), a change of speed will have a strong effect on chatter. The effect of spindle speed is summarized in Fig. 3.24.

In this graph, the "limit depth of cut" b_{lim} is plotted against spindle speed n. Below the border line, cutting is stable and above it chatter arises. In the very low speed range A, process damping suppresses chatter more as the speed is lowered. It is possible to use large depths of cut but M is low anyway because of the correspondingly low feedrate. In the middle range B, the limit depth of cut is lowest and is indicated as the "critical" depth of cut b_{cr}. M is poor because of the low value of b. In range C of high speeds, zones of increased limit depth of cut occur. The highest one is at resonance, speed n_1.

For illustration purposes, the actual speeds are noted for two cases:

1. An end mill with $f_n = 1880$ Hz and $m = 4$ teeth; resonance occurs at 27 000 rpm, smaller stability pockets are found also at 13 500 and 9000 rpm. Process damping acts below 1400 rpm.

2. A face mill with $f_n = 600$ Hz and $m = 10$ teeth; resonance at 3600 rpm, lower stability pockets are at 1800 and 1200 rpm and process damping would act below 114 rpm.

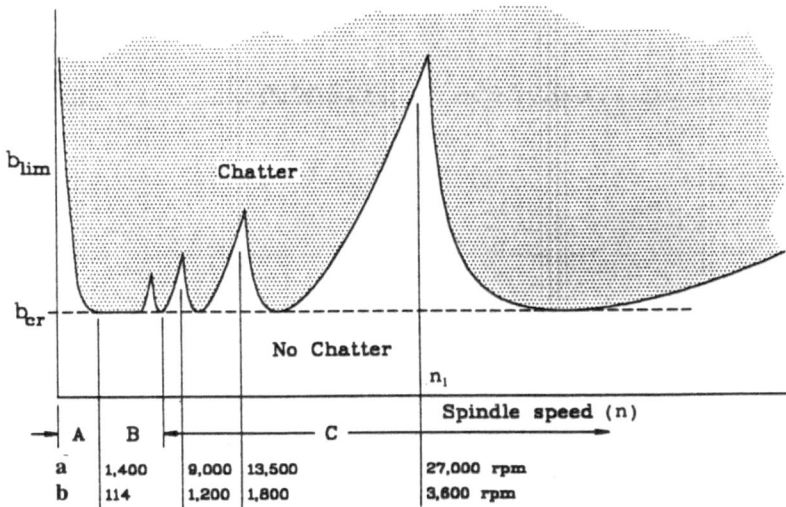

Fig. 3.24. Stability diagram: a end mill with f_n = 1880 Hz and m = 4; b face mill with f_n = 600 Hz and m = 10.

The CRAC system will detect chatter almost immediately, stop the feed, and automatically regulate spindle speed into the highest available stability zone. The system uses a microphone as the primary transducer to detect the sound generated by the milling operations. A spindle-mounted encoder is also used to provide speed information to the system.

The microphone and encoder data are processed using a DSP board which is hosted by an 80486 ISA bus computer that supports the operation of the DSP board, performs display and console functions, and communicates speed, feed and feed-state commands to the CNC controller of the machine tool. The system receives commands from the part program as well as its local console to activate and reset the spindle speed regulation function.

The sound signal is first passed through a filter which removes all periodic components that occur one, two, three times etc. per revolution; these are caused by the run-out of the cutter and by the periodicity of the teeth in a normal chatter-free operation. The onset of chatter is indicated by an increasing periodic signal whose frequency is not an integer multiple of the spindle speed.

The basic algorithm of the system is such that it will select a spindle speed n (rev/s) at which the tooth frequency ($n \times m$), where m is the number of cutter teeth, will be equal to the chatter frequency just calculated. This is the speed n_1 of the highest stability pockets in Fig. 3.24. If this speed is too high and exceeds the speed range available, one-half or one-third of that speed is chosen.

The system acts as a supervisory co-processor to the CNC controller, feeding it optimally stable speeds as they become available. When the system detects chatter, it stops the machine tool axis feeds, and commands the new optimal spindle speed to the CNC controller. When the spindle settles at the new speed, axis feeds are resumed and the system continues to listen for subsequent occurrences of chatter.

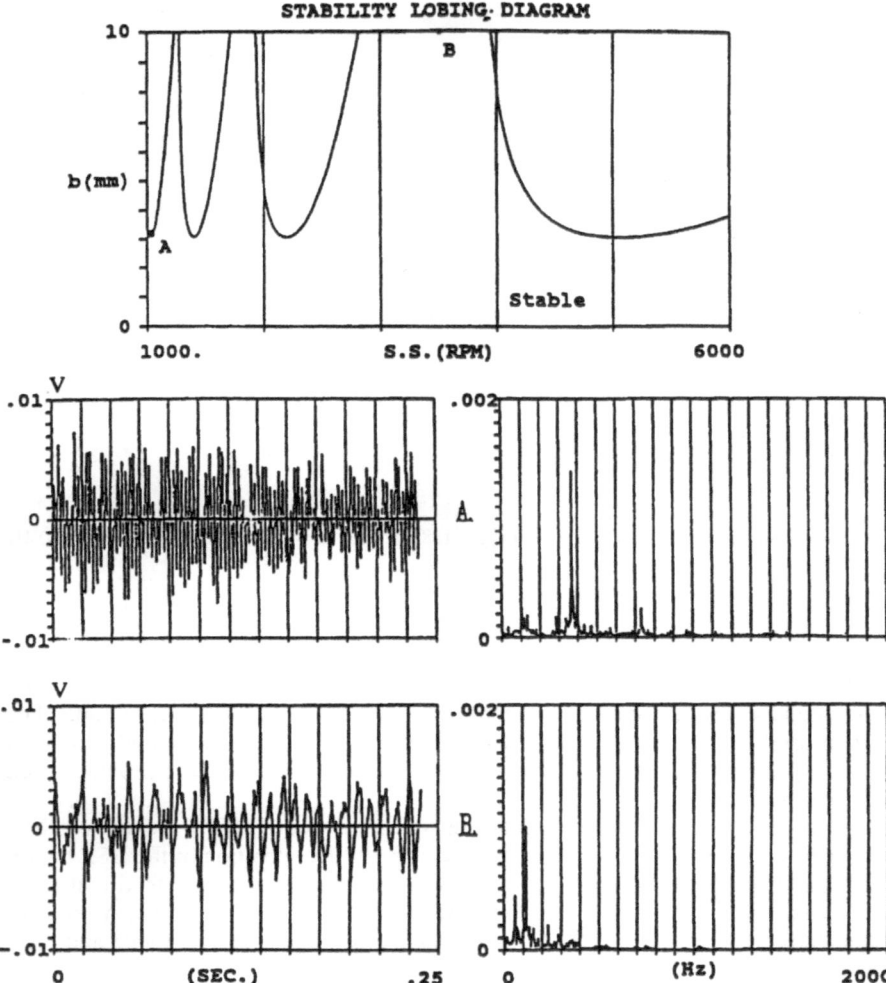

Fig. 3.25. Time and frequency audio records; 100 mm diameter, 6-tooth face mill; 7075–T6 aluminium, $a = 79$ mm.

We will use several examples to illustrate the functioning and performance of the system. In the first case, a face mill of 100 mm diameter and with 6 inserts was used to mill aluminium. For a better understanding, a b_{lim} versus speed diagram was constructed from the measurements of the dynamics on the cutter and this is shown at the top of Fig. 3.25. Bear in mind that the system does not use this information. The starting cut was taken at 1000 rpm and the depth of cut $b = 3$ mm, see point A, and chatter occurred. The record of the sound (left) and its frequency spectrum (right) are shown in plots A. The chatter frequency of 365 Hz is distinguished in the spectrum. The CRAC system regulated to 3475 rpm and the cut was very quiet (not

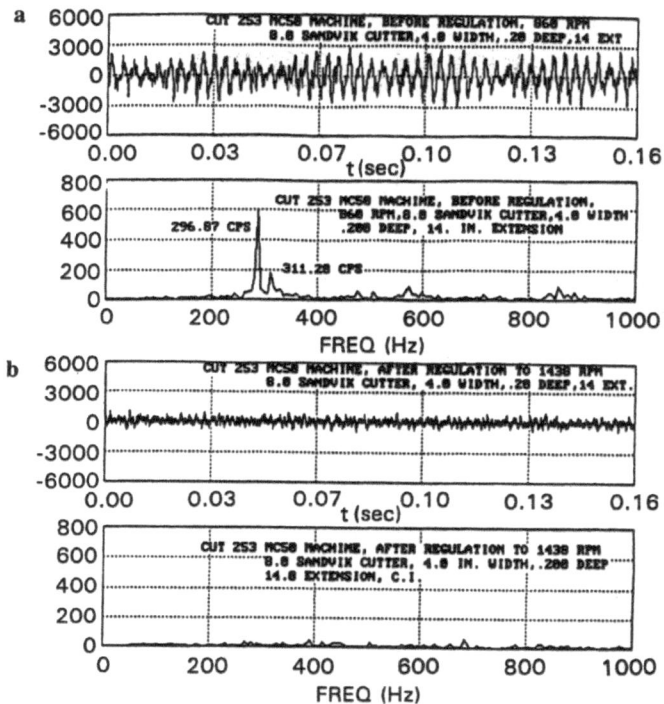

Fig. 3.26. Time and frequency audio records; 100 mm diameter, 12-tooth face mill, a = 100 mm, b = 5 mm, 350 mm spindle extension.

shown). Then, at this speed the depth of cut was increased to b = 10 mm, see point B and also plots B. There is still no chatter. The spectrum contains the run-out component of 3475/60 = 58 Hz and its second harmonic 116 Hz, but no chatter.

The second example concerns the milling of cast iron with Si_3N_4 tools inserted in a face mill of 100 mm diameter and with 12 teeth, attached to the end of a 125 mm diameter boring spindle of a horizontal boring and milling machine, extended by 350 mm. The records of a chattering cut 100 mm wide, 5 mm deep are shown in Fig. 3.26a (sound versus time and below it the spectrum). The chatter frequency of 297 Hz was rather low as it corresponded to the large mass of the cutter and to the long spindle extension. The CRAC system regulated to 1438 rpm – see Fig. 3.26b; vibration was almost completely eliminated.

The third example concerns the high-speed milling of aluminium with a 12.5 mm diameter, 38 mm long end mill with 4 flutes. We will document it in some detail. The transfer functions measured on the tool in the X and Y directions are shown in Fig. 3.27. They indicate a dominant mode at 3250 Hz. This information was used to run a simulation program and to plot peak-to-peak vibration values for spindle speeds between 10 000 and 30 000 rpm versus depths of cuts between 0.25 and 1.1 mm – see Fig. 3.28. The individual lines correspond to depths of cut increasing in steps of 0.05 mm, for a full diameter width of cut. It is clearly seen that at the optimal speed of 25 000 rpm, there is no vibration even at the largest depth of cut, 1.1 mm. In contrast, at 19 490 rpm chatter has already started at 0.3 mm. Other good

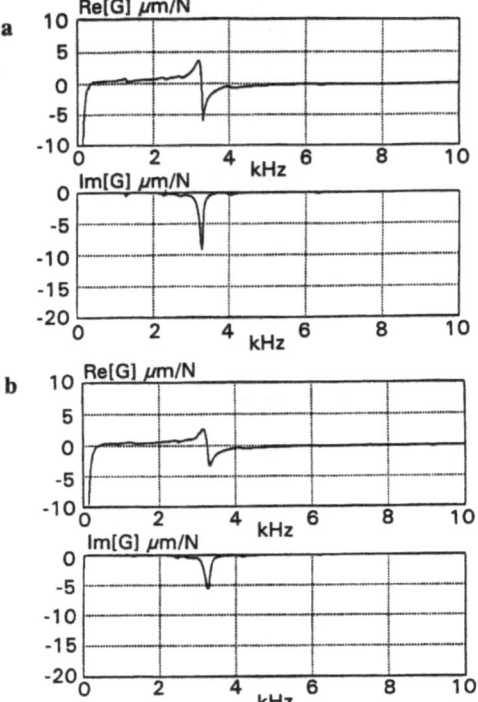

Fig. 3.27. Transfer functions on 12.5 mm diameter, 4-flute end mill: **a** X direction; **b** Y direction.

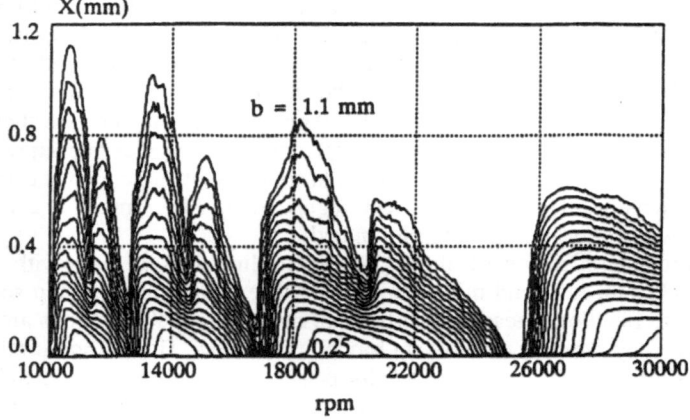

Fig. 3.28. Peak-to-peak vibration for 12.5 mm diameter, 4-flute tool in slotting cut.

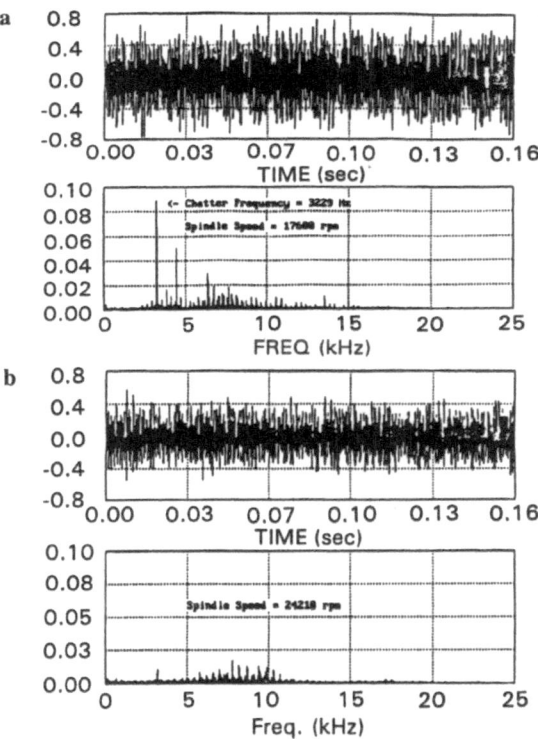

Fig. 3.29. Time and frequency audio records; 12.5 mm diameter, 4-flute tool in slot.

speeds are found at 16 830 rpm and 12 630 rpm; at these speeds, the vibrations are also small. This type of analysis would permit us to recommend the speed of 25 000 rpm. However, without the need for this kind of analysis, the CRAC system detected chatter in a cut at 17 600 rpm (Fig. 3.29a). The chatter frequency of 3229 Hz was detected, and CRAC regulated the spindle speed to 24 210 rpm. Chatter was completely eliminated; the only sound recorded in Fig. 3.29b is the noise produced by the running of the spindle at this rather high speed. Subsequently, it was found possible to increase the depth of cut at this speed by 2.8 times.

3.3.5   Resonant Forced Vibrations in Finish Milling

In finish milling thin-walled structures, such as scrolls for scroll pumps or aircraft panels for fighter aircraft, strong resonant forced vibrations are encountered. These vibrations cannot be suppressed by the algorithm used for the CRAC system for chatter. The problem arises from a combination of strong multiple harmonics of the variable cutting force with the existence of minimally damped modes of vibration of the thin ribs and webs of these integral (one-piece) structures.

In these operations the radial depth of cut a is very low, of the order of 0.1 mm or less while the axial depth of cut b is rather high, of the order of 20–50 mm. The end mills used have diameters of 12.5 mm or less. In some instances, such as those

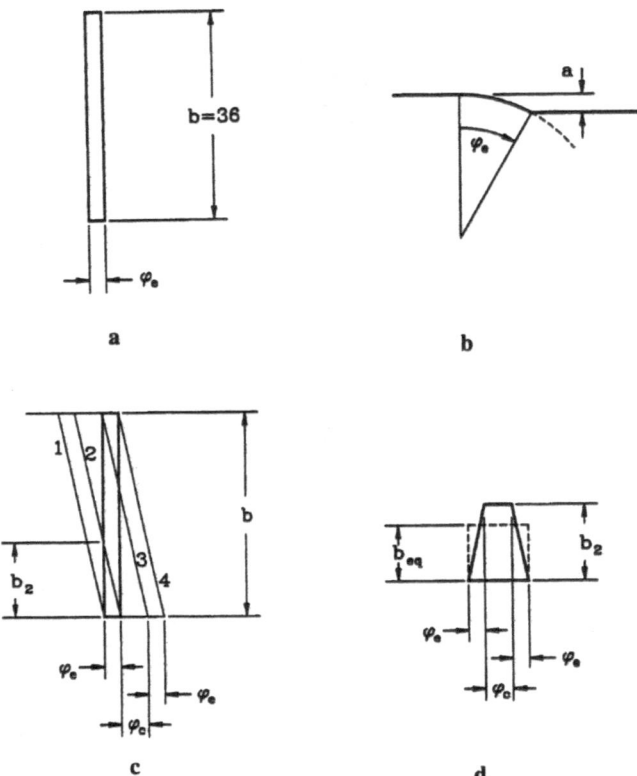

Fig. 3.30. Cutting force variation in fine-finish milling.

where CBN tool edges are used, the tool may have straight edges or edges with a
small helix angle. The tool engagement is explained in Fig. 3.30. In diagrams (a)
and (b) the unrolled area of engagement is shown for a cutter with straight teeth. The
cutting edge on a helical tooth enters in position 1 of diagram (c). As soon as
position 2 is reached the axial depth of the edge engagement reaches b_2. This value
remains constant between 2 and 3, and then decreases and drops to zero at 4. The
effective depth of cut is shown in diagram (d). Assuming a cutter with two teeth,
these pulse-like forces are separated between tooth engagements. For a cutter with
a helix angle of 2°, tool diameter of 12 mm and $a = 0.08$ mm, the trapezoidal pulse
is defined by $\phi_e = 9.4°$, $\phi_c = 3.7°$ and $b_2 = 15$ mm.

The cutting force will have a rather low DC component and a number of variable
harmonic components. The basic harmonic component has the tooth frequency f_t
and is shown as the first spectral line in Fig. 3.31a which applies to the cutter
with two straight teeth at $n = 1250$ rpm. The horizontal scale is in Hz, the vertical
scale is arbitrary. For milling of cast iron, the first harmonic would be about 90 N.
The graph shows that the first six harmonics are still rather strong. A similar graph,

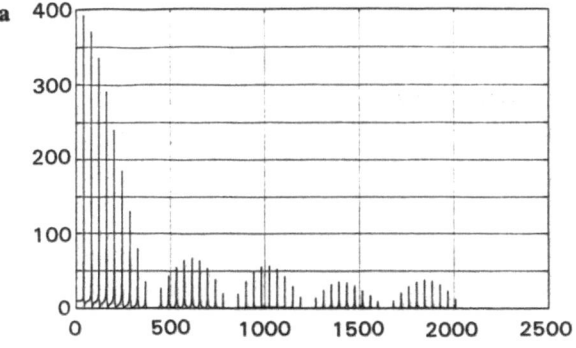

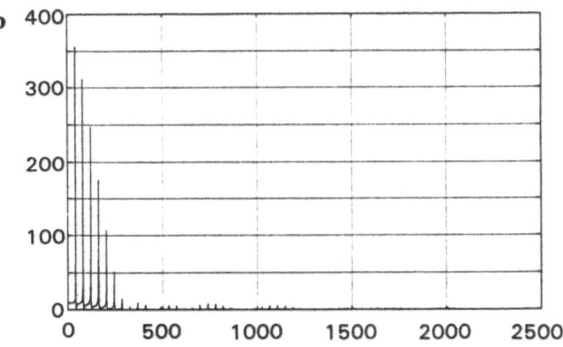

Fig. 3.31. Harmonic contents of the milling force.

Fig. 3.31b, corresponds to the teeth with a helix angle of 2°; however, the first four harmonics are strong.

The mode shapes of a thin-walled rib, 30 mm deep, 1.5 mm thick and 88 mm long, are shown in Fig. 3.32, and transfer functions measured on a similar rib are presented in Fig. 3.33. It is seen that there are a number of flexible modes, with very low damping, about $\xi = 0.002$. Obviously, resonant forced vibrations between the many force harmonics and the rib modes are often encountered.

A record of the sound from a similar operation is presented in Fig. 3.34. The little bubbles on the individual spectral lines indicate integer multiples of spindle speed, $n = 15\,270$ rpm $= 254.5$ rev s^{-1}. The cutter had four teeth. Every fourth bubble is marked by an x. It is seen that there were resonant vibrations at the third and fifth harmonic of f_t.

The control strategy described in Section 3.3.4 for chatter vibrations would backfire in these operations because the system actually filters out speed harmonics and becomes blind to them. A different strategy is in the initial stages of development and cannot yet be discussed. Even so, we thought it useful to outline here the main features of the problem.

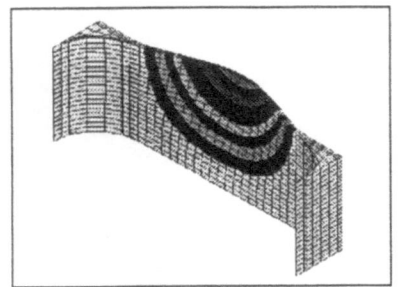

First mode: 2942.737 Hz.

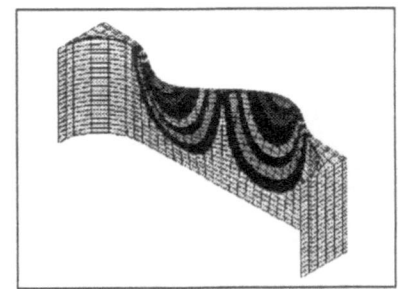

Second mode: 6322.033 Hz.

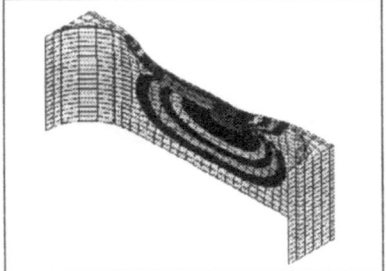

Third mode: 10304.08 Hz.

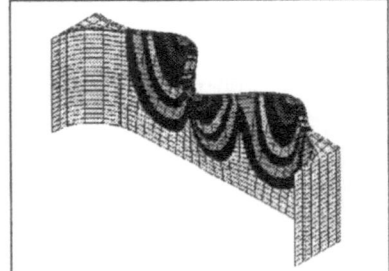

Fourth mode: 11244.82 Hz.

Fig. 3.32. Mode shapes of a thin rib.

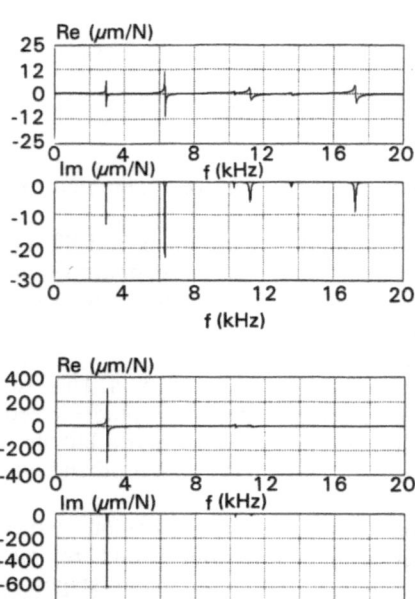

Fig. 3.33. Transfer functions on a thin rib.

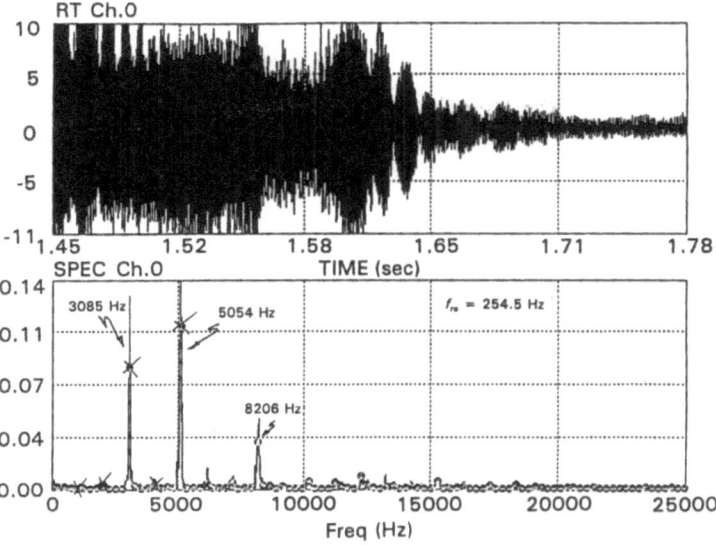

Fig. 3.34. A sound record of milling a thin rib.

References

1. Bishoff B, Hallan M, Moser T, Frohrib D, Ramalingam S. Real time tool condition sensing – fracture detection using a new transducer /failure indentification system. ASME, WAM 1987; PED-26: 69–78
2. Tlusty J, Tyler T. Adaptive control for die milling; criteria and strategies. ASME, WAM 1988; PED-32: 45–60
3. Altintas Y, Yellowley I, Tlusty J. The detection of tool breakage in milling. ASME, WAM 1985; PED-18: 41–48
4. Vierck KC, Tlusty J. Adaptive thresholding for cutter breakage in milling. ASME, WAM 1992; PED-55: 17–32
5. Principe JC, Yoon TW. A new algorithm for detection of tool breakage in milling. Machine tools and manufacture. Pergamon, Oxford, 1991 vol 31-4, pp 443–454
6. Smith S, Delio T. Sensor based control for chatter-free milling by spindle speed selection. ASME J Dynamic Systems, Measurement and Control 1992; 114: 486–492
7. Smith S, Tlusty J. Stabilizing chatter by automatic spindle speed regulation. Annals of the CIRP 1992; 41(1): 433–436

4 Automatic Supervision of Surface Grinding

E. Westkämper

4.1 Introduction

Demands on industrial production change according to the state of technical development and the level of sales. To remain competitive companies have to improve not only their productivity and economic efficiency, but also the flexibility and quality of their manufacturing processes.

In many fields – especially in precision machining – quality is already the determining competitive factor. Not only ever-increasing demands on the precision of workpiece and products are made, but also increased pressures for a more efficient and faster realization of customers' requests.

If these demands are to be met by appropriate manufacturing techniques, it is expected that manufacturing concepts of the future must not only be inexpensive and flexible, but also contain "closed-loop quality-controlled" processes. This chapter deals with the development and application of such quality-controlled systems for precision machining in abrasive processes.

4.2 Quality Parameters and Closed-loop Quality Control of Precision Machining

When modern manufacturing systems are discussed, it is generally assumed that they meet the aims of economic efficiency and productivity. In general, manufacturing times and costs have been reduced as new technological solutions have been applied to solve the problems of manufacturing. Developments in cutting technology, the use of new cutting materials, the application of CNC technology, the realization of developments in machine design and many other advances have contributed to these improvements. However, the potential to make further savings in manufacturing costs and times during precision machining is still high. Firstly, cost reduction depends on the relationship between the work material and the cutting tool material, as well as on the selection of optimal machine-setting parameters by means of closed-loop control systems and external specifications. To some extent, technological developments have resulted in such great improvements that the potential of rationalization is restricted by the demands on workpiece quality.

More and more, grinding is being carried out at the existing limits of technological advances, where although further developments are unlikely, this fact has not caused a lack of demands for increased quality and precision. This has inevitably lead to the growing importance of "closed-loop quality control", i.e. the

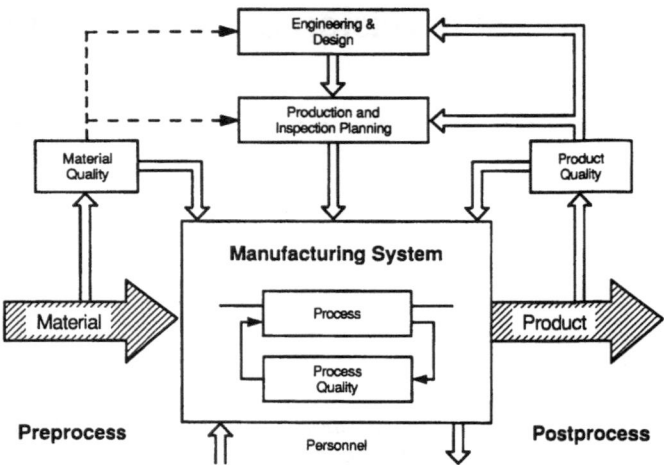

Fig. 4.1. Quality control systems for production process.

optimization of machining parameters according to quality criteria, for process control and supervision.

The aim of industrial development nowadays is to use quality-controlled manufacturing systems which allow optimal use of the technological potential. Closed-loop quality control – basically a limiting control – ensures manufacturing reproducibility. The closer this control can be applied to the process, the higher will be its economic efficiency.

There are a great number of ways to achieve quality-controlled production. Figure 4.1 shows some possible closed-loop quality controls in a very abstract form.

It would be interesting to record only the so-called process quality of an operation, controlling the individual processes by means of parameters so as to ensure that the quality of the machining result is always adequate. Thus we could dispense with any subsequent measurements. Process reliability would be the basis of product quality. This would require the development of a process-integrated quality assurance. If, at the same time, the working ability of the manufacturing system as a whole could be assured, this would be the most efficient method of production.

The most common way of determining manufacturing quality is by means of a variance comparison between those features that are relevant to quality. For example, product quality is determined by measurement. If the dimensions are beyond the range of tolerance, improvement can be attained by a finishing operation. Closed-loop quality control serves as a "post-process control" operation. The loop is closed when correcting variables for the machine, or the process can be derived from the quality deviations.

In practice, a quality test of the surface grinding of a workpiece is only carried out following the machining process and away from the machine-tool working area. Using this method it is possible to detect faulty workpieces only after a time delay. The additional costs of carrying out finishing operations on rejects cannot be prevented.

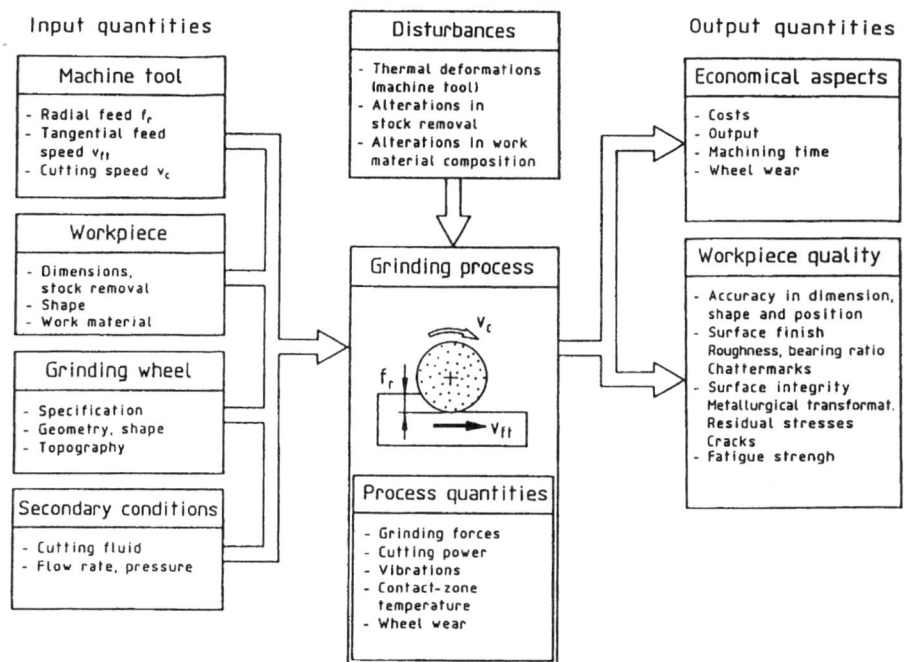

Fig. 4.2. Grinding process as a black-box system.

Naturally, those parameters that need to be changed during manufacture and design can also be derived from the product quality measurements. These so-called large closed-loop controls, which, for example, can influence NC programming, are also "post-process controls". Since they take longer in comparison with the former technique, the times of response must inevitably also be longer. Today, it is conceivable to use these closed-loop controls as the basis of a computer integrated manufacturing (CIM) concept for precision machining, and thus to increase their efficiency.

In some precision machining processes the manufacturing quality also depends on the quality of the raw material and/or the previous machining. For instance, the precision achievable by honing can be improved if machine parameters matched to the initial condition of the workpiece can be determined. Therefore, pre-process measuring is of great importance, especially for precision-machining processes.

To carry out in-process quality control of precision-machining processes, the relationships between the input and output quantities have to be analysed. Saljé has dealt particularly with this subject in numerous scientific papers and experiments. Figure 4.2, for example, shows the parameters and elements of the grinding process as a black-box system.

In a research project being carried out at the Institut für Werkzeugmaschinen und Fertigungstechnik in co-operation with the Institutes of the Technische Universität Hannover, in-process quality control tests and systems of quality control are being developed. These are founded on the fundamental work of Saljé. The priorities of this project are:

1. Measurement of the process parameters as an indirect quality test and of the quality parameters as direct quality testing. The process parameters are determined by process-integrated measuring; the quality parameters are determined mainly by in-process measurements taken inside the working area of the grinding machine.
2. Relationships between the parameters of input, process and quality are determined by means of grinding tests.
3. Parameters are laid down in a model designed to be as universally applicable as possible by way of mathematical functions or algorithms. For instance, cracks and structure transformations can be prevented by relating the setting parameters, the process parameters (forces) and the contact-zone temperatures to each other, in accordance with the model.

Such a quality model makes it possible to design optimal grinding processes during the scheduling of operations, and to carry out appropriate correcting measures if deviations from the required quality occur. Thus, expensive rejects and finishing operations are avoided.

During examinations, input quantities such as machine adjustments and workpiece and grinding-wheel parameters are varied. The grinding and dressing forces are determined by process-integrated measuring, wheel wear by in-process measurements. Parameters of quality, such as the accuracy of dimensions, form and position, as well as the workpiece roughness, are also determined by in-process measurements. In the future, it will be possible to determine grinding temperature by process-integrated measuring, and to detect grinding cracks by in-process measurements.

Undoubtedly the most important quality parameters of the workpiece are deviations from shape and dimensions, in addition to surface quality. These parameters can be used to design a "post-process" closed-loop quality control system. Process parameters such as grinding forces, vibrations and temperatures are suitable not only for indirect quality testing, but also for process-integrated quality control. Examples will be given later in this chapter of the development of sensor and actuator technology for process-integrated testing, in-process testing and supervision of precision-machining processes.

4.3 High-accuracy Closed-loop Position Control for the Feed Axis of a CNC Surface-grinding Machine

Thermally induced deformations in surface-grinding machines have an important effect on the accuracy of dimensions and the shape of the workpiece. In practice, these deformations are compensated by means of statistical process control/testing (SPC). Workpieces taken at random are measured subsequent to the grinding process, away from the machine tool. Although this technique results essentially in improvement of the accuracy of dimensions and shape of the workpiece, a certain number of rejects cannot be avoided.

Figure 4.3 shows the thermally induced deformations in a typical CNC surface-grinding machine in the Y and Z directions which were measured despite the

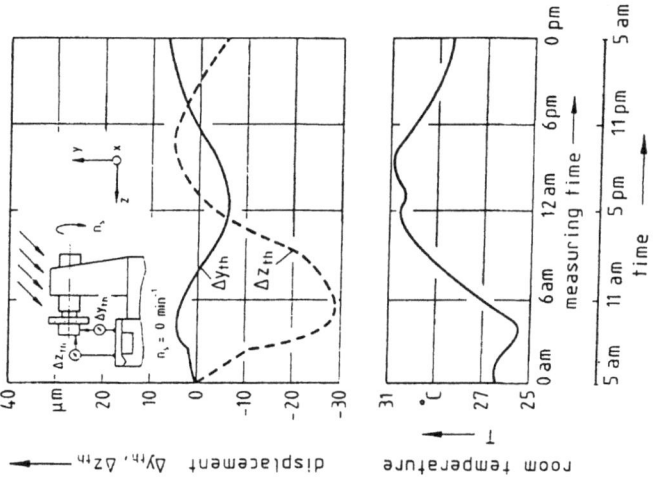

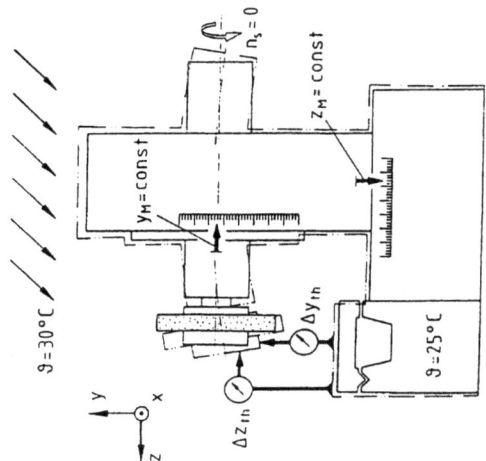

Fig. 4.3. Thermally induced deformations in a CNC surface-grinding machine by external heat sources.

position controllers between the grinding wheel and the supporting table being switched on. The thermally induced deformations due to indirect sunlight amount to about 11 μm in the Y direction and about 35 μm in the Z direction. Without special measures for compensating these deformations, the dimension and shape tolerances of 5 μm cannot be met.

In order to avoid machining errors due to thermally induced machine deformations, the internal position control of a CNC surface-grinding machine is extended along the Y-feed axis by a high-accuracy closed-loop position control. This comprises a new type of piezo-electric measuring system, a piezo-electric measuring interface and a variance comparator in the CNC controller (Fig. 4.4).

By means of the highly temperature-resistant piezo-electric measuring system, the actual position of the grinding wheel is measured against the machine table prior to finish grinding. This system, which is equipped with a diamond stylus, enables contact to be made with the rotating grinding wheel, which is wetted with coolant. Contact between the stylus and the rotating wheel is recognized within some microseconds by an acoustic emission sensor which forms part of a particular processing unit. Immediately afterwards, the stylus is reset from the grinding wheel at a maximum acceleration of 20 000 m s^{-2}. The actual quantity to be measured is determined when the capacitive position sensor of the piezo-electric system contacts the grinding wheel (see Fig. 4.5). If the quantity measured diverges from the desired workpiece dimension, the corrected command signal is passed by the CNC controller to the internal position control, and the grinding spindle is re-adjusted in Y direction. The high-accuracy closed-loop position control remains unaffected by temperature fluctuations within the machine frame and the incremental glass scale.

In the above case, the accuracy of the closed-loop position control is limited to 1 μm because of the resolution of the machine scale needed for re-adjustment. The capacitive position sensor of the piezo-electric measuring system, however, shows a measuring inaccuracy of less than 0.05 μm.

Figure 4.5 shows the schematic set-up of a piezo-electric measuring system with a capacitive position sensor. The measuring range of the system is limited by the maximum stroke of the piezo-electric actuator to about 20 μm.

The transverse acceleration a_x occurring when the diamond stylus contacts the rotating grinding wheel is detected by means of a piezo-electric acoustic emission sensor in the direct proximity of the diamond stylus.

In order to protect the piezo-electric measuring system from mechanical overload across the stoke direction, the diamond stylus is guided by a radially pre-stressed sheet-metal membrane.

The precision capacitor of the internal capacitive position sensor consists of two concentric ring electrodes placed near the stylus. The top electrode is permanently joined to the flexible diamond stylus; the bottom electrode is glued together with the electrode carrier, made of Invar, on to a ceramic insulator (Al_2O_3). By using Invar as the carrier material, the thermal sensibility of the piezo-electric measuring system can be reduced considerably.

The various interface modules of the piezo-electric system intended for contact detection, the processing and linearization of the capacitive position-sensor signal, the control of the piezo-electric high-voltage actuator and communication with the CNC controller are controlled by an interface microcomputer. A special CNC program integrates the fully automatic measuring cycles of the piezo-electric system into the grinding process.

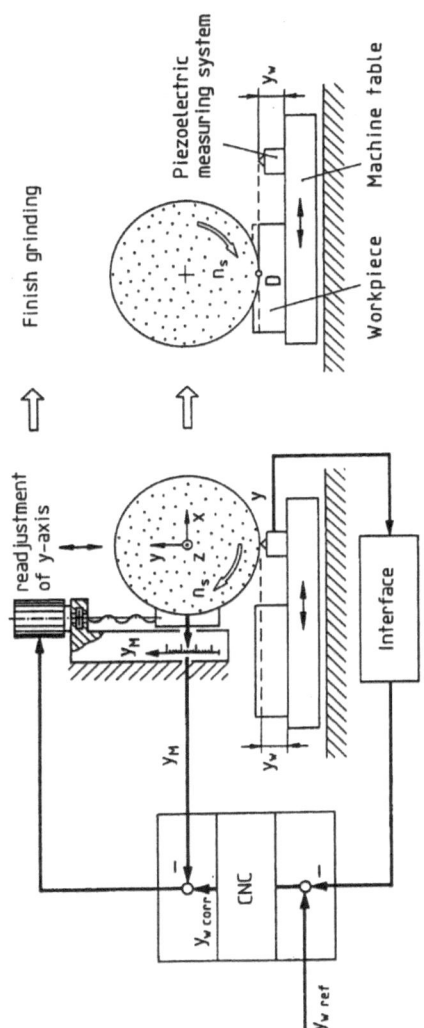

Fig. 4.4. High-accuracy closed-loop position control for increased dimensional accuracy of the workpiece.

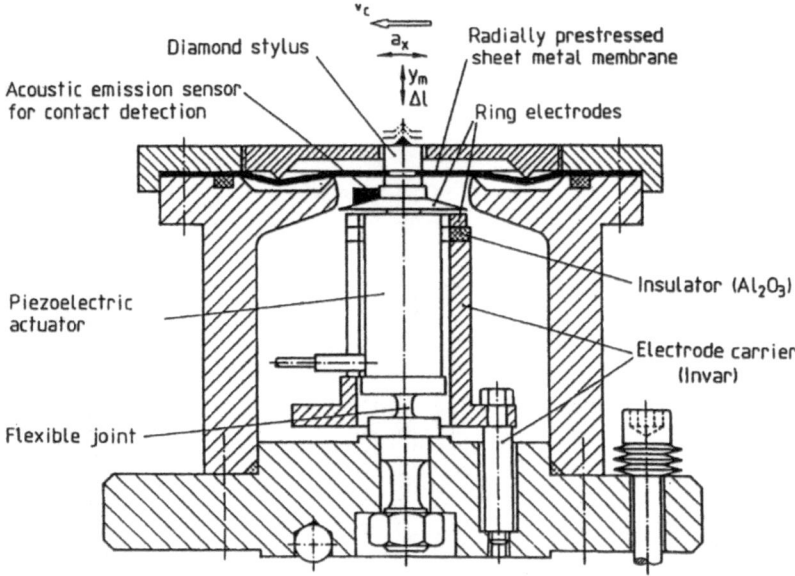

Fig. 4.5. Piezo-electric measuring system to detect the grinding-wheel position.

4.4   In-process Testing of Workpiece Geometry

Whereas the high-accuracy closed-loop position control compensates for the influence of disturbance quantities, such as thermally induced machine deformations, on the grinding process even before workpiece finishing, the measuring process described below is a contribution to process-oriented quality control.

For in-process testing of workpiece geometry, i.e. in the working area of a CNC surface-grinding machine, a two co-ordinate multi-position measuring instrument was developed (Fig. 4.6). With this measuring device, which is constructed as a frame, the longitudinal profile of the workpiece is measured against two reference planes of the machine table immediately after the grinding operation.

Four inductive displacement transducers are used as distance sensors with a measuring range of ±1 mm. They are placed along the Y direction on the workpiece and the reference planes, by means of a stepper motor mounted on the measuring frame. To take measurements, the workpiece is moved by the machine table of the surface-grinding machine in the X direction.

For control of the two co-ordinate multi-position measuring instrument, a logical circuit was developed which integrates an automatic measuring operation into the CNC program of the surface-grinding machine. The acquisition, processing and graphical representation of the measured data as well as the communication with the CNC controller of the grinding machine are carried out by a microcomputer.

Since the bodies of the inductive displacement transducers are connected to each other by a stirrup, differences with respect to the reference planes can be measured very precisely. Thus, first-order measurement errors which are due to thermal

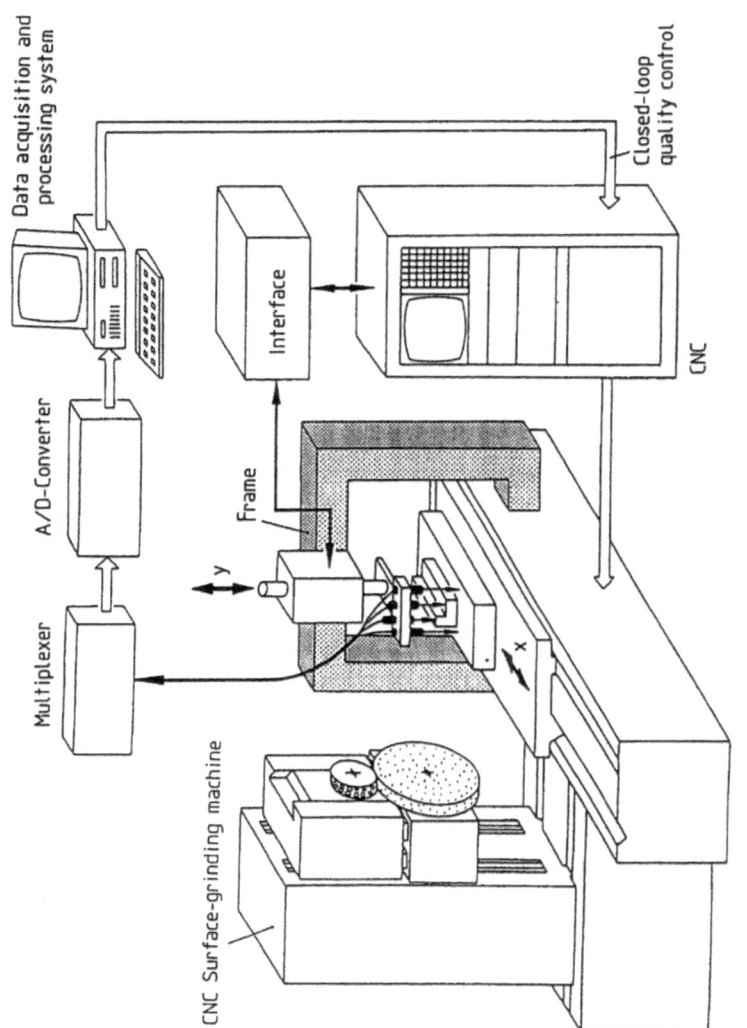

Fig. 4.6. Workpiece-geometry testing inside the working area of a CNC surface-grinding machine.

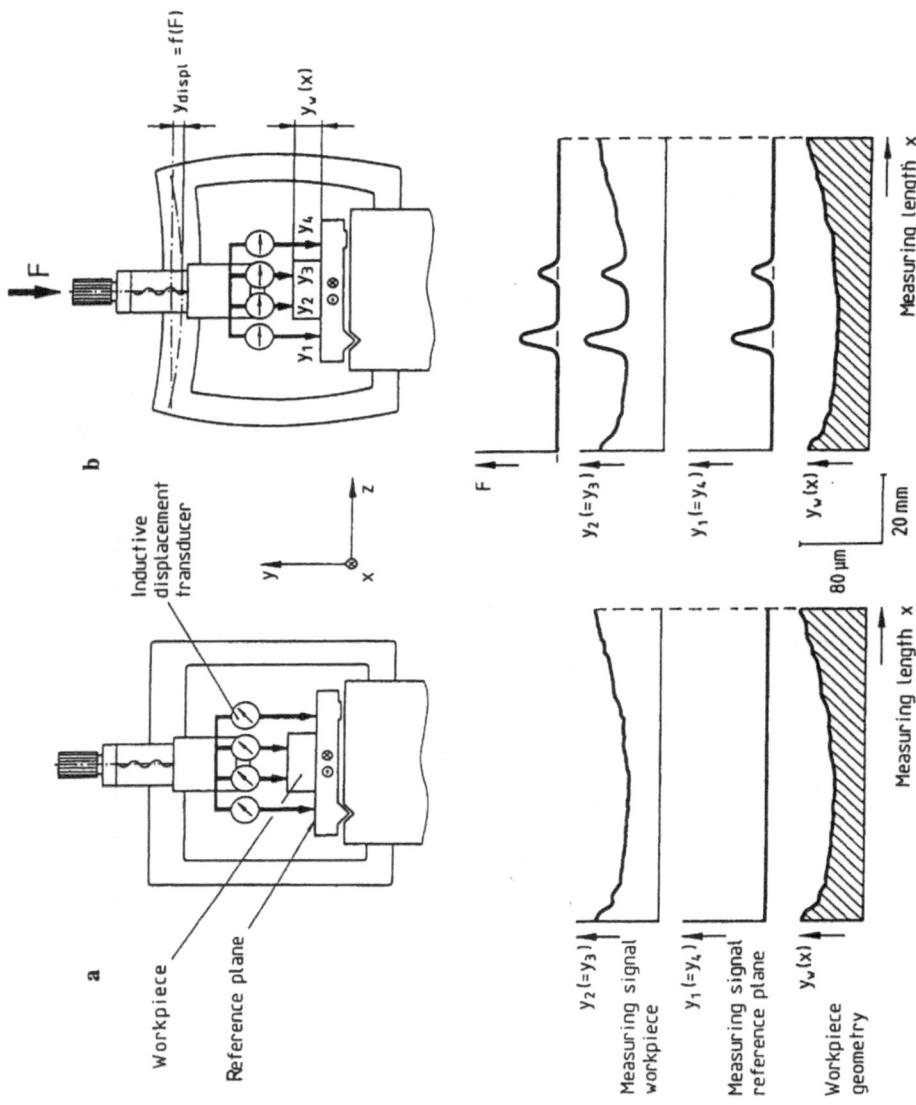

Fig. 4.7. Principle of geometry measuring using reference planes.

deformations in the measuring frame and in the displacement transducers can be eliminated to a large extent.

Figure 4.7 gives an idea of the principle of geometry measuring by means of reference planes. The lower left part of (a) shows the measuring signal caused by the workpiece and the reference plane at a state of normal operation of the measuring frame. After measuring, the workpiece contour is calculated by the computer, which acquires and processes the measuring data as the difference between the values measured for the workpiece and the reference plane. To illustrate disturbance compensation, the measuring frame shown on the right (b) is loaded with a force F. As the load influences the workpiece and the reference-surface signal in the same way, the disturbance can be eliminated by subtraction, and the true workpiece contour calculated.

By means of the processing program, other measuring errors which are caused by surface impurities or abrasive grains embedded in the workpiece surface can be reliably recognized and compensated for using numerical filter algorithms.

With the measuring system described above, a measuring uncertainty of less than $2\mu m$ can be attained even under the rough conditions in the working area of a grinding machine. Compared with three co-ordinate measuring instruments, this level of measuring uncertainty is maintained even without air conditioning.

4.5   In-process Eddy Current Crack Testing

Some workpieces, e.g. jet-engine components made of alloys based on titanium and nickel, are ground only at a low material removal rate, in order to avoid surface damage. This inevitably leads to long and uneconomical grinding times. When the admissible contact-zone temperature is exceeded during the grinding operation because of high material removal rates, micro-cracks are formed on the workpiece surface. For detection of such cracks, the dye penetration test is widely used. Unfortunately, it is a time-consuming and highly polluting method.

At present, the Institut für Werkzeugmaschinen und Fertigungstechnik is working on the development of an in-process eddy current crack-testing system, i.e. within the working area of a CNC surface-grinding machine (Fig. 4.8). Here, the same measuring frame as was developed for workpiece-geometry testing is used to feed the eddy current probes. This testing device is to enable early detection of cracks subsequent to the grinding process. If cracks are occurring, a machine-integrated closed-loop quality control will intervene in the process, for example by reducing the material removal rate.

It is characteristic of the developments shown above that the process parameters, and particularly accuracy, are controlled by in-process acquisition of the values of direct and indirect parameters. These developments have been decisively influenced by the sensor and actuator technology. It can be assumed that other measuring systems will be developed in the future. This same measuring procedure in which quality parameters are quickly acquired can also be employed for highly dynamic control systems.

However, an increasing requirement of monitoring system is not just to assure the quality of the machining results but also to supervise the machine, the periphery and the manufacturing system as a whole, i.e. the working ability of the manufacturing equipment. These aspects of monitoring are examined more closely below.

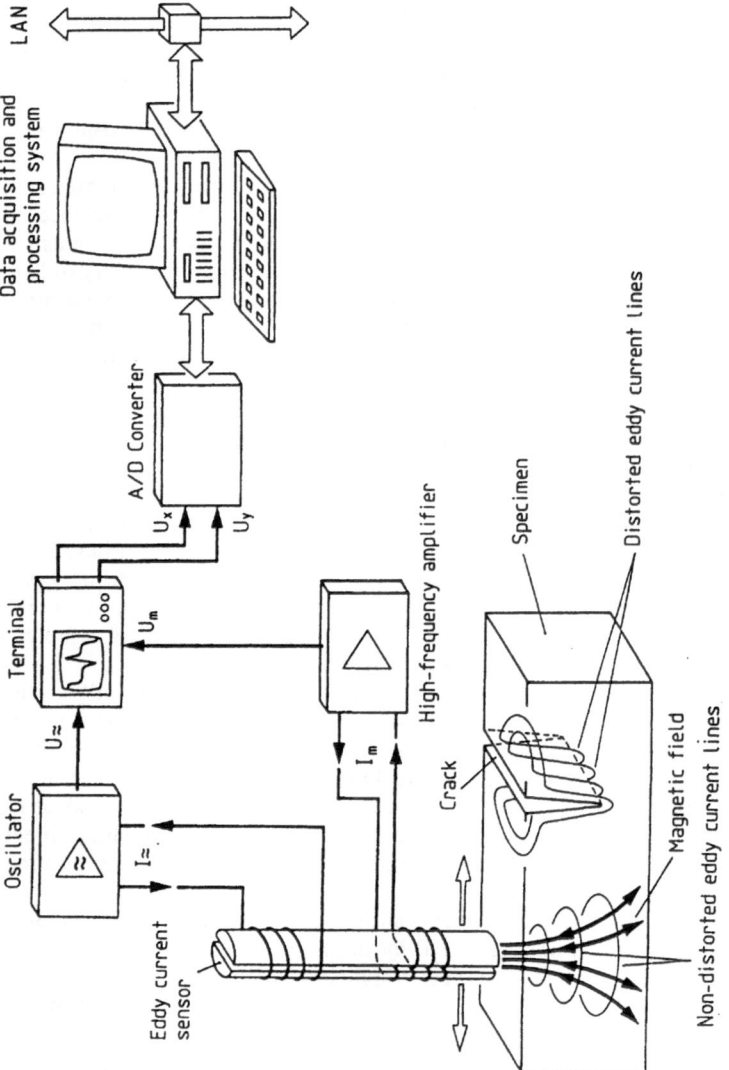

Fig. 4.8. Crack detection by the eddy current technique.

4.6 Diagnosis of Working Ability

There are various possible errors that can affect the working ability of a machine tool. Many of these disturbances cannot be involved in a diagnosis, for economical, technical or technological reasons. In some cases, it would be difficult or even impossible to determine such disturbances and their causes by means of tests. Therefore, a diagnosis system has to be adequately flexible if subsequent extensions or modifications are to be easily brought in at any time.

Figure 4.9 shows the concept of a diagnosing system. For the analysis of the machine status, the chosen quantities are measured by sensors. Following the analysis, the state of the process and the equipment can be estimated. There are certain requirements that apply to the sensors, for example:

No impairment of the machine properties

Little susceptibility to failure

High operational reliability

Wear resistance and low maintenance

Economy

The signals provided by the sensors, i.e. the measured values, have to be prepared for the actual diagnosis. By means of appropriate methods, features characteristic for the identification of the error states of a machine are extracted from the measured values. The signal preparation is subdivided into signal conditioning, data reduction and further procedures which, in previous tests, have proved to be particularly suitable for the detection of the error states.

The signal preparation is followed by the actual diagnosis through so-called classifiers. These classifiers are able to relate certain characteristics, such as the rotational speeds of various components of the grinding machine to certain machine states. These classes of machine state are then passed on by the classifiers to the user interface of the diagnosing system. The user interface translates the classes of machine state into a human-readable form (e.g. bearing failure).

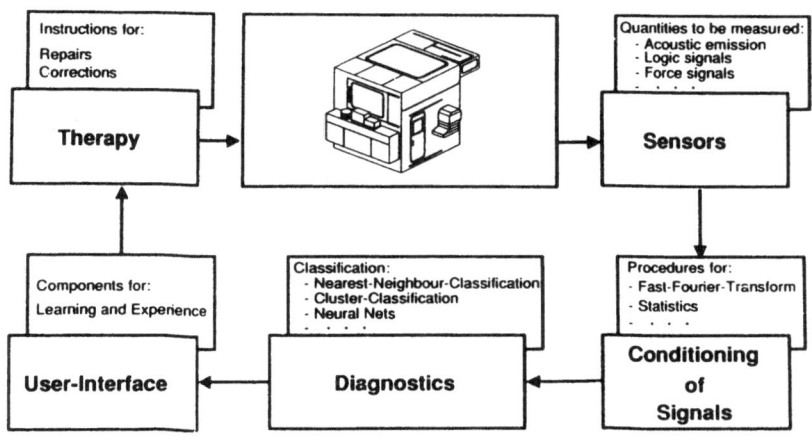

Fig. 4.9. Diagnosing system layout.

Furthermore, the therapy component of the diagnosing system may contain instructions to the operator (e.g. repair procedure) or actuating variables for direct machine-control action.

The Institut für Werkzeugmaschinen und Fertigungstechnik of the Technische Universität Braunschweig has been working for several years on the development of a diagnosis system to be used in grinding machines which is intended to utilize the results from the analysis of structure-borne noise at different points of the machine. At the present state of research, exclusively acoustic signals, detected at a test grinding machine by means of acoustic emission sensors, are being considered as input quantities. In this case, the advantage lies in the fact that the acoustic emission sensors can be attached to the machine without causing problems, particularly without affecting its working ability.

However, a disadvantage has to be seen in the problematic signal processing which often leads to indistinct diagnosis results, i.e. to diagnoses that are not clearly separable from each other (e.g. bearing failure and grinding-wheel unbalance). At present, the classification and interpretation of patterns in the sound level are a subject of intensive research work. However, high-performance computers and new methods of programming and parallel processing would seem to promise some measure of success in the near future.

4.7 Outlook: Manufacturing with Closed-loop Quality Control

Process-integrated quality control, pre-process and post-process control and machine diagnosis are the basic in-process approaches to the technical optimization of manufacturing within a system. In many cases, however, modern manufacturing systems are already part of computerized and integrated manufacturing processes.

If the "process quality" can also be ensured by means of high quality data and information, the whole technical process would be simplified further and, consequently, would become more flexible. The pre-condition would be integration of the preceding manufacturing steps, such as workshop control, production and test planning and designing, into information processing. Within this integration, all aspects of closed-loop quality controls would have to be taken fully into account.

A quality-controlled manufacturing system has to ensure that operational disturbances are avoided and their effects on other system components are minimized. For precision machining, this would mean systematic elimination of the external sources of disturbances. For instance, it has to be guaranteed that NC programs, machine setting parameters and manufacturing facilities etc. have either been tested prior to the beginning of manufacturing or else are available at a sufficient level of quality. The knowledge acquired during the manufacturing process means that product quality has to be assessed at short notice by the planning and control processes.

Such concepts of quality-controlled manufacturing as a whole can only be operated successfully if the specific CIM periphery has been realized.

4.8 Summary

The supervision of precision-machining processes is being encompassed more and more by the closed-loop quality-control system. The aim of this development is to use process parameters to stabilize the process and maintain it even at the extreme ranges of performance. Process parameters are indirect quantities that permit the use of process-integrated closed-loop quality control. Using the example of a high-accuracy closed-loop position control, it was demonstrated how, for example, thermal effects can be compensated for in-process at the machining site.

Direct quality parameters concerning dimensions or the surface can only be determined by "post-process" measurements. Consequently, it is essential to carry out these measurements as near to the actual process as possible and to eliminate disturbing factors. Such closed-loop control systems were illustrated by the examples described above. "Pre-process" measurement and the design of pre-process closed-loop control systems were not dealt with.

Machines for precision machining are complex manufacturing systems – and as such their operation has to be supervised. Machine diagnosis systems equipped with the latest technology help to take measures to avoid error as early as possible.

Further developments will be considerably influenced by the process and the open-loop control technology, as well as by sensor and actuator technology. However, we can anticipate that consideration of these developments for CIM precision machining will have important advantages for the responsiveness of large closed-loop controls in quality assurance.

Further Reading

1. Saljé E. Probleme der Wärmedeformation an Schleifmaschinen und ihre Lösungen. Jahrbuch Schleifen, Honen, Läppen und Polieren, 54 edition. Vulkan-Verlag, Essen, 1987
2. Brinksmeier E, Wulfsberg J-P. Diagnose thermischer Verlagerungen an Schleifmaschinen. Werkstattstechnik 1987; 77
3. Tönshoff HK, Wulfsberg J-P. Meßverfahren zur Diagnose thermischer Verlagerungen in Flachschleifmaschinen. HGF-Kurzbericht, Industrie-Anzeiger 1987; 46/47
4. Schulz H, Mootz A. Berührungslose Verschleißmessung an Profilschleifscheiben. HGF-Kurzbericht, Industrie-Anzeiger 1987; 49/50
5. Autorengruppe: Automatisierte Fertigungsüberwachung. Industrie-Anzeiger 1984; 106
6. Autorengruppe: Maschinendiagnose in der automatisierten Fertigung. Industrie-Anzeiger 1981; 103
7. Weck M. Werkzeugmaschinen, Band 3. VDI-Verlag, 1989
8. Steinhausen D. Clusteranalyse, Einführung in die Methoden und Verfahren. De Gruyter, Berlin 1977
9. Steinhagen H-E, Fuchs S. Objekterkennung. Verlag Technik, Berlin, 1976
10. Schreiber S. Ein Computer lernt buchstabieren. Heise-Verlag ct 1988; 7

5 Automatic Supervision in Physical and Chemical Machining

J.P. Kruth

5.1 Physical and Chemical Machining Processes

The application of new materials (polymers, carbides, composites, ceramics etc.) after the second world war would have been impossible without the simultaneous development of new machining techniques. Those new techniques are often referred to as "non-traditional" in contrast to the "traditional" metal-cutting processes such as turning, milling or grinding. Most of them are characterized by a material-removal process based on physical or chemical phenomena, rather than on mechanical cutting as in traditional machining (Fig. 5.1). Processes such as electro-discharge (EDM), electron beam (EBM), ion beam (IBM), plasma beam (PBM) or laser beam machining (LBM) melt or evaporate the material. Chemical (CHM) and electro-chemical machining (ECM) are based on pure chemical or electro-chemical dissolution of the material. As a result, mechanical strength and hardness no longer limit the machinability. This allows hardened steel in polymer injection dies, fibre reinforced materials, carbides, composites or ceramics to be shaped relatively easily, depending on such physical or chemical characteristics as melting point, specific heat, atomic weight, chemical resistivity etc. Moreover, the absence of contact forces and relative cutting motion between workpiece and tool (electrode or beam) makes it possible to produce intricate shapes even in weak workpieces (foils, tissues etc.).

Some processes, such as ultrasonic (USM), waterjet (WJM) and abrasive jet machining (AJM), still partially rely on mechanical action. However, in these cases too, such physical and chemical phenomena as cavitation, atomic resonance or chemical corrosion play an important role and allow these techniques to be applied in cases where traditional cutting fails. In ultrasonic machining, the tool electrode and hence the abrasive slurry contained in between the tool and workpiece are excited at ultrasonic frequency (20 kHz). Material removal is achieved by impact and cavitation of the slurry.

Today, the designation "non-traditional machining" may be considered obsolete, at least for some of these processes. Electro-discharge machining (EDM), for example, has found such wide applications in mechanical job shops, such as toolmaking companies, that the volume of EDM machining often exceeds conventional milling. The same may be expected in the near future from laser beam machining, e.g. for sheet metal manufacture. The term "physical and chemical machining" (P&C machining) is therefore more appropriate than "non-traditional" machining.

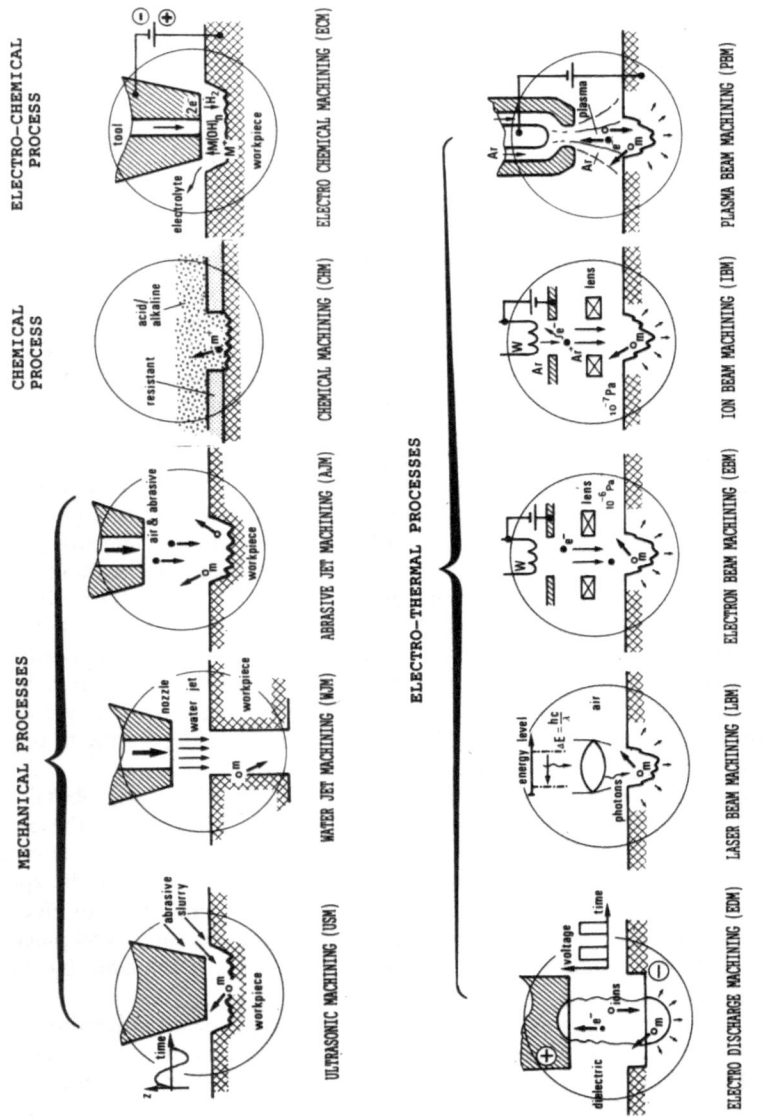

Fig. 5.1. Physical and chemical machining processes.

5.2 Needs for Automatic Control in P&C Machining

As already stated, the breakthrough for new materials was largely the result of the discovery of new physical and chemical machining processes. These latter, in turn, are inconceivable without the recent developments in automatic feedback control. This relates to some very specific needs of P&C machining:

1. *Tool feed control.* The absence of mechanical contact between workpiece and tool (electrode or beam gun), and the low predictability of the material removal rate, often require a feedback control that adjusts the feedrate continuously to the momentary removal rate.

2. *Numerical motion control.* Several P&C machining processes (e.g. beam and jet processes, wire-cutting EDM) can only rely on numerical control to generate a desired workpiece shape.

3. *Safety control.* The erratic character of the P&C processes calls upon sophisticated automatic safety actions.

4. *Process optimization.* The non-deterministic and non-stationary nature of the processes prevents an off-line calculation of optimum working conditions. High efficiency can only be met by tracking the optimum on-line.

Those four control tasks require quite different characteristics of the controller. Feed control and numerical control are generally based on servo systems, while in-process optimization often calls on sophisticated adaptive control. Safety control may be purely hardwired or based on adaptive control. However, the bandwidth of the safety control and optimizing control is quite different. In electro-discharge machining, for example, the safety control should be able to react at about the frequency of the electric sparks used for machining the workpiece, i.e. up to 1 MHz or more. In-process optimization, in contrast, should include a waiting time after each control action in order to allow the process to adapt to the adjustments of the control parameters before evaluating the change. The process settling time may have an order of magnitude of several milliseconds or seconds, resulting in control frequency of a few hertz to less than one hertz.

Therefore, the various control functions will normally be allocated to different controllers with different characteristics. Figure 5.2 [1, 2] depicts a multi-processor control system for electro-discharge machining. CPU1 is specially designed for rapid response and takes care of the servo feed control ("controller") and of safety control based on short-circuit detection. CPU2 takes care of the long-term control action, i.e. adaptive control optimization. No NC controller is involved in this example. Although different controllers are involved, Fig. 5.2 clearly shows that a lot of interaction will normally be needed between the various controllers.

In the most general case, process control in P&C machining will involve four different automatic control loops:

1. Feed servo control loop
2. Safety control loop
3. Optimization control loop
4. Numerical control loop

This is already the case for most electro-discharge machines (Fig. 5.3). This results from the fact that EDM has been, until now, the most successful P&C machining process, justifying a lot of R&D effort. Other P&C machining processes

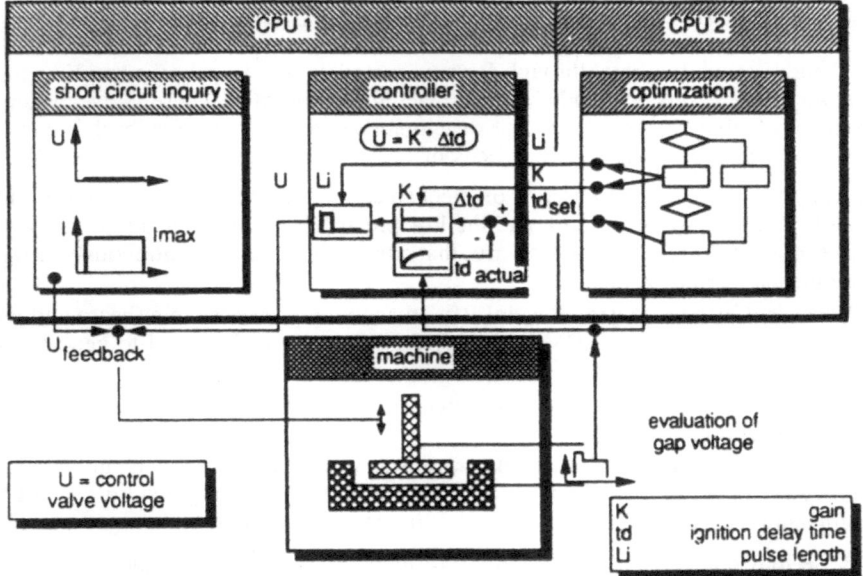

Fig. 5.2. Multi-processor control system for EDM [2].

are lagging somewhat behind and have fewer or simpler control systems. However, the basic principle of on-line supervision, control and optimization is unchanged.

The complexity of most physical and chemical machining processes, the large number of control parameters, the lack of skilled operators, the high equipment cost, the relatively low removal rate and long machining times, and the relative low cost of additional control devices are important additional incitements to automation of P&C processes. However, complexity is also the reason why automatic control may fail and why human operators cannot always be eliminated totally. This is why a fifth feedback loop, representing a human operator, is shown on Fig. 5.3. Automatic sensing devices with visual displays are often provided in order to help the operator in monitoring and controlling the complex process.

The next section will discuss the various control loops.

5.3   Control Loops in P&C Machining

5.3.1   Tool Feed Systems

In traditional machining, a constant feedrate can be applied. This feed determines the material-removal rate. However, material-removal rate is no longer deterministic for many physical and chemical machining processes. As a consequence, feedrate should constantly be adapted to the rate of material removal. This generally calls for a servo-controlled feed system in which feedrate is proportional to some indirect measurement of material removal.

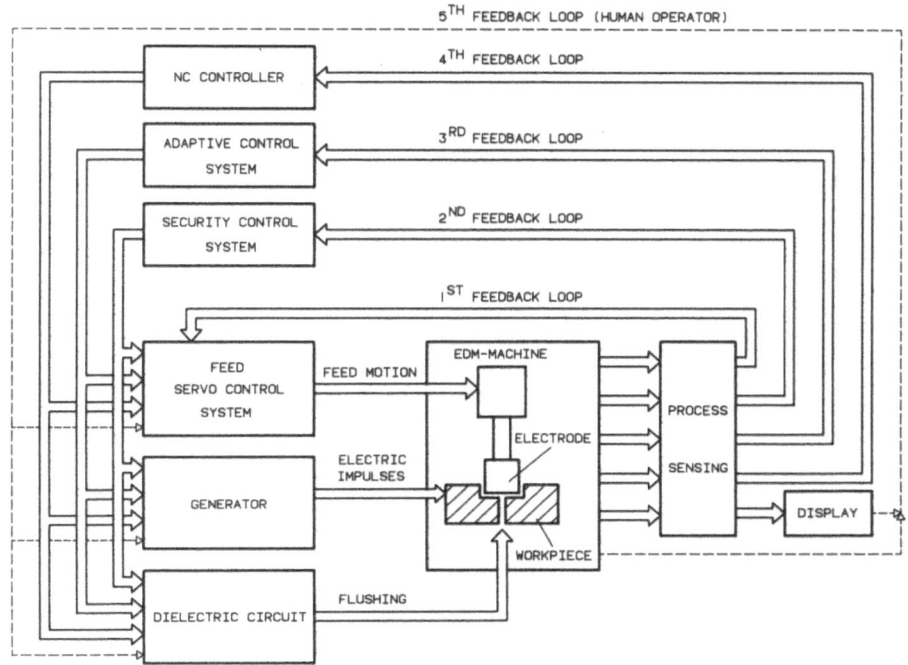

Fig. 5.3. Control loops in EDM.

The problem is further complicated because no contact exists between the tool and the workpiece. For example, in EDM the tool electrode should be kept at a constant distance from the workpiece (0.01–0.1 mm) in order to allow for erosive spark discharges. If the feed is too fast, the distance decreases until short circuits occur between tool and workpiece. No further spark discharge can then occur and machining stops. If the feed is too slow, the gap increases until no discharge is possible. All EDM machines are therefore equipped with a servo-controlled feed system which compares the actual gap to a reference value (Fig. 5.4a). The actual gap size is measured indirectly by recording a sensing parameter such as the average voltage between tool and workpiece (U – Fig. 5.4) or the mean ignition delay of the spark discharges (t_d – Fig. 5.2).

In cases of very accurate machining (e.g. drilling micro holes of diameter 20 μm $\times$ 0.5 mm) double hierarchical servo actuators may be used: for example, a piezo-electric actuator for accurate feed control and a stepping motor for larger displacements.

Adaptive feed control is even needed in electro-chemical machining, where an equilibrium gap occurs between the tool and the workpiece, and where a constant feedrate may theoretically be applied. A feedback control system is at least required to detect and remedy short circuits occurring when material removal unexpectedly drops (Fig. 5.5). The latter often happens as a result of irregular distribution of the electrolytic current flowing between tool and workpiece (temperature or flowrate variation of the electrolyte, passivation layer formed on electrode surface, electric

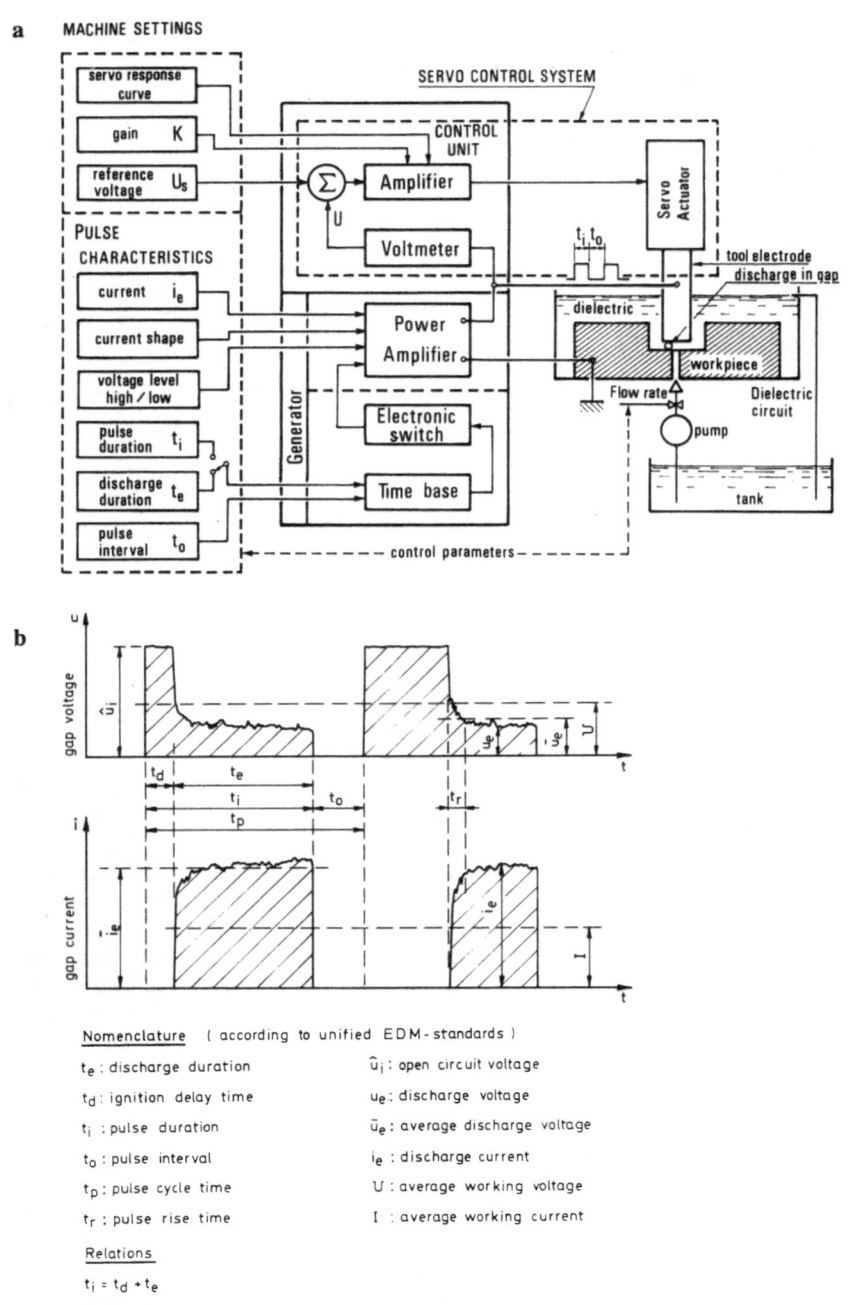

a   MACHINE SETTINGS

Fig. 5.4. Control parameters in EDM: **a** main elements of EDM equipment; **b** voltage and current characteristics of spark.

Within the figure (panel a):

- servo response curve
- gain   K
- reference voltage U_s
- PULSE CHARACTERISTICS
- current i_e
- current shape
- voltage level high / low
- pulse duration t_i
- discharge duration t_e
- pulse interval t_o
- SERVO CONTROL SYSTEM
- CONTROL UNIT
- Amplifier
- Voltmeter
- Power Amplifier
- Electronic switch
- Time base
- Generator
- Servo Actuator
- tool electrode
- discharge in gap
- dielectric
- workpiece
- Flow rate
- Dielectric circuit
- pump
- tank
- control parameters

Within the figure (panel b):

- gap voltage   u
- gap current   i
- t_d t_e t_i t_o t_p t_r
- $\hat{u}_i$ $\bar{u}_e$ U   I i_e

Nomenclature (according to unified EDM-standards)

t_e : discharge duration

t_d : ignition delay time

t_i : pulse duration

t_o : pulse interval

t_p : pulse cycle time

t_r : pulse rise time

$\hat{u}_i$: open circuit voltage

u_e : discharge voltage

$\bar{u}_e$: average discharge voltage

i_e : discharge current

U : average working voltage

I : average working current

Relations

$t_i = t_d + t_e$

$t_p = t_i + t_o$

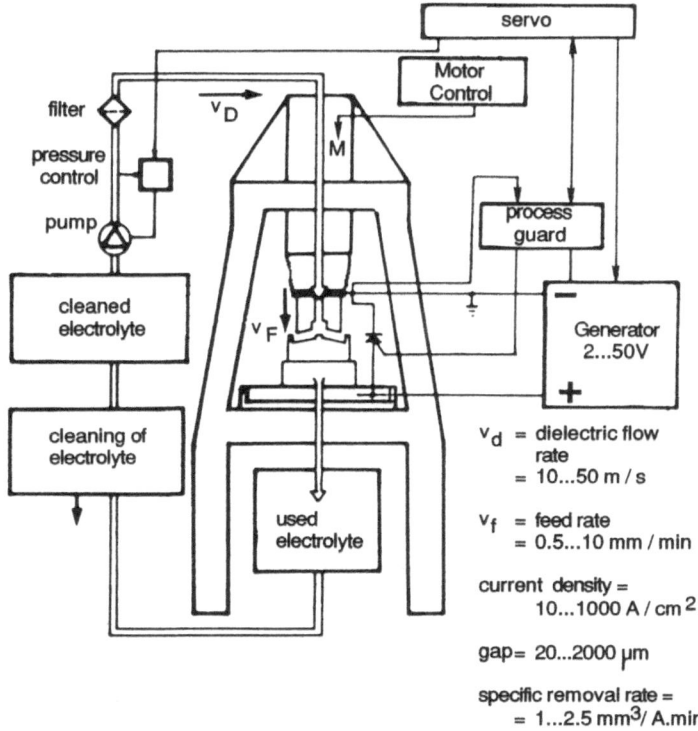

filter

pressure
control

pump

cleaned
electrolyte

cleaning of
electrolyte

used
electrolyte

servo

Motor
Control

v_D

M

process
guard

Generator
2...50V

v_F

v_d = dielectric flow
        rate
     = 10...50 m / s

v_f = feed rate
     = 0.5...10 mm / min

current density =
     10...1000 A / cm^2

gap= 20...2000 μm

specific removal rate =
     = 1...2.5 mm^3/ A.min

Fig. 5.5. Main elements of an ECM equipment.

field concentration at edges etc.). Upon occurrence of a short circuit, the electric current should be interrupted immediately and the electrode retracted. Many ECM machines are now equipped with real servo or adaptive control systems tending to a constant gap, rather than a constant feedrate. Several such systems are described in the literature:

1. Some systems measure the conductivity [3], the pressure [4] or the flow [5] of the electrolyte in order to adjust the feedrate for constant gap size.

2. Other systems adjust the feedrate in order to maintain a constant ratio of gap voltage over feedrate, under controlled electrolyte conductivity.

3. Others impose a high-frequency vibration on the tool electrode and compare the size of the ripple current produced to the ripple on the rectified AC current used for machining. In this way the gap size can be compared with the amplitude of vibration, and an error signal derived to drive a servo motor to maintain a constant gap [3].

In ultrasonic machining, material-removal rate is as unpredictable as in other P&C processes. The feed force should be such that it will not expel all abrasive

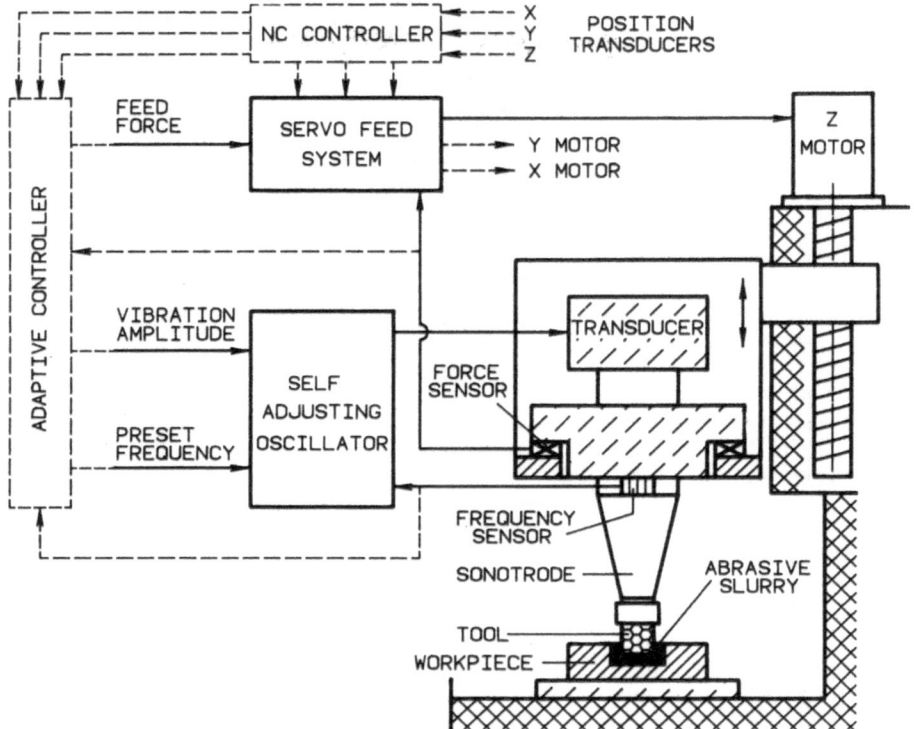

Fig. 5.6. Main elements of a USM equipment.

slurry between tool and workpiece. Most machines, therefore, apply a constant feed force, rather than a constant feedrate. The feed force is generally controlled by a servo system using piezo-electric sensors (Fig. 5.6). Another basic servo control system available on every ultrasonic machine uses a frequency sensor in order to adjust the generator's frequency to the natural frequency of the sonotrode concentrator and tool (Fig. 5.6).

In most of those systems, the machining performance largely depends on the reference value applied to the servo controller. Figure 5.12, for example, illustrates that the material removal rate V_w exhibits a clear maximum against the reference setting of the feed servo system in EDM. Similar parabolic curves are obtained when plotting the removal rate in USM against the feed force. This calls for additional adaptive control optimization loops, which adjust the reference value, servo gain etc. This is discussed below.

The need for rapid tool retraction in cases of process disturbances (such as short circuits, DC arcs etc.), on the one hand, and accurate down feed in order to position the tool at a small gap distance from the workpiece, on the other hand, often calls for special non-linear and asymmetric servo characteristics [1, 2, 6–8]. Several systems also allow adjustment of the overall servo gain (Fig. 5.4a), in order to cope with the large difference in gap between roughing (about one millimetre gap) and super finishing (gap of a few micrometres) [1, 9–11].

5.3.2   Numerical Motion Control

The workpiece shape in physical and chemical machining is controlled:

Either by the tool geometry (die sinking EDM, ECM, USM) in a similar way to that in conventional sheet product fabrication (punching, blanking),

Or by controlled motion of a thin straight tool (wire or beam) through the workpiece (wire-cutting EDM, LBM, IBM, PBM, WJM, AJM).

As a result, all those latter processes were only possible after the development of numerically controlled drive systems or beam deflection systems.

In some cases, the numerical control systems are of a more complex type than those available on conventional metal-cutting machines, because of the special needs of the feed system. This is, for example, the case in wire-cutting EDM, where a wire-electrode is used to cut out a profile in a metal workpiece. Each NC-controlled axis should be completed by a servo feed system adapting the feedrate as described above. Upon occurrence of short circuits, the wire tool should be retracted along the programmed contour, i.e. within the already machine groove. This should be done while maintaining a perfect synchronization between the various NC-controlled axes (i.e. up to five axes in wire-cutting EDM). It requires the NC control unit to be able to step back in the program and execute NC commands in the reverse sense (e.g. circular interpolation commands). Some recent NC controllers for wire EDM machines are also looking ahead in the NC program in order to anticipate when edges have to be cut, for instance. This is done in order to step down the discharge energy a few micrometres before reaching the edge, to guarantee a higher precision [12].

Today, numerical control is commonly encountered in P&C processes which initially did not require it. This has allowed broadening of the process capabilities. Examples are planetary EDM [13, 14] and EDM or USM contouring and pocketing [15], which combine EDM or USM sinking with complex multi-axis NC control and allow intricate shapes to be produced with simple, inexpensive electrodes.

Hybrid NC control systems are used sometimes to combine high positioning accuracy with large displacement ranges. Figure 5.7 shows an electron beam machine which combines a magnetic beam deflection system with an NC-controlled workpiece displacement [16].

5.3.3   Process Supervision: Safety Control and Autonomous Machining

Safety problems, generally related to the non-deterministic or stochastic character of P&C processes, often call for continuous process supervision. Human supervision and interaction is normally too slow to avoid unrecoverable damage. Consider for example the damage caused by a short circuit of 10 000 A lasting for 1 second in ECM, or that of a million arc or short-circuit discharges occurring in EDM in a few seconds: both could induce irrecoverable burn marks on the metallic workpiece. Lasting over longer periods, they could produce a fire and even destruction of the whole plant. Therefore, automatic process supervision and fast safety actions are indispensable.

Besides the short-term safety actions, automatic process supervision systems also aim at larger autonomy of the equipment. Indeed, the low material-removal rates

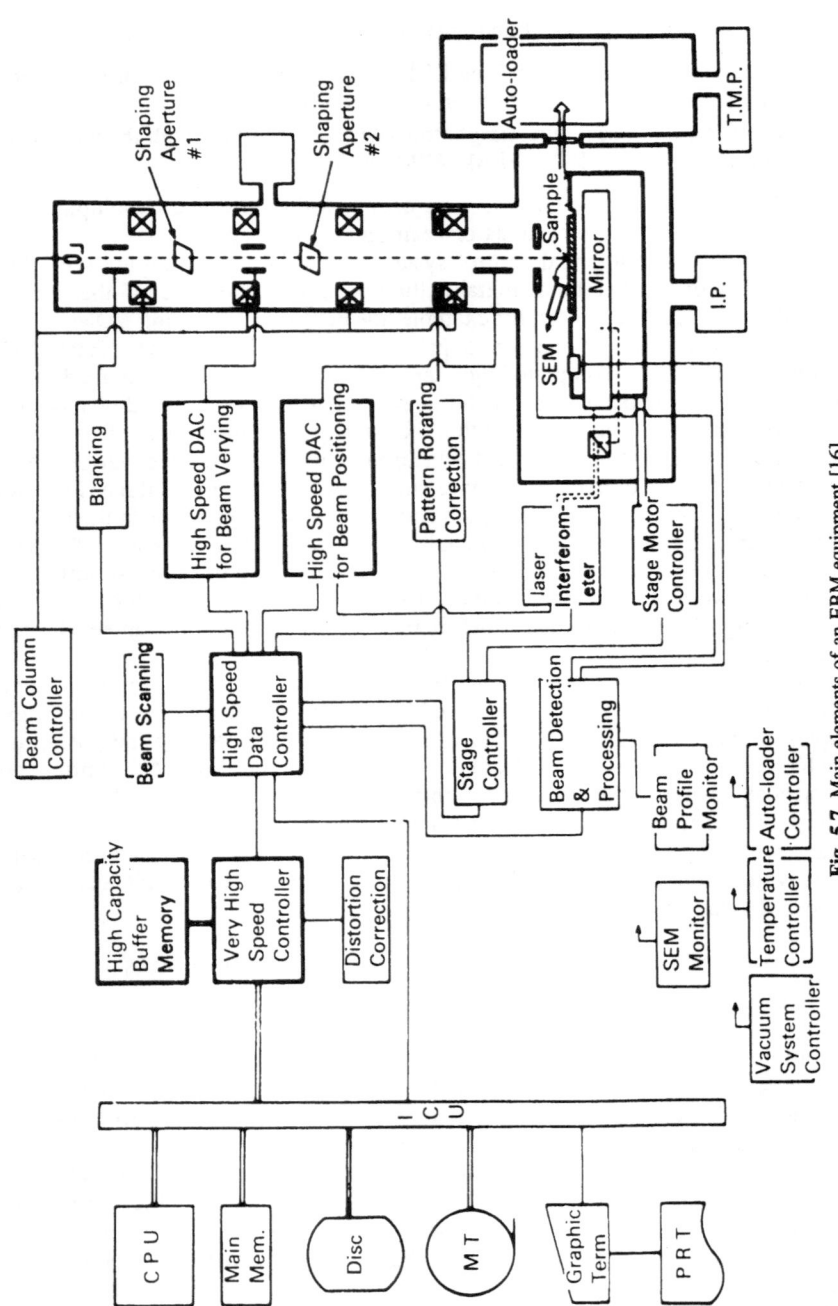

Fig. 5.7. Main elements of an EBM equipment [16].

and long machining times occurring with most P&C processes are major inducements to guarantee reliable operation of the machine without the need for human supervision, even over long periods (e.g. week-ends). Human supervision is even not appropriate during working hours: nobody can justify, either economically or in human terms, putting an operator for hours at the side of a machine where the feed system and contour motion control are fully automated and all the operator has to do is watch the machine and adjust a knob every hour or so.

The actions to be taken by these control systems vary according to the type of process. The range of actions may go from gentle long-range actions, such as switching on or off some cooling device for the electrolyte (ECM) or dielectric (EDM), through short-range actions, such as decreasing/increasing the current amplitude or discharge time, to the complete shut-down of the equipment.

Many safety control systems are described in the literature. A full list is impossible to give, but a few examples are:

Protection against short circuits and steady arcing in EDM (Fig. 5.2) [17–20]

Protection against short circuits in ECM (Fig. 5.5) [21, 22]

Protection against random sparking in electro-chemical machining and electro-chemical grinding [22, 23]

Protection against wire breakage in wire-cutting EDM [24–27]

Protection against tool breakage in USM

Protection against electrolyte boiling in ECM

Protection against excessive dielectric contamination in EDM

Protection against overheating in EBM, IBM, PBM and LBM [28, 29]

These process supervision control systems are sometimes integrated in adaptive control systems. Adaptive control systems are discussed in Section 5.4.

5.3.4   Process Optimization

The undeterministic character of all physical and chemical machining processes makes an off-line selection of optimum machining parameters impossible. Optimum parameter settings can be obtained by:

1. *Performing preliminary trial and error tests.* This solution has the disadvantage of being very time-consuming and totally impracticable for the small series or one-off jobs often encountered in P&C machining. Moreover, the optimum parameter value often varies while machining. In sinking EDM for example, flushing away the workpiece chips out of the gap between tool and workpiece becomes more difficult as the tool penetrates deeper into the workpiece. As a consequence, the reference gap (feed servo setting U_s or t_d) and pulse interval time (waiting time t_0 between successive discharges) have to be increased as machining progresses, and it is impossible to define a single optimum setting for those parameters. Moreover, optimum servo-setting and pulse interval time are totally different after short circuit or arcing conditions have occurred.

2. *In-process optimization by a human operator regularly adjusting the control parameters.* This control strategy allows for the non-stationary character of the optimum parameter values. Major problems in the case of manual optimization are that:

(a) It may be preferable not to tie up an operator during the long machining times involved in P&C processes, mainly because the number of manual interventions may remain low.

(b) The optimization problem may be difficult because of the large number of parameters influencing the process and because of interference between those parameters. As an example, Fig. 5.4 lists some of the control parameters of the EDM process.

(c) It is very hard for a human operator to identify in-process when the optimum machining is reached. For example, in USM or sinking EDM, optimum machining may refer to a maximum material-removal rate in conjunction with a low tool electrode wear. Neither of the latter parameters can be directly evaluated in-process. Indeed, measuring the actual feedrate is equivalent to measuring the sum of both, and adjusting the machine for a maximum feedrate may result in increasing the tool wear, rather than the workpiece removal rate. Manual optimization will in most cases require some hardware sensing or monitoring devices in order to assist the operator to evaluate the process (Fig. 5.3).

3. *Adaptive control.* The long machining times, the high equipment cost, the need for hardware process monitoring for purposes of safety or optimization, the non-stationary character of optimum machine settings and the lack of skilled operators all favour the use of adaptive control in physical and chemical machining. This explains why AC is more frequently applied to P&C machining than to conventional machine tools. Most EDM machines are equipped with some kind of AC system. As will be seen later, various types of such systems are used [30]. Most commonly, adaptive control constraint or self-adjusting control systems and adaptive control optimization or hill-climbing systems are used [18, 31]. However some model reference AC systems [10], learning systems [32] and knowledge-based systems [25, 33, 34] are used as well.

The remainder of this chapter will discuss in more detail the adaptive control techniques for process supervision and optimization in physical and chemical machining.

5.4   Adaptive Control in P&C Machining

5.4.1   Short Historical Overview

The first attempt to apply adaptive control to P&C machining processes goes back to the late 1960s. In 1969, Colwell [35] reported an adaptive control optimization system for electrochemical grinding (ECG). ECG is a combined ECM and mechanical grinding process. Similar work was carried out in the early 1970s at Technion-Israel, and several adaptive control optimization (ACO) and adaptive control constraint (ACC) systems for ECG were developed [36–38]. AC systems have also been developed for classical electro-chemical machining [3, 23, 39], but the limited breakthrough of this process hampered large-scale developments.

At the same time, EDM grew to become the most popular P&C process and consumed the largest research efforts. This gave rise to a large number of commercial AC systems. Extensive lists of early commercial and academic AC systems for EDM are given in [31]. The parameters controlled by those systems and the strategy applied are reported in survey tables.

Adaptive control is used in both sinking EDM, in which a three-dimensional electrode is sunk into the workpiece, and for wire cutting EDM, where a thin wire electrode is used to cut out a prismatic part. AC in sinking EDM [1, 8, 10, 11, 18, 40–43] aims at a minimum machining time and limited tool electrode wear. Main objective in wire cutting EDM are machining speed and avoidance of wire breakage. Recently, AC systems have been developed for planetary EDM [13, 14, 43].

The recent growth in importance of laser machining initiated several research projects on AC for LBM [28, 29, 44, 45]. Most of those systems are still at the research level and are not fully operational. But the commercial importance of LBM will undoubtedly stimulate such developments.

Ultrasonic machining is also well suited for adaptive control process optimization (Fig. 5.6). Parameters which may be optimized are the feed force, the vibration amplitude of the tool, the flowrate of the abrasive slurry and the frequency of intermediate tool retractions. Some early developments have been initiated, but no publications or working systems are recorded.

As for other P&C processes, the advances of their control systems will largely depend on their respective market shares.

5.4.2   Sensing Problems and Sensing Parameters

One of the major problems when applying adaptive control to P&C machining is the impossibility of measuring in-process the output parameters to be optimized, i.e. the material removal rate, the tool wear, the workpiece dimension and accuracy, the workpiece quality, the surface roughness etc. Hence these output parameters or machining results will have to be evaluated indirectly by measuring other parameters which correlate well with the aimed output. These parameters are called "sensing parameters". They require the development of specific sensors in order to be measured in-process. A few examples are given here.

The quality of a workpiece machined with laser beam (e.g. LBM hardening, welding or cutting) can be evaluated on-line by monitoring some beam properties: total beam power, beam intensity distribution, beam dimension, focus location etc. [46]. König [29] gives a survey of applied sensing devices – see Fig. 5.8. A sensor developed by his research team for the detection of the beam intensity distribution is depicted in Fig. 5.9. This sensor is used in an AC system, together with a surface roughness sensor and with some "visual" process observation based on high-speed cinematography or thermography (Fig. 5.10) [28, 44]. A similar nail type sensor has been used to adaptively control the beam focusing in electron beam machining. Other researchers use CCD cameras and fast photodiodes for on-line detection of the roughness of laser cuts [47].

Controlling the workpiece accuracy in ECM is rather difficult, because of the difficulties in predicting the size of the gap separating the workpiece from the tool electrode. Bignon and Bedrin [48] have developed a sensor which estimates the gap by using eddy currents.

In ECM the material removal rate can be identified rather accurately by measuring the feedrate of the ram holding the tool electrode. Indeed, the absence of tool wear allows the establishment of a relationship between ram position, ram speed and material removal rate [3]. Hence, ram position is often used as one of the input signals in an adaptive control system. This no longer applies to sinking EDM, where ram speed equals the sum of tool wear rate and workpiece removal rate.

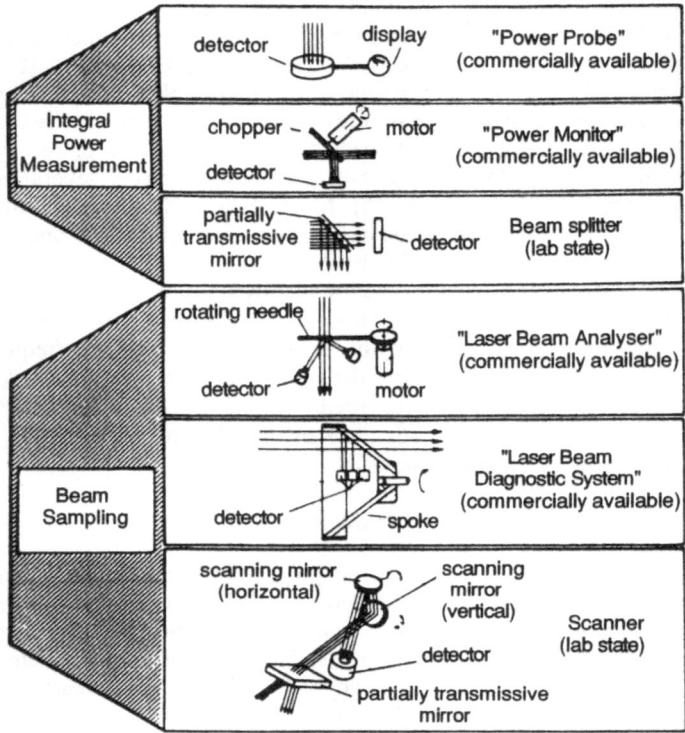

Fig. 5.8. Sensing devices for LBM [29].

Much research has been devoted to finding suitable sensing parameters which allow workpiece removal and tool electrode wear during electro-discharge machining to be evaluated [8, 17, 19, 43, 49]. An extensive survey is given in [31]. Most of the parameters for EDM are based on the recognition of different types of discharges [1, 8, 9, 10, 18, 24, 25, 40–42]. By analysing the time evolution of the gap voltage (voltage measured between tool and workpiece) and gap current (current flowing between tool and workpiece), four basic types of pulses can be recognized (Fig. 5.11):

Open-circuit pulses having a constant high voltage, i.e. no discharge breakdown occurs, hence there is no voltage drop or current rise

Normal discharges showing a breakdown (voltage drop until about 20V and current rise to adjustable value) after a certain ignition delay

Abnormal discharges (arc discharges), having no ignition delay (voltage does not reach high level)

Short-circuit pulses depicting a nearly zero voltage

Some research workers have extended this pulse classification to over 16 pulse types

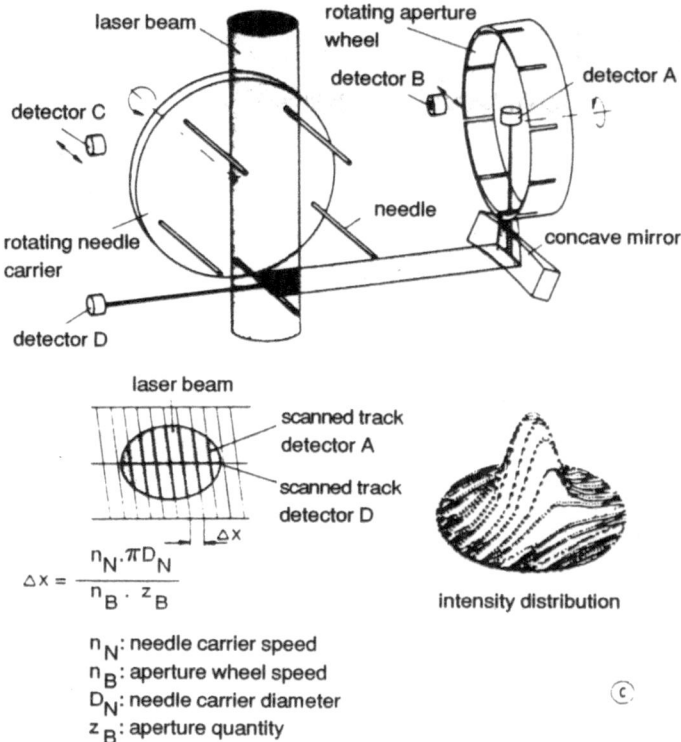

Fig. 5.9. Laser beam sensor developed at the TH Aachen [29].

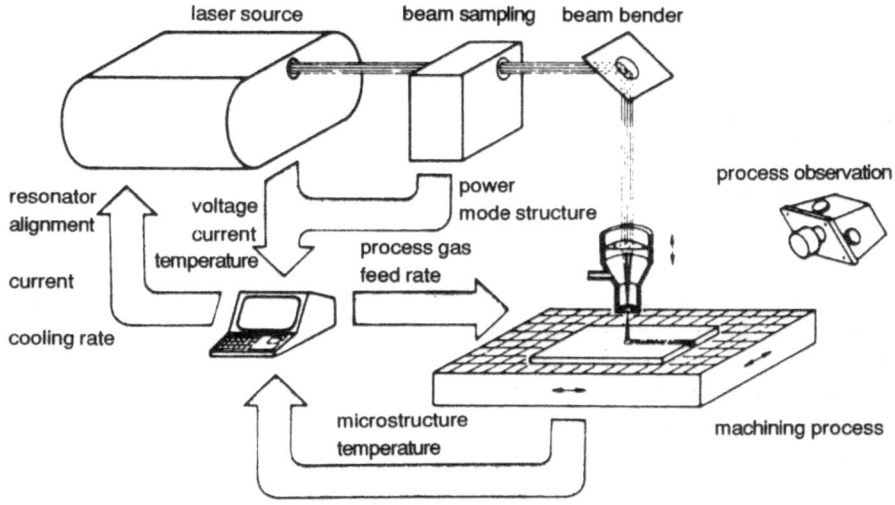

Fig. 5.10. Adaptive control system for LBM [29].

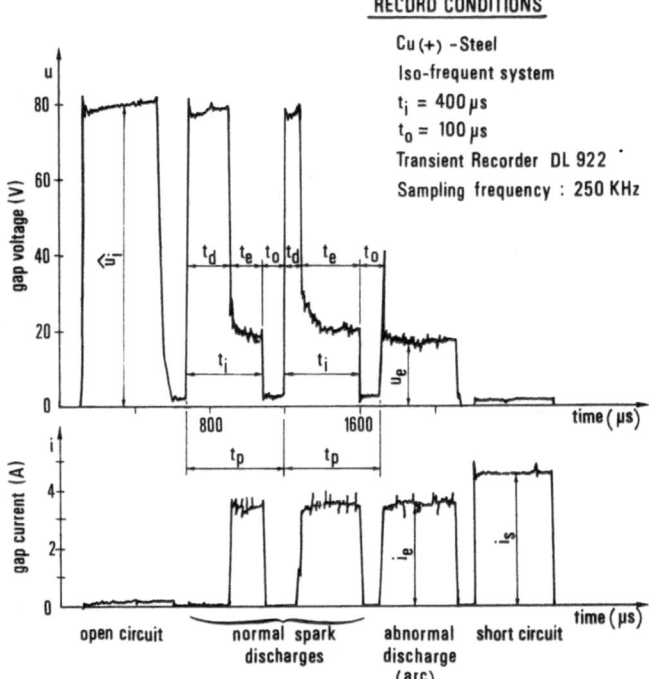

RECORD CONDITIONS

Cu(+) -Steel
Iso-frequent system
$t_i = 400\,\mu s$
$t_0 = 100\,\mu s$
Transient Recorder DL 922
Sampling frequency : 250 KHz

Fig. 5.11. Basic pulse types in EDM.

and involved pattern recognition techniques for this purpose [8]. Other EDM sensing parameters are based on specific pulse characteristics (Fig. 5.4b), such as ignition delay time, discharge duration time or pulse cycle time, noise on voltage signal, pulse rise time, or on emitted radio frequency signals [49]. Some of those sensing parameters will be discussed in the next section.

5.4.3 Sensing Parameters and Type of AC Control

Process evaluation in adaptive control systems is based on a "performance index", which is defined as being one or a combination of several sensing parameters. The type of the performance index or sensing parameter determines the adaptive control strategies to be applied. This is illustrated in Fig. 5.12. The figure applies to the EDM process. It allows strategies to be identified which maximize the material removal rate V_w while limiting the tool electrode wear υ by optimizing the servo reference voltage (reference gap size).

Sensing parameters are classified in two groups:

Odd sensing parameters depicting no absolute maximum or minimum within the considered range of the control parameters (Fig. 5.12a). They require an adaptive

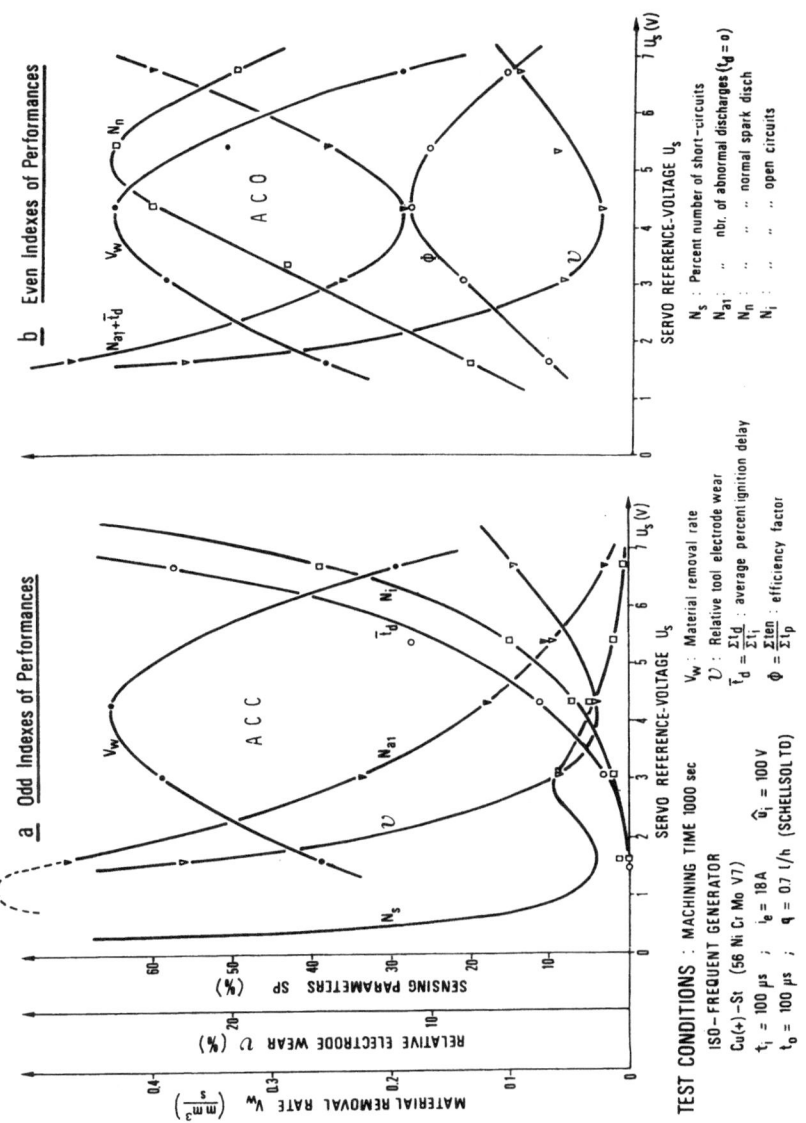

Fig. 5.12. Even and odd sensing parameters in EDM.

control constraint strategy (ACC) which tends to keep the sensing parameter at a constant predefined value.*

Even sensing parameters with a clear maximum or minimum within the considered range of the control parameters (Fig. 5.12b). These are suitable for an adaptive control optimization strategy (ACO) which continuously tends to enlarge (or reduce) the value of the sensing parameter.

Both types of sensing parameters and AC strategies have been applied to physical and chemical machining processes.

ACC systems have been developed for normal electro-chemical machining in order to adjust the supply voltage for a constant electrolyte flowrate, a constant current or a constant electrolyte temperature [3]. Many commercial EDM machines are equipped with ACC systems that adjust the servo reference voltage or the pulse interval time (Fig. 5.4). Those systems apply different sensing parameters: constrained number of specific types of discharge (short circuits, abnormal discharges, contamination discharges, open circuits) [42, 50, 51], aimed dielectric strength or gap impedance [11], constant ignition delay [11, 42], minimum level for electric noise signal on discharge voltage [20], minimum level of radio frequency emitted by the EDM process [19, 52] etc. An extensive survey with references is given in [31].

ACC systems face two major difficulties as compared to ACO:

1. It requires a calibration procedure to define the constraint value to be aimed at. This calibration may have to be repeated for each combination of machining conditions. In EDM, for instance, the constraint value may change with tool electrode material or shape, workpiece material, penetration depth into workpiece, dielectric type or flowrate, electrode polarity, discharge current, discharge time, pulse interval time, generator voltage etc.
2. Multi-parameter optimization with ACC requires the combination of several constraint controls using different odd performance indexes, whereas only a single even index is needed with ACO.

This latter is clarified by comparing Fig. 5.13 and Fig. 5.14. Consider that the reference voltage of the feed servo system U_s and the pulse interval time t_0 of an EDM machine have to be adjusted for a maximum material removal rate V_w. Figure 5.13 shows that an ACC strategy aiming at a single constraint value, for instance 10% average ignition delay ($t_d = 10\%$), will not necessarily end up as the absolute optimum (U_s^{**}, t_0^{**}). This ACC constraint may be used to optimize the servo setting, but a second constraint is needed in order to optimize t_0. In contrast, an ACO strategy aiming at a maximum efficiency ϕ (i.e. ratio of effective discharge time to total sensing time) will lead to the right combination of optimum servo setting U_s^{**} and pulse interval time t_0^{**} (Fig. 5.14). The latter efficiency factor ϕ has proven to be one of the best sensing parameters for multi-parameter optimization in EDM in terms of the maximum material-removal rate [8, 53].

---

* The adjectives "odd" and "even" may be misleading. In mathematics the function is *even* when its graph is symmetric with respect to the y-axis, that is, the functional value $f(x)$ does not change when x is replaced by $(-x)$ [e.g. $y = \cos(x)$]. The function is *odd* when its graph is symmetric with respect to the origin, that is, we receive the negative of $f(x)$ when x is replaced by $(-x)$ [e.g. $y = \sin(x)$]. ACC may be used (see Chapter 1) when the quality index (performance index) of the process is a *monotonic* function in the working range of the controlled variable – Fig. 15.12a may be misleading. (V_w and v are not performance indexes.) [Editor]

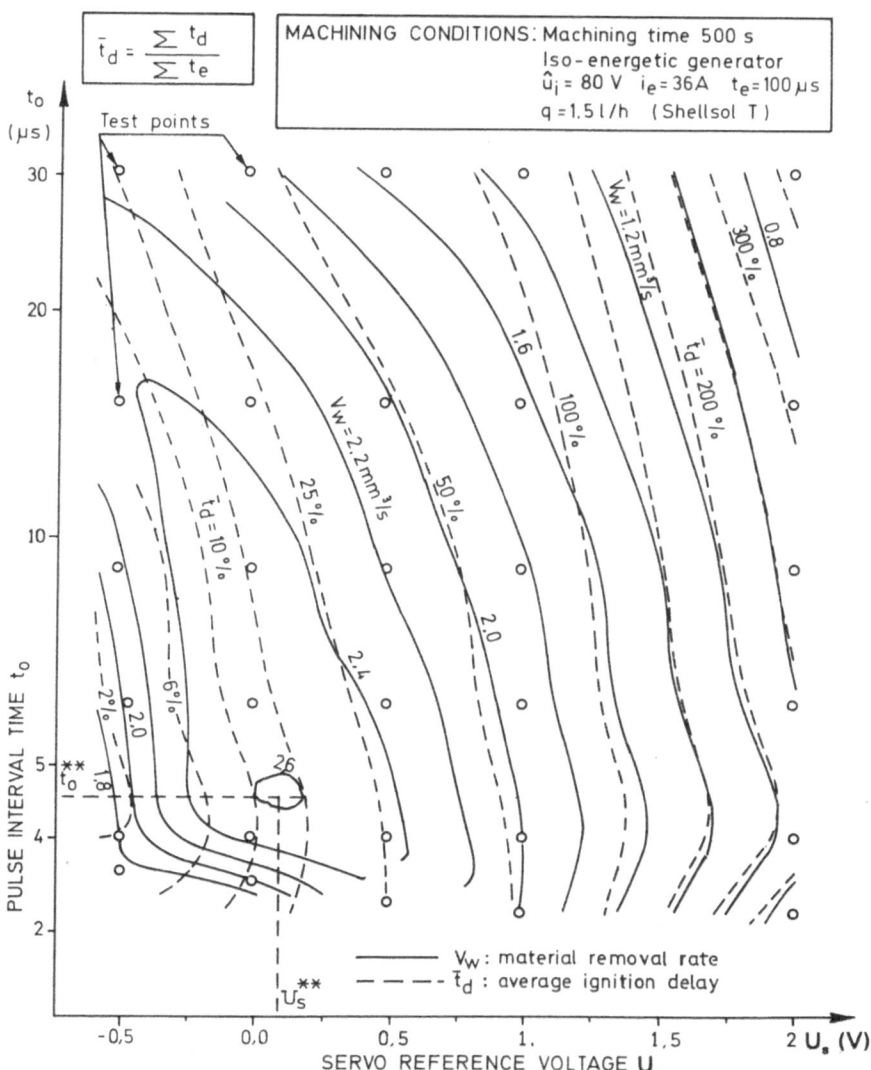

Fig. 5.13. Correlation between average ignition delay t_d and material removal rate V_w in EDM.

ACO systems have been developed for other P&C processes. As long ago as 1969 Colwell reported an ACO system that adjusted the feed force and the gap voltage in electro-chemical grinding [35]. Similar work was carried out at the Technion [36, 37].

Recent developments at the Technion however applied ACC to multi-parameter optimization in electro-chemical grinding (ECG) [54]. ECG combines classical abrasive grinding with electro-chemical machining (Fig. 5.15a). This is achieved by using a metallic bound grinding wheel and applying a DC current between this grinding wheel (cathode) and the workpiece (anode). The process has two dominant control parameters: the feedrate f and the supplied voltage U_m (Fig. 5.15b):

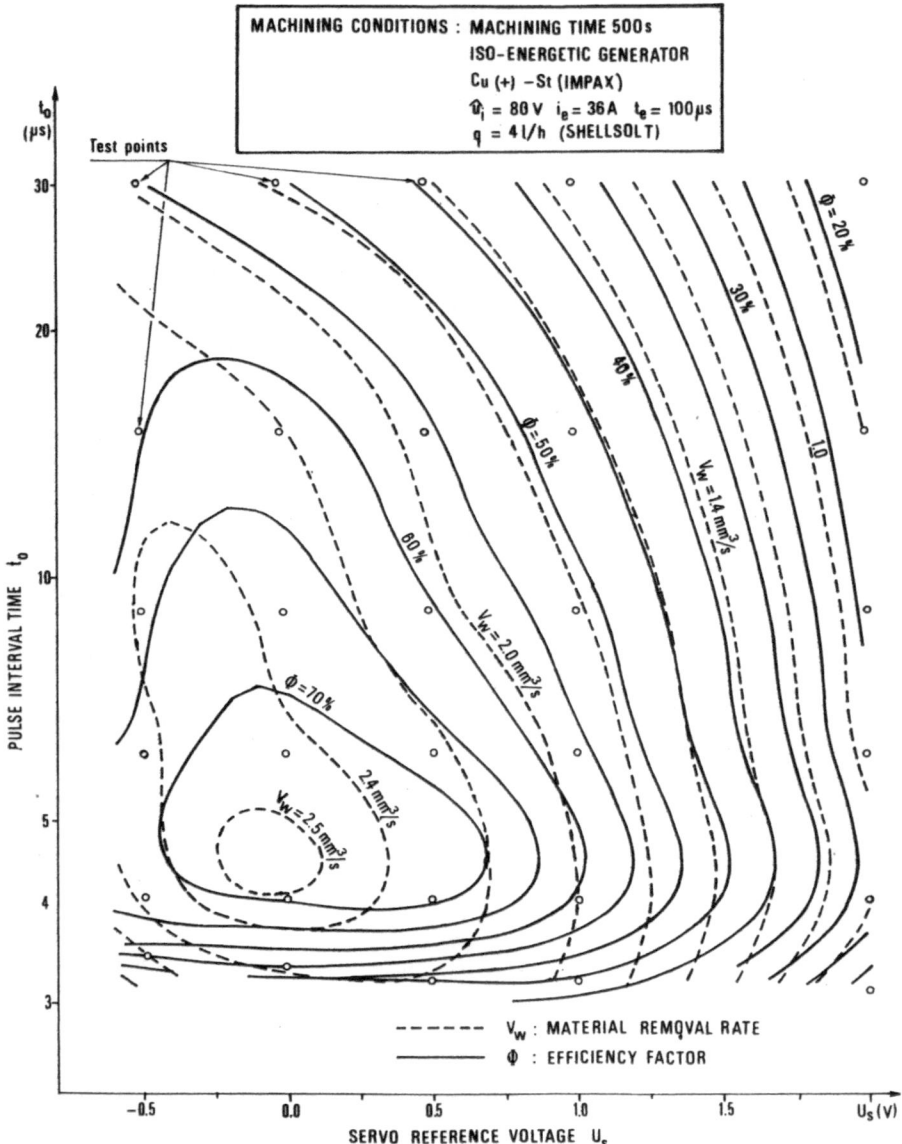

Fig. 5.14. Correlation between efficiency factor φ and material removal rate V_w in EDM.

1. *Feedrate f.* Larger feedrates increase the rate of abrasive machining. Too large a feedrate may however induce sparking (i.e. electric discharges occurring between the grinding wheel and the workpiece) and damage the workpiece. Figure 5.15b therefore depicts a sparking limit: the maximum allowable feedrate is obviously smaller when larger voltages are applied. The sparking limit is not known in advance and has to be identified on-line by the ACC system.

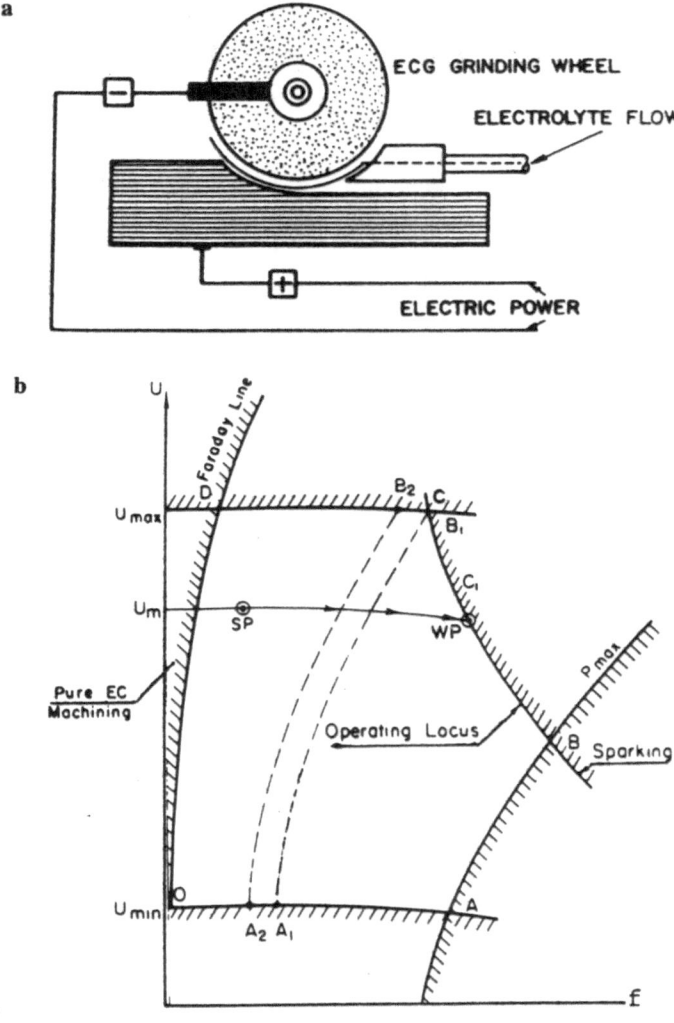

Fig. 5.15. ACC of the ECG process: **a** layout of the ECG process; **b** control constraints.

2. *Supply voltage U_m.* Larger voltages increase the rate of electro-chemical machining, but augment the overcut and reduce the accuracy. Overcut occurs because electro-chemical machining continues after the grinding wheel has passed along. The actual voltage across the grinding wheel and workpiece U decreases slightly with increasing feed (Fig. 5.15b, line U_m–SP–WP or line U_{min}–A). The supplied voltage U_m is the one coinciding with $f = 0$ (i.e. the U-axis). The decrease of U with f is not known in advance and has to be detected on-line. It can be used as a way to identify the sparking limit [54].

Figure 5.15b shows the control space $U\text{–}f$ with five constraint curves:

Minimum supply voltage (line $U_{min}\text{–}A$)
Maximum supply voltage (line $U_{max}\text{–}C$)
Minimum required voltage U to initiate ECM machining (Faraday line O–D)
Sparking limit (line B–C)
Maximum grinding wheel power (line A–B)

The overcut constraint is not represented in Fig. 5.15b.

The control system developed by Shpitalni combines two separate ACC strategies in order to optimize f and U_m. The first one adjusts the feedrate in order to reach the sparking constraint or the power limit: see line SP–WP (starting point–working point) in Fig. 5.15. This is done while machining a workpiece with a constant supply voltage U_m. The second ACC strategy adjusts the voltage U_m in between two workpieces: it adjusts the supply voltage U_m for the next workpiece based on the overcut size measured on the previous part. Figure 5.16 displays a control sequence. It relates to a surface profile grinding operation: the grinding wheel entered the part at one side and exited it at the opposite side. The two upper plots show how U and f changed as functions of time. The lower plot shows how the working point moved in the $U\text{–}f$ control space.

A similar dual ACC control system has been developed for EDM machining at the WZL laboratory in Aachen [43]. A first control loop adjusts the pulse interval time t_0 between successive discharges as a function of the measured ignition delay t_d of the previous discharge (Fig. 5.14). If this control fails to keep t_0 within prescribed limits (t_{0min}, t_{0max}), a second ACC loop adjusts the reference voltage of the servo feed system (U_s).

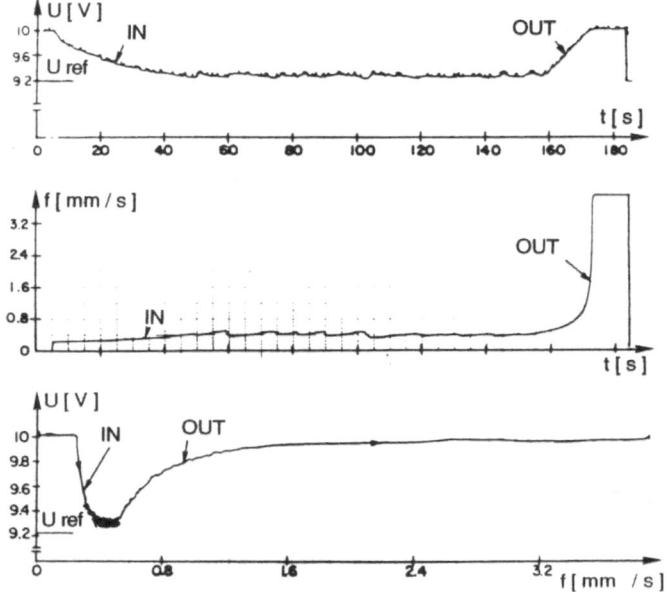

Fig. 5.16. Evolution of the control parameters in time.

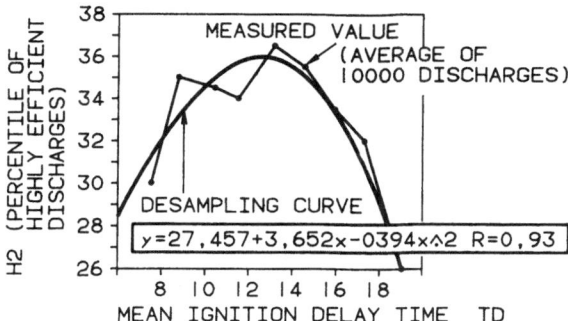

Fig. 5.17. Approximation of the sensing function by means of parabolic regression [1].

5.4.4 Optimum Search Algorithms for ACO

5.4.4.1 Process Variability

The search for the extremum of an even-sensing parameter in P&C machining is often hampered by the large process variability. Indeed, even under constant working conditions the sensing parameters may fluctuate widely around a mean value. Hence the sensing parameter needs to be evaluated and averaged over a sufficiently long period of time. This is in contradiction with the need for rapid action on process disturbances. As a consequence, safety control loops and optimization control loops are generally totally separated in P&C machining, and each one acts at a different response speed.

Even when averaging over a long period of time (e.g. 10 000 discharges in EDM, that is, over 1 second) the variation on the sensing parameter may still largely deform the general outlook of the sensing function as plotted versus the control parameter (Fig. 5.17). This may lead to a wrong estimation of the optimum. Weck *et al.* apply parabolic regression in order to identify the optimum: Fig. 5.17 [1, 2]. Figure 5.17 applies to an AC system adjusting the reference signal of the feed servo system (in this case, the reference ignition delay t_d) in order to achieve a maximum amount of highly efficient discharges (H2). The AC system first enters a test phase in which the percentage of efficient discharges are measured for 5 to 7 different adjustments of the servo reference values. The measured results are interpolated numerically by a parabolic curve. If the correlation coefficient is too low, the measurements are discarded and a new measuring cycle is started.

Performing trial tests with different settings of the control parameter is not always possible. In EDM, for example, it may not be applied when optimizing the pulse interval time t_0, since lowering t_0 below the optimum will induce, almost instantaneously, arcing and destabilize irremediably the EDM process. Optimizing the pulse interval time in EDM therefore requires more sophisticated strategies. A control system, using some kind of self-learning ACO algorithm, that adjusts automatically some constraint values and the controller's gain has been developed by Kruth [18, 53].

Rajurkar suggests using even more sophisticated stochastic analysis methods (i.e. DDS – data dependent systems) in order to cope with the process variability in EDM [49] or ECM control [23].

5.4.4.2   Hill-climbing Algorithms

Different ACO strategies for multi-parameter optimization have been studied at the University of Leuven [18]. They relate to 'hill climbing' optimization strategies, where an extremum of a performance index (i.e. sensing parameter) is traced. The study compares so-called sectioning strategies (Fig. 5.18) and steepest-ascent strategies (Fig. 5.19). The examples shown in Fig. 5.18 to Fig. 5.21 apply once more to the optimization of the servo reference voltage U_s and the pulse interval time t_0 in terms of a maximum efficiency ϕ. The left-hand side plots of Fig. 5.20 and Fig. 5.21 show the evolution of the two control parameters and the resulting performance index (efficiency) versus time. The right-hand side plots display the evolution of the working point within the U_s–t_0 control space.

A sectioning strategy (Fig. 5.18) optimizes the various control parameters separately, on a cyclic basis. Each control parameter is optimized individually by searching for an extremum of the performance index, while the other parameters are kept constant; once the local optimum is reached (in this case, an optimum along a t_0 or U_s section – Fig. 5.18) or after a certain time, the considered control parameter is set to its previous best value while the next parameter is optimized. In Fig. 5.18, the servo setting U_s is optimized first, while t_0 is kept at 25 μs. After seven adjustments a local optimum was found for the servo ($U_s = 0.3$ V). The servo was left unchanged, while the pulse interval was optimized during five successive adjustments of t_0 and so on.

A steepest-ascent strategy (Fig. 5.19) tries to optimize the various parameters simultaneously, in such a way as to move towards the top of the hill (extremum of the performance index) along the steepest path. This steepest path represents the shortest way towards high values of the performance index (i.e. efficiency ϕ). In the particular case of Fig. 5.19, the steepest path is found by adjusting alternately each control parameter and making each adjustment step of a particular parameter proportional to the local partial derivative of the performance index of that parameter. This partial derivative is evaluated using the results of the previous adjustment of that particular parameter, that is

$$\text{derivative: } \frac{d\phi}{dU_s} = \frac{(\phi_o - \phi_{-1})}{(U_{so} - U_{s-1})}$$

$$\text{adjustment step for } U_s: \ U_{s+1} = U_{so} + \Delta U_s \text{ with } \Delta U_s = K \frac{d\phi}{dU_s}$$

where U_{s-1}, U_{so} and U_{s+1} are the previous, present and next servo setting, respectively. This technique allows elimination of the exploration steps often used in steepest-ascent optimization for evaluating the partial derivatives [55].

Figure 5.20 shows that the steepest-ascent strategy results in a faster move to the optimum. Figure 5.20a relates to a sectioning strategy, while Fig. 5.20b resulted from a steepest-ascent search. Comparing the two right-hand side diagrams indicates that only twelve adjustments were needed to reach the optimum with the steepest-ascent strategy, whereas 27 steps were required with the sectioning strategy. Full ϕ contour lines could however not be plotted in those diagrams, because the total control ranges of U_s and t_0 were not explored (compare to Fig. 5.18 and Fig. 5.19 for contour plots).

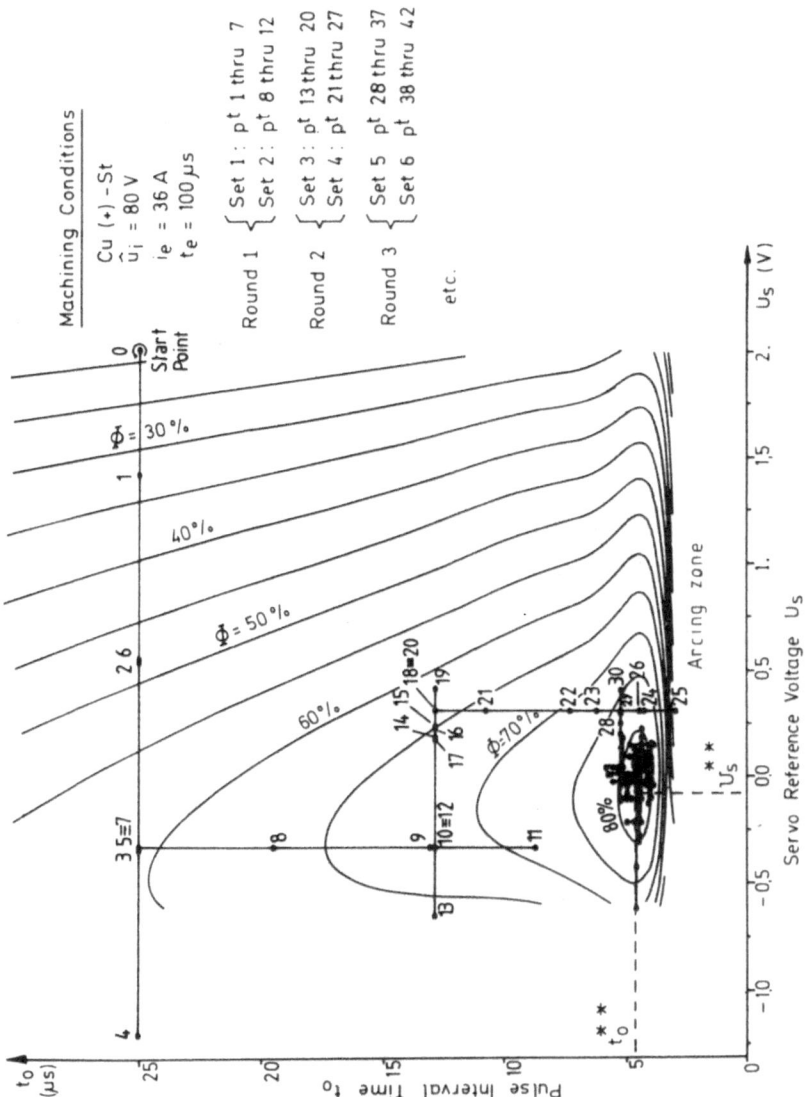

Fig. 5.18. Organization of sectioning search.

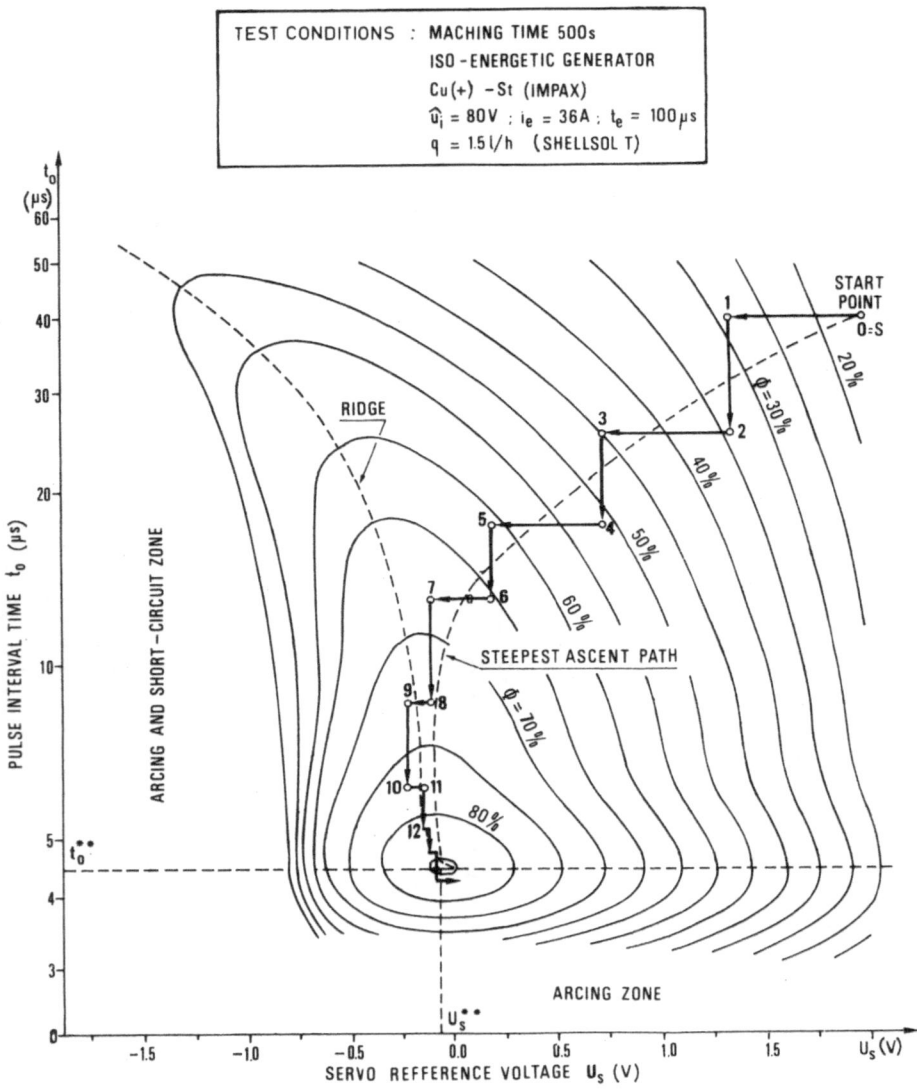

Fig. 5.19. Organization of steepest-ascent search.

The left-hand side diagrams of Fig. 5.20 show that the maximum efficiency was reached after about 1 minute for the steepest-ascent algorithm, compared with about 3 minutes for the sectioning strategy.

5.4.4.3   Ridges

Ridges often occur on the surface representing the performance index versus various control parameters. This is due to interaction between the control parameters [10, 18]. In the case of EDM, for example, the optimum pulse interval time is larger for small gaps (i.e. small servo reference voltages), than for large ones. This explains the ridge visible on Fig. 5.19. A similar ridge can be seen in Fig. 5.13 along the contour line of 10% t_d.

Sharp ridges on the hill to be climbed may cause the optimization strategy to stick on the ridge at a point lower than the absolute maximum, because this point appears as a local optimum of function U_s as well as of t_0. Even a steepest-ascent strategy may have difficulties in detecting that still higher efficiencies may be reached by moving along the ridge. This may result in a slow convergence to the top, as shown in Fig. 5.21b. The situation was worsened in this figure by taking a starting point ($t_0 = 60$ µs) very far from the optimum. However, the steepest-ascent strategy reached a high efficiency after about 1 minute (see Fig. 5.21b, left diagram), while this was only achieved after 4 minutes with the sectioning strategy (Fig. 5.21a).

Some authors describe techniques to accelerate the search to an extremum in the case of ridges or ways that follow ridges [56–58]. To our knowledge, none of those techniques have yet been applied to physical and chemical machining. In the examples of Fig. 5.18 to Fig. 5.21, the pernicious influence of the ridge was reduced by transforming the control space by taking the logarithm of t_0: using "log t_0" instead of t_0 as control parameter reduces the elongation of the ridge. Rajurkar achieved a similar result with a servo system that avoided interaction between servo gap control and the pulse interval time. This servo system is included in a model reference AC system [10].

5.4.5   Knowledge-based Adaptive Control in P&C Machining

A lot of human experience and knowledge is necessary to achieve optimum results with P&C machining. Many P&C machines provided with adaptive control still need human supervision and operator intervention when machining deteriorates because of events not foreseen or allowed for within the AC controller. For example in wire-EDM, the wire often breaks suddenly. This rupture can have many reasons: too much power dissipated in the wire, too much axial tension applied to the wire, unstable process, bad surface conditions etc. An adaptive control system cannot anticipate wire rupture in every situation. However, an experienced operator identifies dangerous machining conditions and can prevent rupture in most cases.

The latest evolution in automation tries to integrate the operator's knowledge in a "knowledge-based system" or "expert system". The ultimate goal of the expert system is to control the whole machining process without any intervention from a human operator.

A knowledge-based system for process control in sinking-EDM was presented in 1989 [34]. A similar system for wire-EDM has been developed by Dekeyser (Fig. 5.22) [25, 33]. This system is written in TURBO-PROLOG on an IBM-PC. The knowledge is represented by some kind of 'IF...THEN...' rules: i.e. IF (situation or

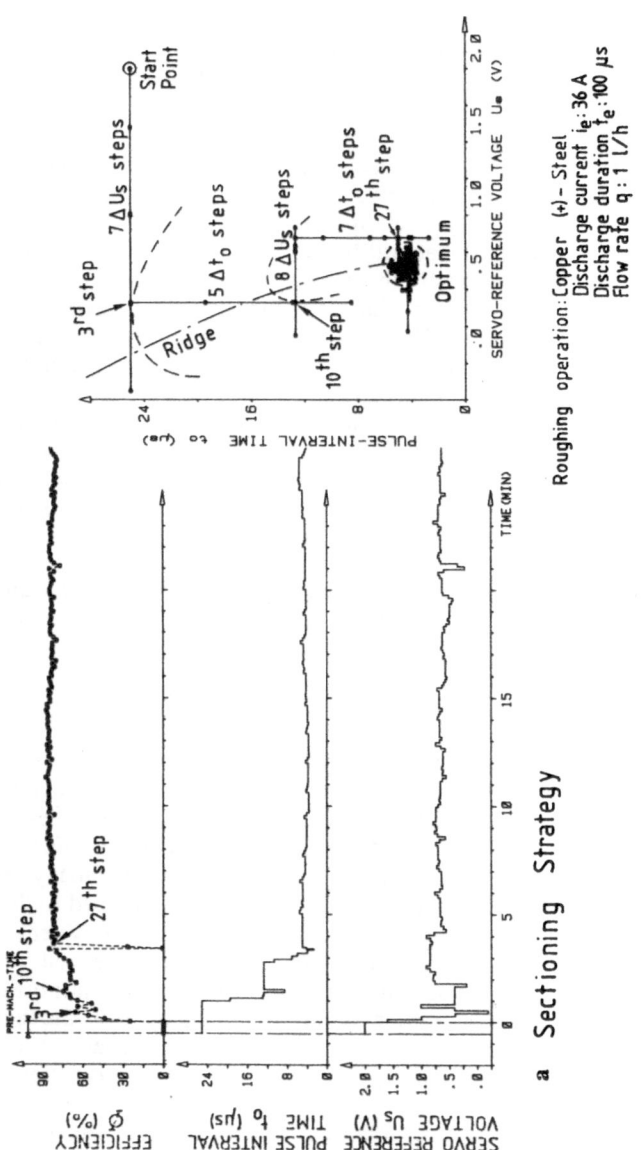

Roughing operation: Copper (+) - Steel
Discharge current i_e: 36 A
Discharge duration t_e: 100 μs
Flow rate q: 1 l/h

a   Sectioning   Strategy

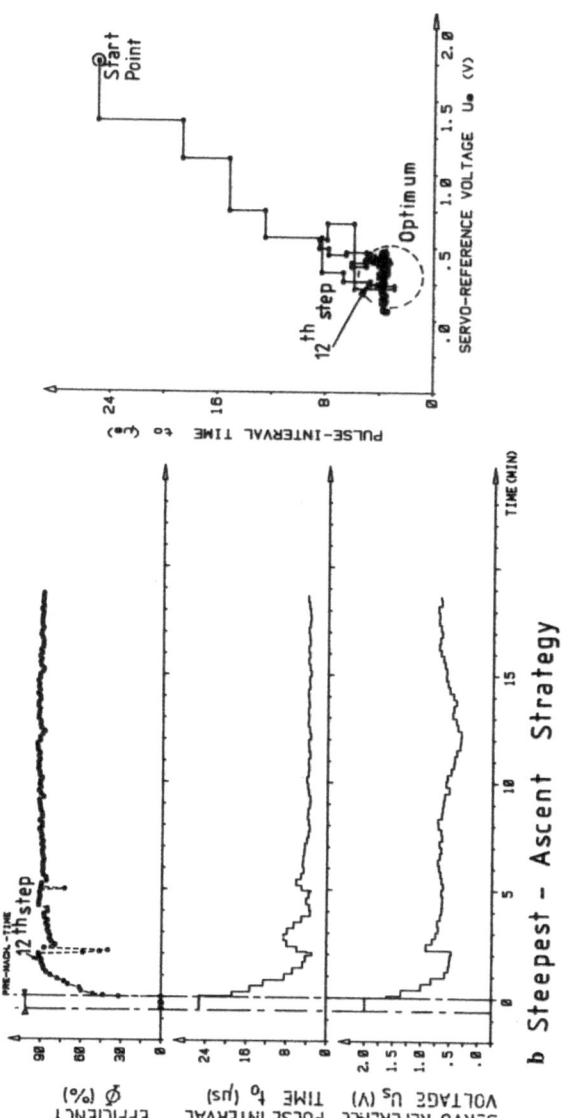

Fig. 5.20. Comparison of sectioning and steepest-ascent strategy.

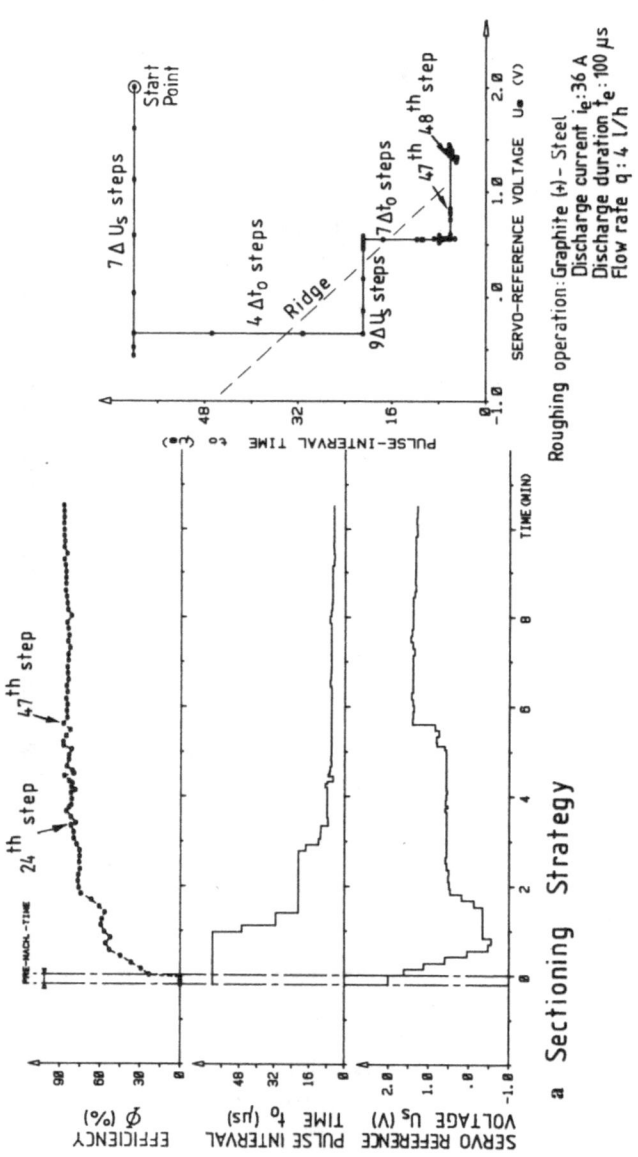

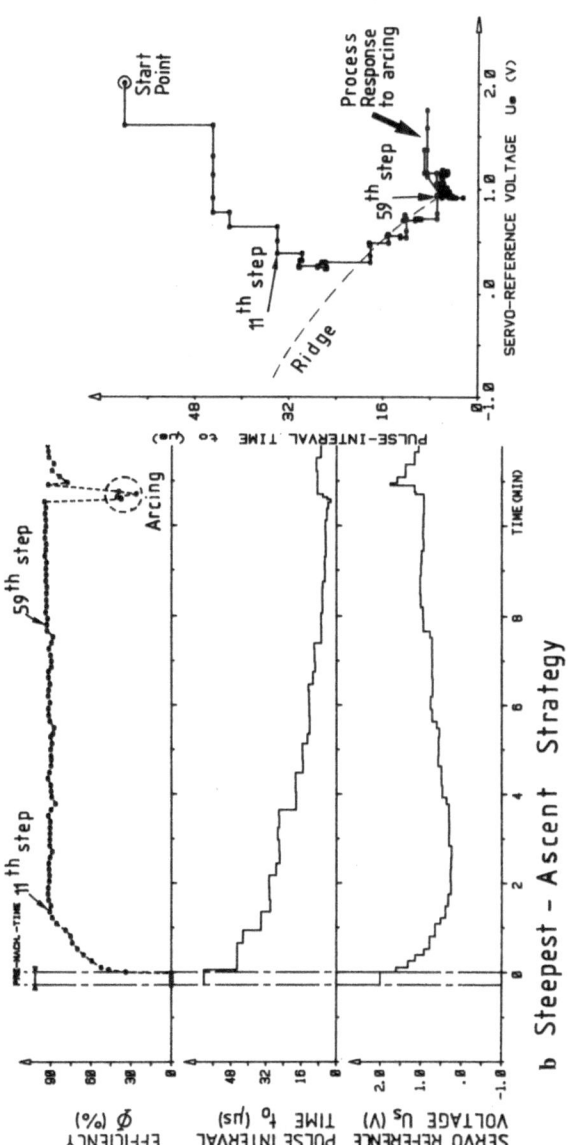

Fig. 5.21. Second comparison of sectioning and steepest-ascent strategy.

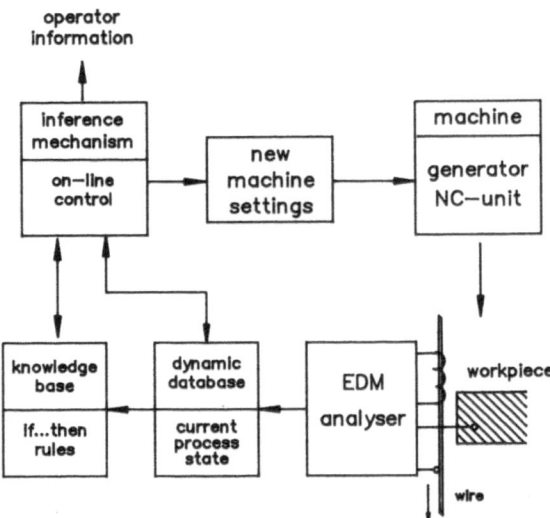

Fig. 5.22. Knowledge-based AC system for EDM.

event) THEN (suggestion or action). The control system evaluates the process (detection of situation or event) and uses the knowledge base to find the right action. A typical AC rule is, for example: IF too many short circuits THEN decrease frequency of discharges and feed servo setting.

Two kinds of knowledge are distinguished:

1. *Heuristic knowledge.* This is operator's knowledge obtained through training and experience. Some of those heuristic rules are represented in the trouble–cause diagram of Fig. 5.23. It relates to the possible cause of wire rupture. Not all the rules yield automatic process adjustment. For example the following two rules:
   IF_THEN_SUGGEST("the wire is broken", "the axial force is too large")
   IF_THEN_TAKE("the axial force is too large", "adjust the wire force, the correct value is ‹value› Newton")
   will only generate a message to the operator, because no automatic adjustment of the wire force is provided for.
2. *Theoretical knowledge.* This represents knowledge based on a physico-mathematical model of the thermal load of the wire electrode. This model allows identification of an adaptive control constraint strategy which limits the heat power dissipated in the wire.

The knowledge comprises rules and facts. It is stored in two different databases (Fig. 5.22). The dynamic database contains the measured sensing parameters used for evaluating the actual process state: i.e. facts such as number of short-circuit pulses, performance index. The content of this database changes at each process evaluation. The knowledge database contains the 'IF...THEN...' rules representing the heuristic and theoretical knowledge and the permanent facts (e.g. type of operation, workpiece material, ACC constraint for the number of short circuits or the dissipated power etc.). The inference mechanism or control system combines the

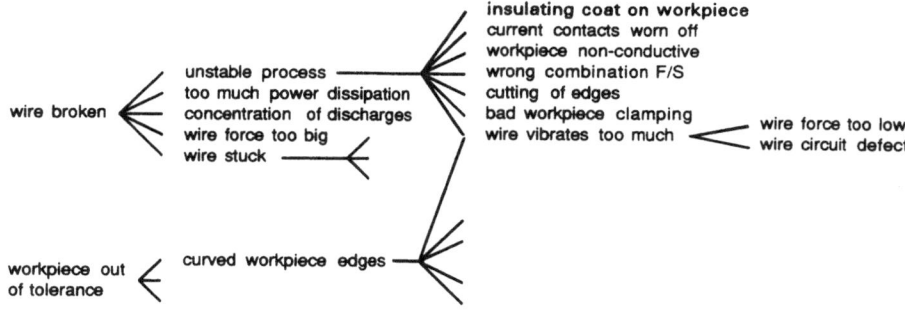

Fig. 5.23. Heuristic trouble-to-cause rules for wire-EDM.

actual process data with the knowledge rules and calculates the new machine settings.

Experiments with this expert system showed an improvement of the cutting speed by 20%, with reduced risk of wire rupture in dangerous machining conditions (e.g. when cutting corners). Morita *et al.* [34] even report improvements of 45% in sinking-EDM compared with using a machining speed with constant control parameters, and some 4% compared with operation by a skilled human.

5.4.6   Results Obtained with Adaptive Control in P&C machining

Many authors report significant progress resulting from the application of adaptive control to P&C machining. Figure 5.24 compares material-removal rates achieved for identical surface roughness and electrode wear with and without AC. A gain of 100% is often achieved, with even 400% for one particular roughing condition. Staelens and Kruth [14] developed a control system optimizing planetary EDM. Time savings of 45% were noticed with circular, and 64% with rectangular electrodes. Many other authors have reported quantified results which combine several benefits for sinking or planetary EDM, such as higher removal rates, lower electrode wear, better surface roughness or a smaller thermal influenced zone [1, 8, 10, 11, 18, 42, 53, 59].

Few papers report an increase of material removal rate in ECM, ECG or LBM. This is probably because adaptive control for those processes aims rather to increase workpiece quality or accuracy. Shpitalni [54] gives convincing photographs to show increased accuracy in adaptive controlled electro-chemical grinding. This mainly relates to better control of the overcut. Krampitz and Bruckmann [39] obtained similar results in electro-chemical machining. Adaptive control in LBM mainly aims at improved workpiece quality in terms of thermal damage, depth of hardening, defects in coatings etc. [28, 29]. These latter developments are quite recent and it may take some time before significant results are obtained.

5.5   Conclusion

The long machining times resulting from low material-removal rate and high risks of disturbance arising from the erratic character of P&C machining processes need automatic control if reliable and safe working conditions are to be established. Many

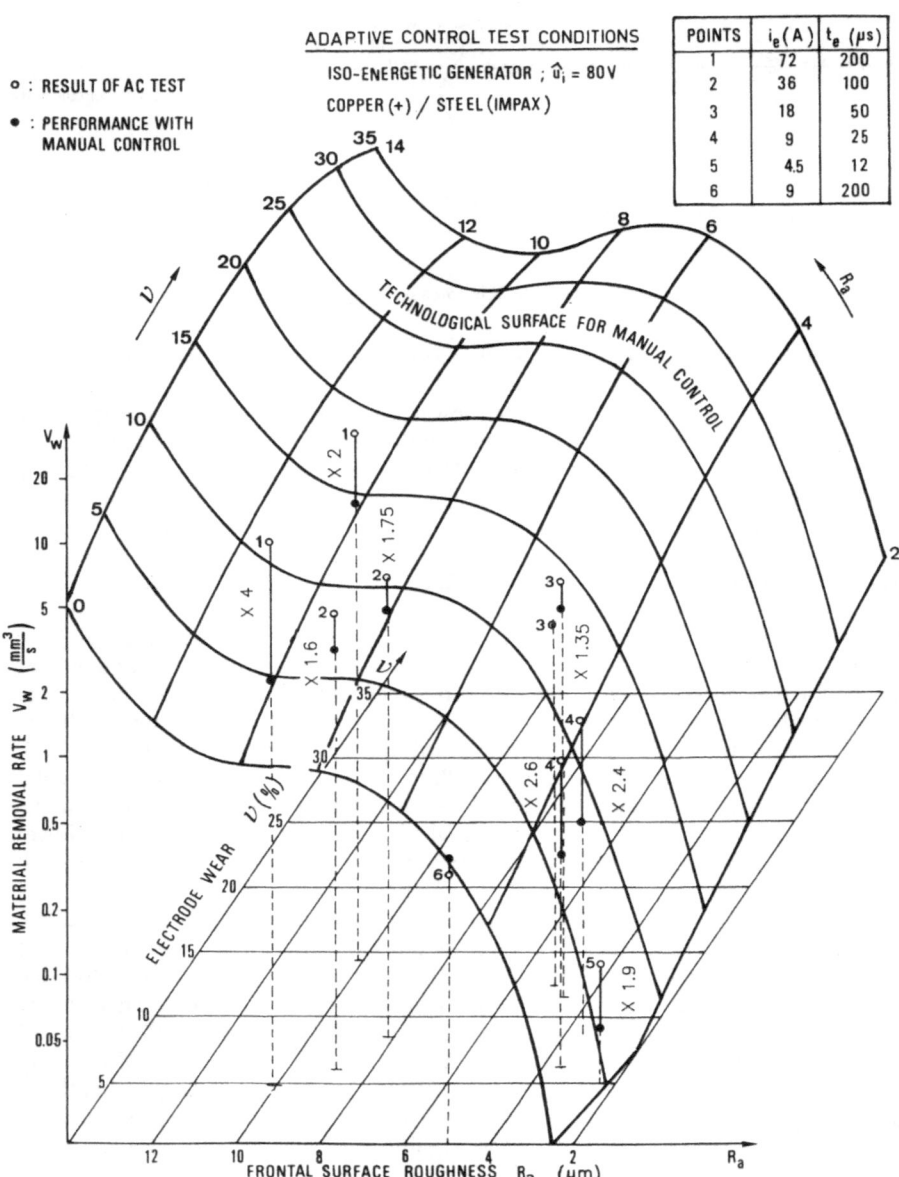

Fig. 5.24. Comparison of material removal rate with and without adaptive control in EDM.

developments have demonstrated that adaptive control not only allows process autonomy and safety to be increased, but also that substantial gain in process efficiency can be achieved. Adaptive control is an answer to the lack of experienced operators able to handle these complex processes. This is the reason why adaptive control has found much wider applications in physical and chemical machining than in traditional machining processes. Today, all EDM machines are equipped with automatic control systems. Industrial applications are also known for electro-chemical machining, electro-chemical grinding and electron beam machining. It is believed that these applications will extend rapidly to many other physical and chemical machining processes, such as laser beam machining and ultrasonic machining.

References

1. Weck M, Dehmer JM. Digitale adaptive Regelung des Funkenerosionsprozesses. VDI-Z 1989; 131(2): 39–44
2. Dehmer J. Adaptive control optimization of EDM gap control. In: Integrated European Course in Mechatronics, Aachen, Germany, 10–14 April 1989
3. Larsson CN, Zhu S. Micro-computer control of an electrochemical machine. In: 22nd Intl Machine Tool Design and Research Conf, 1981, pp 394–399
4. Freer HE, Martin HP. A pressure sensitive adaptive control system for ECM. In: Proc 6th Intl Symp Electro-Machining (ISEM-6), 1980, pp 264–297
5. Larsson CN. Adaptive control of ECM based on flow measurement. In: Proc 6th Intl Symp Electro-Machining (ISEM-6), 1980, pp 293–300
6. Kremer D, Lebrun JL, Moisan A. Effects of ultrasonic vibrations on the performances in EDM. Annals of the CIRP 1989; 38(1): 199–202
7. Behrens A, Odensass Ph. Increasing efficiency of spark-erosion – sinking machines by adapting control using digital processors. In: Proc 10th Intl Symp Electro-Machining (ISEM-X), Magdeburg, 6–8 May 1992, pp 112–120
8. Dauw DF. On-line identification and optimization of electro-discharge machining. PhD thesis, KU Leuven, 1985
9. Heuvelman CJ et al. Microcomputer controlled spark-erosion. EDM Digest 1980; II(5): 24–28
10. Rajurkar KP, Wang WM. A new model reference adaptive control of EDM. Annals of the CIRP 1989; 38(1): 183–186
11. König W, Weck M, Ennig H-J, Peuler H. Electro-discharge sinking – development of an AC system composed of subcontrol loops. In: Proc 20th Intl Machine Tools Design and Research Conf, vol III, Electrical Processes, 10–14 September 1979, pp 575–581
12. Kruth JP, Snoeys R, Lauwers B, Juwet M. A generalized post-processor and process-planner for five-axes wire EDM-machines. Annals of the CIRP 1988; 37(1): 203–208
13. Weck M, Dehmer JM. Adaptive Optimierungs der Planetärerosion. Industrie-Anzeiger 1989; 3(4): 18–26
14. Staelens F, Kruth JP. A computer integrated machining strategy for planetary EDM. Annals of the CIRP 1989; 38(1): 187–190
15. Kruth JP, Lauwers B, Clappaert W. A study of EDM pocketing. In: Proc 10th Intl Symp Electro-Machining (ISEM-X), Magdeburg, 6–8 May 1992, pp 121–135
16. Tarui Y. Lithography systems for VLSI. Bull Japan Society of Precision Engineering 1984; 18(2)
17. El-Menshawy MF, Bhattacharyya SK. The use of acoustic techniques for monitoring and controlling the EDM process. In: Proc 19th Intl Machine Tool Design and Research Conf, UMIST, Manchester, 1978, pp 559–566
18. Kruth JP. Adaptive control optimization of electro-discharge machining. PhD thesis, KU Leuven, 1979
19. Hon KK, Razavi ES. A contribution to the monitoring and control of EDM based on high frequency methods. SME Manufacturing Technology Review 1987; 2, 413–417
20. Shaw TW, Lee LC, Crookall JR. Automation of the EDM process. In: Proc 20th Intl Machine Tool Design and Research Conf, vol III, Electrical Processes, 10–14 September 1979, pp 591–598
21. Van Osenbruggen C, de Regt C. Electro-chemical micromachining. Philips Technical Review 1984; 42(1): 22–23

22. Matthes HG. Zum Schutz der Elektroden, Leistungselektronik verhindert Lichtbögen beim elektrochemischen Senken. VDI Nachrichten 1982; 5(29): 16–17

23. Rajurkar KP, Schnacker CL (submitted by Lindsay RP). Some aspects of ECM performances and control. Annals of the CIRP 1988; 37(1): 183–186

24. Jennes M. Adaptive control of EDM-wire cutting based on real time pulse analysis. PhD thesis, KU Leuven, 1988

25. Dekeyser W. Knowledge-based system for wire-EDM. PhD thesis, KU Leuven, 1988

26. Tanimura T, Heuvelman CJ, Veenstra PC. Properties of the servo gap sensor with wire spark erosion machining. Annals of the CIRP, 1977; 25(1): 59–63

27. Kinoshita N, Fukui M, Gamo G. Control of wire-EDM: preventing electrode from breaking. Annals of the CIRP 1982; 32(1): 111–114

28. König W, Herziger G, Willerscheid H, Wissenbach K. Temperaturmessungen beim Laserstrahlhärten. Laser Magazin 1988; 1: 16–22

29. König W, Meis FU, Willersheid H, Schmitz-Justen C. Process monitoring of high power CO_2-lasers in manufacturing. In: Proc LIM-2, Birmingham, 1985, pp 129–140

30. M'Pherson PK. Systems and manufacture: an introductory study. In: Proc CIRP Seminars on Manufacturing Systems 1972; 1(3): 209–264

31. Snoeys R, Dauw DF, Kruth JP. Survey of adaptive control in electro discharge machining. J Manufacturing Systems 1983; 2(2): 147–164

32. Mueller PA. Trainable adaptive control for automated machining. SME technical paper, MS 72–132, 1972

33. Snoeys R, Van Brussel H, Dekeyser W. Knowledge-based system for fault diagnosis and process-control in wire-EDM. In: Proc ASME Winter Annual Meeting, Research and Technological Developments in Nontraditional Machining, PED Vol 34, 1988, pp 117–133

34. Morita A, Imai Y, Noda A, Maruyama H, Kobayashi K. Fuzzy controller for EDM. In: Proc 9th Intl Symp Electro-Machining (ISEM-9), Nagoya, 1989, pp 236–239

35. Colwell LV. Automatic adaptive control for electro-chemical grinding. Annals of CIRP 1969; 18(1): 577–584

36. Lenz E, Levy GN. An approach to optimization of electro-chemical grinding. In: Manufacturing Engineering Transactions, vol III. SME, 1975, pp 941–949

37. Levy GN, Lenz E. An Optimizer for ECG. In: Proc 3rd North American Metalworking Research Conf (NAMRC III), Carnegie-Mellon Univ, Pittsburg, Pennsylvania, May 1975, pp 941–954

38. Shpitalni M, Koren Y, Lenz E. Adaptive control of the ECG process. In: Proc CIRP Seminar Manufacturing Systems 1978; 7(3): 187–196

39. Krampitz R, Bruckmann G. Rationalization through microcomputer-controlled electrochemical machining in parts production. In: Proc 7th Intl Symp Electro-Machining (ISEM-7), Birmingham 1983, pp 434–442

40. Tseng MM. A systematic approach to the adaptive control of the electro-discharge machining process. ASME J Engineering for Industry 1978; 100(August): 303–310

41. Lascoe OD. Minicomputer for on-line process optimization. EDM Digest 1981; III(1): 14–25

42. Barz E. Strategie für die selbstständige Optimierung des funkenerosieven Senkens. Dissertation, TH Aachen, 1976

43. Weck M, Slomka M. Adaptive Regelung des Senkerodierens. VDI-Z, 1985; 9: 319–323

44. Willerscheid H. Prozesskontrolle in der Lasermaterialbearbeitung. In: Sem Lasermaterialbearbeitung, Fraunhofer Institut für Produktionstechnologie, IPT, 1988

45. Olsen FO, Andersen KE, Raben N, Thomassen FB. Investigations in methods for adaptive control of laser cutting. In: Proc Conf Laser Materials Processing (LAMP), Osaka, Japan, 1987, pp 267–272

46. Gregersenn O, Olsen FO. Beam analyzing system for CO_2 lasers. In: Proc ICALEO, 1990, pp 27–35

47. Jorgensen H, Kechemair D, Olsen FO. On-line monitoring of the laser welding process. DVS-Berichte 1991; 135: 1–6

48. Bignon C, Bédrin C. (submitted by Weill R). Application of eddy currents to the in-process measurement of the gap in ECM. Annals of the CIRP 1982; 31(1): 115–118

49. Rajurkar KP, Pandit SM, Wittig WH. Pulse current signal as a sensor for on-line computer control of EDM. In: Proc NAMRC-II, SME Manufacturing Engineering Trans, 1983, pp 379–385

50. Croockall JR, Shaw TW. Improving the operation and machine performance in EDM. EDM Digest 1982; IV(3): 18–19

51. Saito N, Kobayashi K. A method for adaptive control in EDM process. In: Proc 3rd Intl Symp Electro-Machining (ISEM3), Vienna, 12–15 October 1970

52. Bhattacharyya K et al. An adaptive control system for EDM process. In: Proc 8th North American Metalworking Research Conf, 1980, pp 229–234

53. Kruth JP, Snoeys R, Van Brussel H. An adaptive control solution to the problem of automation and optimization of EDM. In: Proc 7th North American Metalworking Research Conf. NAMRC-VII, 1979, pp 307–314

54. Shpitalni M. A double loop adaptive control system for the ECG process. SME Manufacturing Technology Review, 1983, pp 29–35

55. Eveleigh VW. Adaptive control and optimization techniques. McGraw-Hill, New York, 1967

56. Porter B, Richardson EJ. Search strategies for the self-optimizing computer control of the metal-cutting process. Proc CIRP Seminar Manufacturing Systems 1978; 7(3)

57. Wilde DJ, Breightler CS. Foundations of optimization. Prentice-Hall, Englewood Cliffs, New Jersey, 1967

58. Kazakevich VV, Mochalov IA. Sequential algorithms of accelerated hill climbing in inertial plants to be optimized. Automation and Remote Control 1977; 38(1): Part 1 32–40

59. Rajurkar KP, Ahmed MS, Royo GF. Effect of RF control and orbital motion on EDM performance. In: Proc 9th Intl Symp Electro-Machining (ISEM-9), Nagoya, 1989, pp 30–33

6 Automatic Supervision in Metal-forming

K. Chodnikiewicz and L. Olejnik

6.1 Introduction

In metal-forming processes, workpieces are most commonly formed by using at least two tools (a punch and a die) which move linearly with respect to each other. One part of the tooling is generally fixed to a slide of the press, and the other to the press table.

This chapter deals only with those metal-forming processes performed using presses. However, nearly all metal-forming processes (e.g. upsetting, forging, blanking, deep drawing, bending, extrusion etc.) can be performed in this way.

The working cycle of the press can be divided into the following stages:

Press ram stationary at its upper position

Press ram moves towards the workpiece

Working stroke, during which the workpiece is plastically deformed

Return stroke of the press ram

The following three functions have to be performed in each working cycle:

1. Feeding of the sheet, billet etc. into the working space of the press, as well as correct placing of these with respect to the tools.
2. Plastic deformation of the workpiece.
3. Removal of the finished workpiece from the working space.

A disturbance of any of the above functions decreases the quality index of the whole process [1]. Such disturbances as cracking of the tool, inappropriate feeding or placing of the sheet or billet to be formed, plastic instability of the material etc. will decrease the quality index suddenly. Others, such as tool wear, press element wear etc. influence the quality very slowly. In the case of a disturbance of the first kind, a supervisory system must stop the process as soon as possible. In the case of the second kind of disturbances, the process should be stopped when the workpiece quality falls below a certain, *a priori* determined value.

The functional requirements which should be fulfilled by the supervisory system of the metal-forming process can be stated as follows:

1. The feeding of sheets or billets should be stopped if the finished workpiece has not been removed from the working space of the press.
2. The press should be stopped not later than during the approaching movement of the press ram if certain parameters of the metal sheet or billet (dimensions, mass,

temperature, hardness etc.) are outside given tolerances. Similarly, the press should be stopped following incorrect placing of the metal sheet or billet with respect to the tools.

3. The press and all auxiliary devices should be stopped in the case of:
   a) failure of the tools
   b) incorrect shape or cracking of the workpiece
   c) galling of the tool.
4. The wear on the tools should be continuously inspected and estimated. If changes in the shape as a result of wear exceed *a priori* determined limits, the process should be stopped.
5. All parameters that influence the proper working of the press and its auxiliary devices should be continuously inspected and estimated. These include, for example, the pressure of the working medium, the temperature of bearings, the lubrication quality, the press load etc.

It is worth taking notice of the fact that none of the requirements listed above are involved with adaptive control, as this type of control applies only to a very limited number of metal-forming processes.

The requirements given above refer equally to automated and non-automated metal-forming processes. Thus, it seems reasonable to consider the following question:

How should the modern meaning of the term "Automatic Supervision in Metal Forming" be understood?

To answer this question we would first like to recall the definition of a supervisory system given by Szafarczyk and Chisholm [1]. This definition is very broad and embraces supervision system devices both based on micro-switches and controlled by computers. According to this definition, the combination of monitoring and control is the essence of all supervisory systems. Szafarczyk and Chisholm have distinguished automated supervision of a manufacturing process from the narrower meaning of the term for example, automated supervision of feeding. It is time to say that automated supervision in metal-forming, as defined by its narrower meaning, has been used for a long time. Various electric and electronic devices are used to inspect the work of the press and stop operation if a malfunction occurs. These devices can be termed a supervisory systems. However, in order to obtain a fuller answer to the question above, we must discuss the typical ways of fulfilling the requirements which have been stated earlier.

The removal of the finished workpiece from the working space of the press (requirement 1) has most frequently been checked by using micro-switches, inductive, optical, acoustic and similar sensors. This inspection includes operation of the press ejector as well as displacement of the workpiece itself. The workpiece, while being removed from the working space, must activate the specific sensor. The same sensors can be used to avoid the feeding of two strips of sheets instead of only one and to verify the correct placement of the sheet or billet with respect to the tools etc. The inspection of the billet temperature and such parameters as the oil and/or air pressure, the bearing and guide lubrication etc. can be carried out in the same way as for other machines.

Producers of presses aim at fulfilling requirements 1, 2 and 5, being aware of the serious consequences of improper feeding, of not removing the finished workpiece from the working space of the press, and similar disturbances of the press working

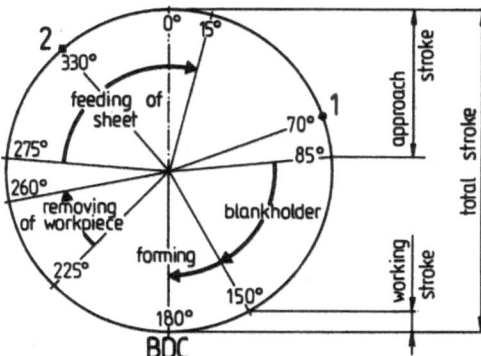

Fig. 6.1. Cyclogram of crank press equipped with feeding device: 1 – checking of feeding; 2 – checking of workpiece removal.

cycle (Fig. 6.1). All these disturbances can cause press overloading and ultimately failure of the press. Thus, presses, especially mechanical ones, are equipped with various overload safety devices. These have been designed to ensure safe operation rather than to supervise the manufacturing process. According to the definition discussed above, overload safety devices fulfil the functions of supervisory systems, however in a way which is far from perfect. The simplest overload safety devices (shear plates, shear pins, slip clutches etc.) are not adjustable to any setting, or at least such adjustments are very difficult to make. Therefore, the overload safety devices under discussion are able to stop the press when tool failure occurs, but only if this disturbance is connected with overloading of the press. If tool failure simply causes the press load to decrease, the press will not be stopped.

Next we come to requirements 3 and 4, and to the role played by the person who operates and supervises the press and its auxiliary devices. Most frequently the fulfillment of requirements 3b, 3c and 4 consists in inspecting the quality of the workpiece after it has left the press. However, in this case any decision about continuating or interrupting the process is long after the instant in which the workpiece was manufactured. This always involves the risk of losses resulting from unnecessary stoppage of the press, or from the manufacture of workpieces of poor quality.

However, it should be emphasized that an experienced press operator will immediately notice many disturbances in the metal-forming process, for example a tool cracking. This statement may seem obvious but it has an important outcome. The essence of the matter is that an experienced operator inspects the quality of the process on the grounds of a characteristic symptom or a set of symptoms (syndrome) [1], and then takes appropriate supervisory action (most frequently by interrupting the process manually). The modern trend is to replace the experienced operator by a supervisory system which acts in the same way as this person. This implies that just one or a few physical values need be monitored and evaluated by the supervisory system in order to obtain the same supervising results. This conclusion will be discussed later in more detail.

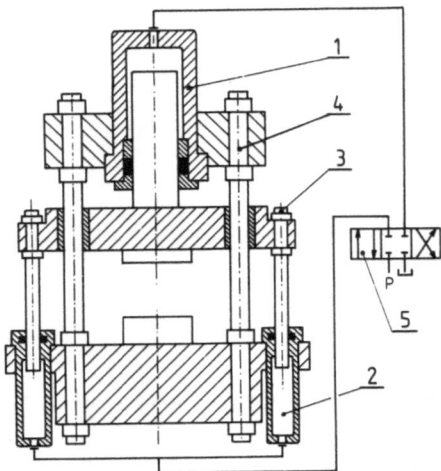

Fig. 6.2. Scheme of hydraulic press:
1 – main cylinder; 2 – return cylinder; 3 – ram;
4 – frame; 5 – directional valve.

6.2   Influence of Press Type on Supervisory Action

Interruption of the process is the most common supervisory action in metal-forming. The possibility of stopping the whole system completely is most frequently dependent on the inertia of the press ram-driving mechanism. Thus, some features of presses which influence supervisory action will be examined below.

Hydraulic, typical screw, crank and clutch-controlled screw presses will be considered. The main differences between them lie in the driving system. In a hydraulic press the main part of the driving system is a hydraulic cylinder with a piston connected to the press ram. In a screw press, a screw mechanism is utilized; in a crank press, a slide–crank mechanism. We will use the name "crank press" for all presses in which the crankshaft is the main part of the driving mechanism.

Hydraulic Press. In order to start the approach movement of the press ram, a directional valve (Fig. 6.2), which connects a pump to a chamber of a hydraulic cylinder, must be switched over. The moment of switch-over does not depend on the previous working cycle of the press. The ram velocity during the approach movement, as well as during the working stroke, is relatively low. The working stroke lasts at least 0.1 s [2]. Normally, the press ram can be easily stopped during any stage of the working cycle, because only switch-over of the valve into the neutral position is needed. The safety valve protects the press against overloading.

Screw Press. There are several types of screw presses, which have different driving systems for the flywheel. In all screw presses the screw is connected to the flywheel, most commonly by means of a slip clutch. Similarly, as in the case of the hydraulic press, in order to start the approach movement of the ram, a control signal is needed. The brake is disengaged and a driving torque is applied to the flywheel (Fig. 6.3), causing its accelerated, rotational motion. If the angular speed of the flywheel reaches a given value, the driving system is disconnected. The ram speed at the end of the approach movement ranges between 0.5 and 1.5 m s^{-1}, depending on the type

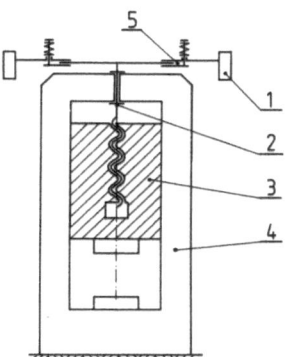

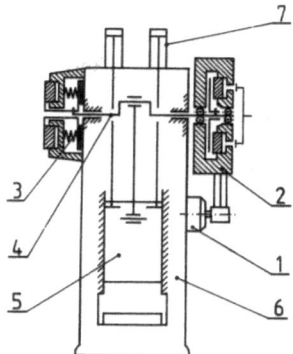

Fig. 6.3. Scheme of typical screw press (driving system not shown):
1 – flywheel; 2 – screw; 3 – ram; 4 – frame; 5 – slip clutch.

Fig. 6.4. Scheme of crank press:
1 – electric motor; 2 – flywheel with clutch; 3 – brake; 4 – crankshaft; 5 – ram; 6 – frame; 7 – counterbalance cylinder.

of press. The time of the working stroke depends on the kinetic energy accumulated in the flywheel, as well as the reaction force of the workpiece being formed. This time in average conditions ranges from 10^{-2} to 10^{-1} s [2]. In the final instant of the working stroke the speed of the ram is zero. Immediately after the working stroke the return movement begins, caused, at first, by elastic deflection and later by the driving system. The flywheel and the screw rotate in the opposite direction to that of the approach movement. In the second part of the return movement the brake stops the flywheel, the screw and the ram. If necessary, the same brake also stops the ram during the approach movement. The effective inertia of the parts to be stopped (ram, screw and flywheel) is quite large and, for this reason, the efficiency of the brake is rather limited. The braking distance ranges from 1/2 to 2/3 of the total stroke of the press ram. Thus, the ram moving towards the workpiece will be effectively stopped (before the working stroke) only if the braking action begins in the first part of the approach movement. This fact limits the time for inspections, which have to be done before the working stroke starts.

During the working stroke, the braking is inefficient. So, if this stroke proceeds improperly, the supervisory system should stop the press ram in its upper position and so make the next working cycle impossible.

Crank Press. The constant stroke of the press ram is a characteristic feature of all crank presses (Fig. 6.4). In some presses, the stroke of the ram can be adjusted to obtain a given value, but only when the press is stationary. The flywheel of the crank press is continuously driven by the electric motor. In order to obtain the motion of the ram, the crankshaft brake must be disengaged, and the clutch engaged. The clutch connects the flywheel to the crankshaft, and the ram starts to move up and down, travelling from the upper dead centre (UDC) to the bottom dead centre (BDC). A reciprocating motion of the ram continues until the moment when the clutch is disengaged and the brake is engaged. If an intermittent mode of the press action has been set, the ram stops automatically close to its UDC after each working cycle. In the case of continuous work, the ram is stopped by means of the special control signal sent from the control system of the press.

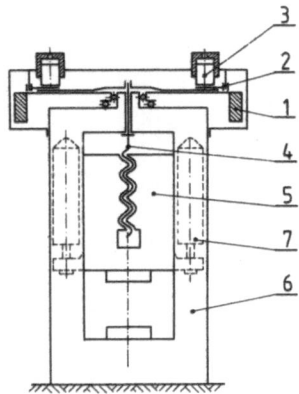

Fig. 6.5. Scheme of clutch-controlled screw press:
1 – flywheel; 2 – clutch; 3 – clutch cylinder; 4 – screw;
5 – ram; 6 – frame 7 – return cylinder.

The time of the approach movement as well as the time of the working stroke of the crank press depend on the number of ram strokes per minute, n, and the ratio x/S where S is the press ram stroke and x is the distance of the ram from its BDC at the instant of contact between the punch and the workpiece. The ratio x/S depends in turn on the type of metal-forming process. Let us assume that $x/S \approx 1/8$, which is the value commonly met with in practice. Under this assumption the approach movement takes about $23.4/n$ seconds. This means, for example, that for $n = 100$ strokes per minute we obtain an approach movement time equal to 0.23 s. However for $n = 600$ strokes per minute, this movement takes only 0.04 s. In the first case, the press ram can be stopped during the approach movement. In the second case, the time is too short to stop the press ram before it contacts the workpiece.

The facts just discussed have found their expression in the safety regulations which are in force in many countries (e.g. [3]). According to the general regulations, it must be possible to stop the ram during the approach movement, irrespective of the type of action (continuous, intermittent) of the press. However, this requirement does not relate to high-speed mechanical presses in which the brake must stop the ram in the vicinity of its UDC despite the fact that a stopping signal has been sent during the approach movement.

If the ratio x/S is equal to 1/8, the working stroke lasts no longer than $6.6/n$ seconds. For $n = 100$ strokes per minute we obtain a time of working stroke equal to 0.066 s. For $n = 600$ strokes per minute this time is only 0.011 s. Thus, it is practically impossible to stop the press ram during the working stroke. Moreover, such action could create self-locking of the slide–crank mechanism. Thus, if the working stroke proceeds improperly, the supervisory system must stop the press ram in the vicinity of its UDC, and make the next working cycle impossible.

Clutch-controlled Screw Press. This is a new type of press in which positive features of screw presses are combined with positive features of crank presses [4]. The flywheel (Fig. 6.5) is driven by the electric motor and rotates continuously, in a similar way to the flywheel of a crank press. In order to connect the flywheel to the screw, the disc clutch is engaged. The effective inertia of the screw and the ram is relatively small, so starting the screw takes a little time. During the working stroke the flywheel is not stopped, but its angular speed is reduced in a similar way to that of the crank press. At the end of the working stroke the clutch is disengaged.

Disengagement of the clutch is controlled by the forming force or by the permissible decrease in angular speed of the flywheel. The return stroke of the press ram and the screw is obtained by means of hydraulic cylinders.

Owing to the short time of disengagement of the clutch (5 to 10 ms), and the small value of the effective inertia of the screw and press ram, the hydraulic cylinders are able to stop the ram during its approach movement. Sometimes the special brake acting on the screw gives an additional braking torque. If the working stroke proceeds improperly, the supervisory system must stop the press ram at its upper position and so make the next working cycle impossible.

6.3 Working Cycle Supervision

6.3.1 The Press–Tooling–Workpiece System and its Disturbances

It is widely known that in the metal-forming operation the yield stress σ_f depends on the equivalent strain ϵ and the strain rate $\dot{\epsilon}$. This statement can be expressed as

$$\sigma_f = \sigma_f (\epsilon, \dot{\epsilon}, ...)$$

where the dots . . . represent other factors that influence the yield stress σ_f. The reaction force F_t of the workpiece being plastically deformed is a function of σ_t, whereas the strain ϵ and the strain rate $\dot{\epsilon}$ can be expressed by the relative displacement of the working surfaces of the tools, x_t, and its velocity, $\dot{x}_t$. Thus one can write

$$F_t = F_t (x_t, \dot{x}_t, ...)$$

The press–tooling–workpiece system is represented diagrammatically in Fig. 6.6. The displacement x_t and the velocity $\dot{x}_t$ can be considered as the input values of the subsystem "workpiece". The force F_t is the output value of this subsystem.

Similarly the displacement x of the ram and its velocity $\dot{x}$ are the input values of the subsystem "tooling". The force F applied to the ram and to the table of the press is the output value of this subsystem. The force F can be written as

$$F = F_t + F_a$$

where F_a is the sum of all the additional forces acting on the press ram. The force F loads the main elements of the press. However, owing to inertia, the dynamic stresses in these elements (e.g. frame) can be much bigger than the static ones which correspond to the force F.

The foundation of the press is loaded by the force F_f. The thermal power N_t and the acoustic power N_a are emitted by the tooling and the workpiece to the environment. The power N_e is supplied to the electric motor. All these physical quantities will be called the system parameters [1].

In Fig. 6.6 the symbols Δs, ..., Δp have been used in order to indicate the most commonly occurring disturbances suffered by the tooling and the workpiece. Under ideal working conditions disturbances do not occur, and all system parameters versus time relationships ($F_t(t)$, $x_t(t)$, $x(t)$ etc.) are also ideal. Any disturbance changes all the time relationships of the system parameters. Obviously, the work of the supervisory system should be based on such a system parameter whose values change considerably after a given disturbance. The choice of a suitable parameter depends on the type of metal-forming process involved and also on the type of machine being used to perform the process.

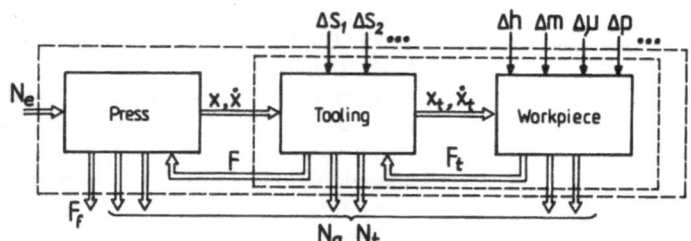

Fig. 6.6. Scheme of press–tooling–workpiece system. Meaning of symbols: Δs_1 – tool failure: Δs_2 – tool wear; Δh – improper initial shape of workpiece; Δm – improper material, $\Delta \mu$ – disturbance of lubrication; Δp – improper placement of workpiece; other symbols explained in text.

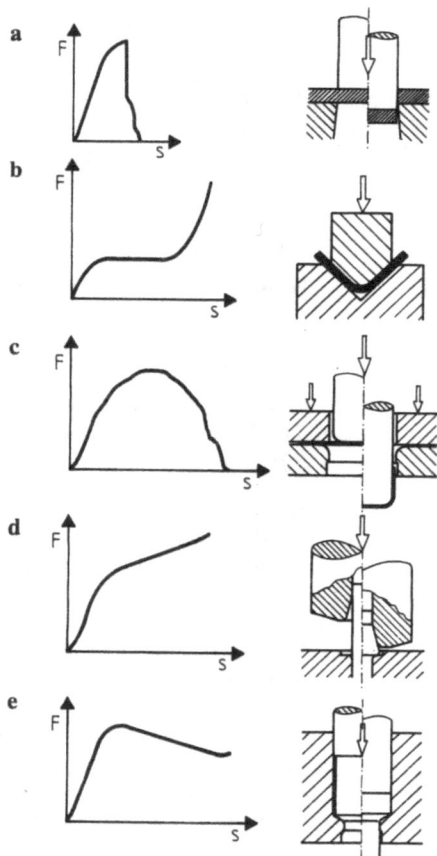

Fig. 6.7. Typical force–travel diagrams for chosen metal-forming processes: **a** sheet cutting; **b** bending; **c** deep drawing; **d** upsetting; **e** forward extrusion.

The force–travel diagram (Fig. 6.7) has been recognized as the best description of a given metal-forming process [5–7]. As it is very difficult to measure the displacement x_t, a measurement of x is taken instead. The relationship between these two variables can be written in the following form:

$$x = x_t + F_t/C_t$$

where C_t is the stiffness of the tooling. (For the sake of simplicity we assume that the force F_t, the displacement x_t, as well as an axis of the press all act along the same direction.) Similarly, the force F or even the stress in the frame of the press is measured instead of the force F_t.

It is widely known that in some supervisory systems used in metal-forming processes only the maximum value of the force F_t is monitored. In other systems, the relationships between force and time, or between the force and an angle of rotation of the crankshaft, are monitored.

Next we will consider which information needs to be obtained, and which can be ignored, in supervisory systems based on the measurement of different physical quantities.

6.3.2 Choice of Supervisory Parameter

As the forming force is changed by any disturbance in the metal-forming process, it seems advisable to measure that force. The force F_t, or F, can be measured by means of strain gauges, piezo-quartz or piezo-ceramic sensing elements. To measure stresses in the frame, strain gauges are commonly used. The influence of two typical disturbances of the bulk metal-forming process on the value of the maximum forming force, F_M, is shown in Fig. 6.8. The same disturbance results in different values of the force increase, ΔF_M, depending on the type of press. It can easily be seen that ΔF_M is not sufficient by itself to distinguish the type of disturbance involved. This distinction can only be made considering the complete relationship $F(x)$ or fragments of it called "windows". In this case the essence of performance of the supervisory system lies in checking whether the inequality

$$F_{1i} < F(x_i) < F_{2i}$$

has been fulfilled (Fig. 6.9a). In this inequality F_{1i} and F_{2i} are the lower and the upper permissible values of the forming force for any given position of the press ram denoted as x_i. Sometimes it may be more advantageous to recognize the position of the press ram as the supervisory parameter. In such a case the following inequality

$$x_{1i} < x(F_i) < x_{2i}$$

has to be checked (Fig. 6.9b). Such a replacement is not possible if the relationship $F(x)$ in the given interval has the form of a constant function (Fig. 6.10a) but is necessary if the relationship $F(x)$ has the form of a step function (Fig. 6.10b).

Let us examine the consequences of choosing various supervisory functions, taking as an example the bulk forming operation carried out by means of a crank press, which can, in the simplest form, be modelled as shown in Fig. 6.11. The spring C represents the elasticity of the press. The simplified elastic characteristic of the press is given in Fig. 6.12a, whereas the diagram of the bulk forming operation is shown in Fig. 6.12b. The point $x = 0$ corresponds to the BDC of the unloaded ram

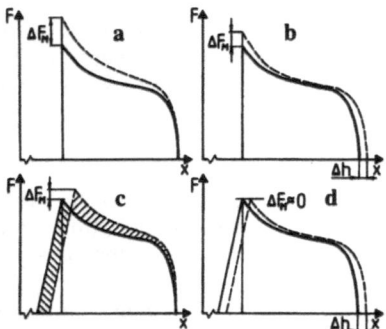
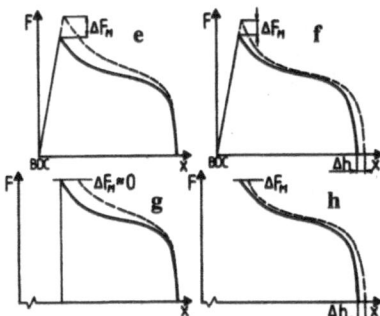

Fig. 6.8. Influence of two different disturbances on bulk metal-forming process carried out with different presses: a,c,e,g – material too hard; b,d,f,h – billet too high; a,b – hydraulic press; c,d – typical screw press; e,f – crank press; g,h – clutch-controlled screw press (the end of working stroke controlled by forming force).

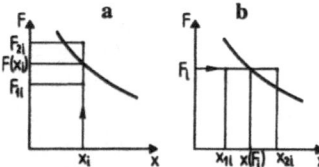

Fig. 6.9. Relation $F(x)$: a F – supervised parameter checked for the given x_i; b vice versa.

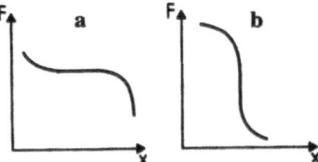

Fig. 6.10. Specific cases of $F(x)$ relationship: **a** the quantity F must not be checked for the given x_i; **b** the quantity x must not be checked for the given F_i.

of the press. The operation starts at the point x_s (compare Fig. 6.11a). As the press is being elastically deformed, the plastic deformation of the workpiece represented by the line SM (Fig. 6.12b) runs simultaneously with the elastic deformation of the press represented by the line OM. At the instant of termination of the loading stage the maximum force, F_M, is reached; the plastic deformation of the workpiece is represented by the segment x_s-x_M, and the elastic deformation of the press by $x_M = f_M$ (compare Fig. 6.11b). Next, the unloading stage of the bulk forming operation starts. The elastic deformation of the press decreases according to the line MO (Fig. 6.12b) whereas the press ram stays almost at a standstill until the press is fully unloaded (compare Fig. 6.11c). Thus, during the unloading stage the angle of rotation of the crankshaft α changes from α_M to α_O, whereas the position of the press ram is almost unchanged.

For this reason one can safely state that only the loading stage of the operation may be supervised on the grounds of the relationship $F(x)$. It can be seen from Fig. 6.12c that the relationship $F(\alpha)$ supports supervision of both the loading and the

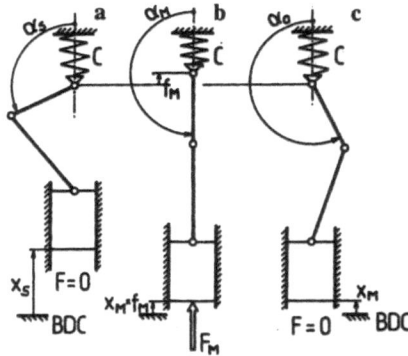

Fig. 6.11. Model of crank press: a the punch meets a workpiece; b the BDC; c the end of unloading stage of the bulk forming operation.

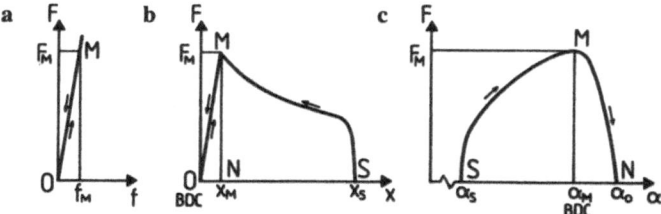

Fig. 6.12. The forming force versus: a elastic deformation of the press; b displacement of the press ram; c angle of rotation of the crankshaft.

unloading stages of the bulk forming operation: the loading stage is represented by the curve SM and the unloading stage by the curve MN. It is worth noting that

$$(\alpha_O - \alpha_M)/(\alpha_M - \alpha_S) \approx 1/4$$

This means that about 20% of all angle measurements fall within the unloading stage. This makes supervision of the tool damage easier, for in most cases damage occurs when the forming force is close to its maximum. Thus, supervision of both the loading and the unloading stages of the bulk metal-forming process, based on the relationship $F(\alpha)$, seems to be a much better prospect than that control based on the relationship $F(x)$.

The time history of the forming force, $F(t)$ is similar to that of $F(\alpha)$. So, from the point of view of monitoring of the working stroke, both these relationships can be equally well used by a supervisory system. It should be noted that the relationship $F(t)$ is influenced not only by disturbances of the process but also by those occurring in the press. This feature of the relationship $F(t)$ can be recognized as an advantage, because the complete system (press–tooling–workpiece) is under inspection. On the other hand, a diagnosis of press disturbance based on the time history of the force seems to be difficult. Different disturbances can influence the force variation in the same way. In sheet metal-forming processes the force F decreases before the ram reaches the BDC position. This means that the relationship $F(x)$ is unique during the whole working stroke. There is also an additional argument for the relationship $F(x)$

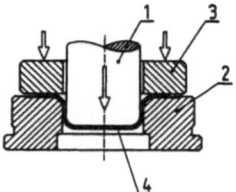

Fig. 6.13. Scheme of deep-drawing process: 1 – punch; 2 – die; 3 – blankholder; 4 – workpiece.

when it is used to supervise blanking and piercing processes. Knowing the force–travel diagram, one can calculate the deformation work, which is widely accepted to be a good measure of tool wear [2].

Three tools are used in deep drawing (Fig. 6.13). The third tool, a blankholder, exerts a force on other regions of the sheet in order to prevent wrinkles. Single-action presses used for deep drawing must be equipped with an additional fixture which performs the function of a blankholder ram. This fixture is a mechanical, pneumatic or hydraulic die cushion, most frequently situated under the table of the press. In double-action presses the punch is fixed to the inner ram, whereas the blankholder is attached to the outer ram of the press. The quality of a deep drawn component depends on the correct choice of blankholder force, F_b. Cracks near the bottom of the component are caused by too high a force F_b. Too low values of F_b result in wrinkles in the collar or the walls of the deep drawn component. Every crack produces a sudden drop in the forming force F_t. However, force drops resulting from short lengths of crack are very small. In such cases, the acceleration of the punch or the press ram has been recognized as a more suitable supervisory parameter than the forces F_t or F [6].

Any increase in the gap between the sheet and the blankholder means the beginning of wrinkle formation. Thus this gap should be the next supervised parameter of the deep drawing process. Gap supervision is described in detail by Siegert *et al.* [6].

In some cases the choice of the supervisory relationship is dependent not only on the type of metal-forming process but also on the press type. Consider, as an example, a bulk metal-forming process carried out with a typical screw press. The time history of the angular flywheel (screw) velocity ω describes perfectly the work of the press during the approach and return stroke. Thus the relationship $\omega(t)$ seems to be very useful for supervising both these strokes. Next, knowing the relation between F and ω, it is easy to monitor the correctness of the working stroke. Indeed, the velocity ω and the force F are connected by the law of conservation of energy [2].

When describing the block scheme of the press–tooling–workpiece system (Fig. 6.6) it was stated that only under static conditions were the stresses in the press frame a good measure of the forming force F_t. Very often the press operates in conditions that are far from static. It is widely known that rapid contact of the punch with the workpiece, as well as rapid decrease in the forming force, are the two main reasons for press vibrations. For example, in high-speed crank presses, the time history of stresses is influenced by both the forming force and the dynamic features of the press. This is the reason why the time history of stresses in the frame of the press dramatically changes when the number of strokes per minute of the press

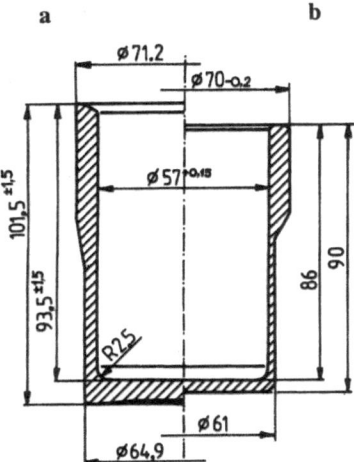

Fig. 6.14. Container shape: **a** after sizing; **b** after machining.

increases. At the same time, the time history of the forming force does not change [7]. Thus, replacement of the forming force by stresses in the press frame should be preceded by a dynamic analysis of the press–tooling–workpiece system.

Coming back to Fig. 6.6, it is necessary to state that the electric power N_e can be measured instead of the forming force but only for those types of hydraulic presses equipped with a constant flowrate pump. In all presses equipped with a flywheel, the time history of the power N_e is quite different from the time history of the forming force.

In some specific cases, the acoustic power N_a, as well as the thermal power N_t, can be monitored by a supervisory system [8].

6.4 Case Study: Backward Extrusion

6.4.1 Process Description

The automatic supervision of the backward extrusion of a laboratory centrifuge container will be described next. This container is of crucial importance to the performance of the centrifuge. High strength and small weight of container are essential. Thus, the aluminium alloy containing 3.8–4.8% Cu, 0.4–1.1% Mg and 0.4–1.0% Mn has been selected. Natural ageing is one of the characteristic features of this alloy.

Several metal-forming operations are needed to obtain the final shape of the workpiece [9]. First, the bar cut is slightly upset. Then, the backward extrusion is applied. Next, three subsequent ironing operations are carried out. Finally, the bottom of the container is sized (Fig. 6.14).

The final quality of the container is mostly dependent on the shape and mechanical properties of the extruded cup. Thus, the backward extrusion was

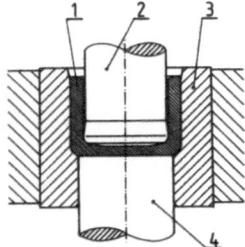

Fig. 6.15. Scheme of backward extrusion process: 1 – workpiece; 2 – punch; 3 – die; 4 – ejector.

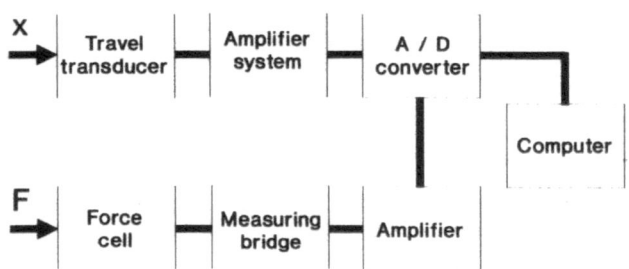

Fig. 6.16. Block diagram of F and x – data acquisition.

chosen to be automatically supervised (Fig. 6.15). This operation was carried out by means of a hydraulic press. The force F acting on the press ram and the relative displacement x of the basic plates of tooling were measured (Fig. 6.16).

6.4.2   Bases of Supervision Strategy

The force–travel diagram is shown in Fig. 6.17. At first the force increases owing to strain hardening. Then a reduction of force is observed, as the lubrication properties of the zinc stearate, used as a lubricant, improve with temperature, which increases as a result of the plastic deformation. The maximum forming force F_M is dependent on the plastic properties of the material to be formed. These are correlated with the mechanical properties of the extruded cup, and finally with the mechanical properties of the container. At the point where the upper tooling plate meets its stop fixed to the lower tooling plate, the minimum value of forming force F_E is obtained. This force depends mainly on the lubrication properties during the working stage, as can be seen from Fig. 6.18. Poor lubrication results in galling of the workpiece surface. This means that force F_E can be considered as a measure of surface quality of the workpiece, especially for the inner surface of the container. Finally, the force acting on the press ram rapidly increases. The final force F_0 (Fig. 6.17) corresponds to the instant when the main press cylinder is being connected to the tank.

According to Lange [2], during the first stage of the backward extrusion process the billet is axially upset and the die cavity is filled. At the end of this stage, full

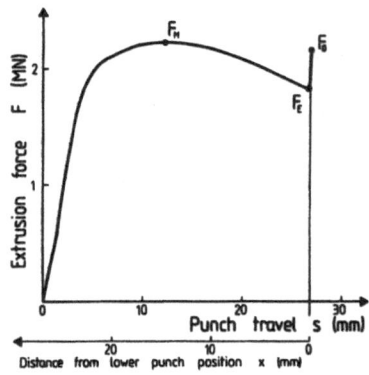

Fig. 6.17. Force–travel diagram for backward cup extrusion.

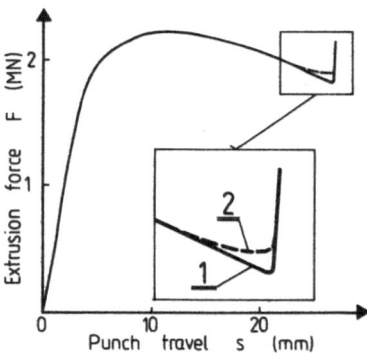

Fig. 6.18. Influence of lubrication on the force–travel diagram: 1 – proper lubrication; 2 – poor lubrication.

contact between the billet and the tools is obtained (Fig. 6.19a). The distance between the punch-face in this position and its lowest position (Fig. 6.19b) is equal to x_{iF}. The following equation can be written

$$0.25 \pi d_p^2 x_{iF} + V_C = V$$

where d_p is the punch diameter, V_C the volume of the body ABCDEF and V the billet volume. As $x_{iF} \approx x_F$, one can rewrite the above equation in the form

$$0.25 \pi d_p^2 x_F + V_C \approx V$$

where x_F is the distance between the upper and lower tooling plates at the end of the first stage of the backward extrusion process. Thus, x_F can be chosen as a measure of the billet volume and/or of the final height of the extruded cup. The force F_F corresponding to the distance x_F can be estimated from the approximate formula

$$F_F = 0.25 \pi d_d^2 C [\ln((x_S + g)/(x_F + g))]^n$$

where d_d is the die diameter, C and n are constants in the equation describing the flow curve, g is the thickness of cup bottom and x_S is the distance defined in Fig. 6.19c.

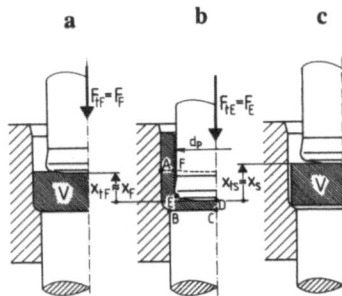

Fig. 6.19. Three characteristic punch positions: **a** die cavity filled; **b** end of extrusion; **c** punch contacts the billet.

6.4.3 Supervision Strategy

The supervision strategy has been developed on the grounds of the considerations discussed in the previous section. The strategy described below has been chosen to inspect specific features of the backward extrusion process as well as the quality of the extruded cup. The "window technique" has been used, which means that at specified instants the supervisory system checks whether specific inequalities (related to supervised parameters) are fulfilled. The supervision strategy is summarised in Table 6.1.

If a given inequality is fulfilled, no control signal is generated by the supervisory system. The signal "+1" means that the upper limit of the supervisory parameter has been exceeded; the signal "−1" informs that the value of supervisory parameter is below the lower limit. Thus, proper supervisory action can be taken and a diagnosis of disturbance can be given.

6.4.4 Teaching Stage

It is known that, if the extrusion process is carried out correctly, the supervisory parameters will lie within specific ranges. Manufacturing tests have to be used to determine the limits of each range. During these tests the process correctness as well as the quality of the extruded cup should be inspected by an experienced engineer. The limit values for the backward extrusion of the container cup have been determined in the way explained below.

Let us start with the values x_{01} and x_{02} (see Table 6.1) which limit the permissible range of the lowest position of the press ram. Assume also that the minimum value of the forming force is equal to $F_{0,min}^i$, where i is the ordinal number of the given manufacturing test (Fig. 6.20). The final force is equal to $F_{0,max}^i$. This force is reached at the instant when the main press cylinder is being connected to a tank. The lower limit F_{01} is equal to the lowest value of $F_{0,min}^i$, which can be written as

$$F_{01} = \min_{i=1,2,\dots m} (F_{0,min}^i)$$

where m is the total number of tests, of which there should be several dozen. Similarly the upper limit

$$F_{02} = \max_{i=1,2,\dots m} (F_{0,max}^i)$$

has been chosen. Certain ram positions, S_{01} and S_{02}, correspond to the forces F_{01} and

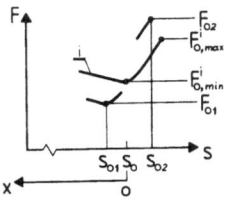

Fig. 6.20. Determination of the lower punch position.

Table 6.1. Supervision strategy of the backward extrusion

Window	Supervisory parameter (SP)	SP checked as function of	Permissible range of SP	Inspected features[a]
S	F	$x > x_S$	$F = 0$	1. The proper position of the billet in the die cavity
F	x	F_F	$x_{F1} < x < x_{F2}$	1. Volume of the billet 2. Final height of the cup
M	F	x_M	$F_{M1} < F < F_{M2}$	1. Plastic properties of the material 2. Mechanical properties of the cup
E	F	x_E	$F_{E1} < F < F_{E2}$	1. Correctness of lubrication 2. Quality of the cup surface
O	x	x_O	$x_{O1} < x < x_{O2}$	1. The lowest ram position 2. Thickness of the cup bottom

[a]1. Process quality features, 2. Cup quality features.
The lower indices mean: S – the punch contacts the billet; F – the die cavity filled; M – the maximum extrusion force; E – the end of extrusion; O – the cylinder connected to the tank.

F_{02} (Fig. 6.20). The arithmetic mean

$$S_0 = (S_{01} + S_{02})/2$$

has been used to calculate the conventional lower ram position. This ram position is chosen as the origin ($x = x_0 = 0$) of the x-axis, which has been used many times in previous sections of this chapter.

The other limit values have been determined in a similar way. It is obvious that the method described above has been given only as an example; it is possible to develop other formulae defining F_{01}, F_{02} and x_0, for example using more advanced statistical techniques.

As tool wear influences die and punch dimensions, the manufacturing tests should be repeated from time to time and perhaps new limiting values for the supervisory parameter may become necessary.

Other very interesting examples of supervision in metal forming have been described by Hellwig [10], Doege and Brendel [11] Doege et al. [12] and Song Zong-hua et al. [13].

References

1. Szafarczyk M, Chisholm AWJ. Automatic supervision in manufacturing systems. Principles, classification and terminology. In: Proc 3rd Intl Conf Automatic Supervision, Monitoring and Adaptive Control in Manufacturing, AC'90, CIRP, Rydzyna, Poland, 1990, vol 1, pp 7-22
2. Lange K (Ed). Handbook of metal forming. McGraw-Hill, New York, 1985
3. VGB-Vorschriften: 11.062 Blatt, Exzenter und verwandte Presse (Order No. VGB7n5.1). Verband der Gewerblichen Berufsgenossenschaften, Bonn, 1987
4. Estreicher M, Scholz G. Kupplungs-Spindelpressen zum produktiven, wirtschaftlichen Schmieden. Werkstatt und Betrieb 1986; 119(9): 802–806
5. Schmoeckel D. Automatic supervision in forming. In: Proc 3rd Intl Conf Automatic Supervision, Monitoring and Adaptive Control in Manufacturing, AC'90, CIRP, Rydzyna, Poland, 1990, vol 1, pp 59–71
6. Siegert K, Klamser M, Straube O. Error identification during deep drawing. Zeitschrift für Wirtschaftliche Fertigung 1990, 85(8): 416–420
7. Brankamp K, Bongartz B. Der moderne Stanzbetrieb. VDI-Verlag, Duesseldorf, 1986
8. Hettig A, Lange K. Werkzeugüberwachung beim Vollvorwärtsfließpressen im Hinblick auf die Ermüdungsrisserkennung. Umformtechnik 1992; 26: 2: 95–97
9. Kuczynski K. Cold extrusion of the cup shaped workpieces. Przeglad Mechaniczny 1988; 4: 10–14 (in Polish)
10. Hellwig W. Sensortechnik an Stanzmaschinen. Schweizer Maschinenmarkt 1986; 28: 29–33
11. Doege E, Brendel T. Verfügbarkeit von Tiefziehpressen verbessern mit Diagnosesystemen. Bänder Bleche Rohre 1991; 32(2): 45–50
12. Doege E, Schomaker K-H, Brendel T. Sensors and diagnostic systems in forming machines. Annals of the CIRP 1992; 41(1)
13. Song Zong-hua et al. Real-time defect detection in flow-forming processes. J Materials Processing Technology 1992; 32: 365–370

7 Assembly Process Supervision within Flexible Automatic Assembly Systems

A. Arnström and M. Onori

7.1 Introduction

7.1.1 Background

A Flexible Automatic Assembly System (FAA) is a system in which different products or variants of a product are assembled automatically. Such a system should also be capable of accepting new products/product variants in as simple a way as possible. The ability to change over automatically from one product assembly to another is also desirable, especially if the system is to be run on several unmanned shifts. Small batch sizes are another goal for an FAA system, in order to fulfil the requirements of JIT (just in time) philosophies.

What then is so special about FAA systems when it comes to process supervision? Why cannot the same principles described elsewhere in this book be applied in this case?

Let us first of all make clear to the reader that "assembly" in itself is a very poorly described operation. "Assembly", "mounting", or "putting together" does not comprise just one type of operation. These terms comprise a *class of operations*. Consider the following "assembly" operations:

Screwing
Snapping
Pressing
Insertion
Glueing
Soldering
"Non-rivet" riveting
Applying gaskets
      O-rings
      seals
      lubricants
      etc.

Obviously, this list could be extended. What do these assembly operations have in common, especially when it comes down to describing the process itself? The answer has to be: not much.

So far we have described the "assembly" section of what is carried out in an FAA system. In addition there are other, "non-assembly" operations. Consider the following of these "non-assembly" operations:

Parts handling
Orientation of parts (feeders, magazines, vision etc.)
Mapping of tools, grippers, fixtures etc.
Materials planning
Program generation for robot(s), feeders, conveyors etc.

It is becoming painfully clear that the relationship between complexity and flexibility is inevitable. If such systems are to be truly flexible, the solutions required will be far from easy. This is especially true since at present there are no *de facto* standard solutions for these different operations. Currently the only way to solve problems is by adopting *ad hoc* solutions.

To summarise, FAA systems are technically very complex and should be able to cope with many different products and product variants in small batches. Furthermore, it should be possible to update such a system to accommodate new products involving new technologies. It is essential that FAA systems are capable of being continuously updated.

7.1.2   Steps to Take – an Overview

Modern manufacturing theories emphasize the need for production that is dictated by consumer demands, not by what is needed to fill the warehouse shelves. This in turn leads to giving priority to short throughput times, lead times, changeover times and small batches. In other words, an FAA system must be highly reliable.

Assembly represents one of the final stages in the production sequence. The product also represents the highest cost or value at this stage. The conclusion from this is that a *high degree of reliability* is of the utmost importance in the assembly system. There are two main ways to achieve this:

1. Make the process 100% reliable, or at least very close;
2. Let the system take care of any problems that arise and return it to its functional status automatically.

One school of thought advocates 1 as the only method to apply from a technical and economical point of view – Redford [1]. However, it must be added that this step incorporates not only the assembly system but also all the parts to be assembled. A conformity of parts of 100% is extremely difficult to achieve. Normally this depends on the functional geometry and its tolerances as specified in the drawings. Remember that in most assembly systems the parts are oriented and gripped relative to certain of their features. Hence one should demand 100% conformity for these part features, which obviously adds substantial cost to the product.

Method 2 presents the problem of adding complexity to an already complicated system. It will, however, give the FAA system the theoretical possibility to regain functionality after an error, * and, as we shall explain in detail later, many errors are

---

* The term "error" in this chapter is equivalent to "failure" or "malfunction" in the rest of the book. [Editor]

of a trivial nature. However, if one does not intervene when these trivial errors occur, fatal errors can eventually result.

Hence, for successful implementation of FAA systems, a balance must be struck between the two methods. There is no single "right" way to solve this problem. A combination of both methods based on a well-projected price/performance plan should be the aim. Obviously one cannot blindly trust an "intelligent error recovery system", and the software and hardware within the system should be made as robust and economically viable as possible. The reader should bear this in mind as the remaining part of this section will investigate the details of automatic supervision – process monitoring, error detection and error recovery – within FAA systems.

7.2 Background Work

Research in the field of automatic process supervision and error recovery has been carried out over several years by various researchers. Let us begin by giving the definition of automatic error recovery according to Fielding: "The process by which a system returns itself to normal operation after conditions which generally hinder the intended purpose of the system".

Obviously, such automatic supervision and error recovery systems are software based and utilize all types of transducers. The simplest way to design these programs is by using "hard programming" techniques. What this means is that the supervision and recovery routines are embedded into the actual robot programs at the robot controller level. This is a rather crude approach and results in a very complex debugging environment. Gini *et al.* [2] concluded that if such an approach is taken, up to 80% of the program code will consist of sensor signals, error recovery routines and "if–then" statements. This was confirmed by Nyström [3] who pointed out that conventional robot controllers also exhibit a limited ability to handle complex data structures. Similar results were also obtained by Giacobbe when investigating a system for packing diskettes (IBM): 90% of the system coding was devoted to automatic error recovery. Basically, the hard programming approach is time and code intensive without solving the main problem of having to deal with complex data structures.

Such an approach also makes any debugging work extremely tedious and enhancement of the software system is very time consuming, if at all possible. Hence there is a real need for alternative solutions.

An obvious conclusion would be to abandon the use of robot-controller-based software and write more generic, user-friendly programs at the cell-controller level. Such systems are based on a variety of software philosophies, from procedural code to rule-based systems and expert systems. At the present time there are two schools of thought on this issue:

1. *Rescheduling*: If one of the stations or pieces of equipment malfunctions, the system checks to see if there is any other station within the FAA system that could temporarily take over the task in progress. Alternatively, it checks whether there is any batch in the assembly queue which does not require the malfunctioning station. In both cases the error recovery system lays out a new assembly plan and reschedules the FAA system *while* looking for a possible solution to the problem [4].

2. *Active recovery*: The supervisory system not only detects that an error has

occurred but also classifies it as a defined error. One or more pre-programmed recovery routines are linked to this error in a specific database. According to special algorithms, a given recovery routine is chosen, downloaded to the robot (or machine) and executed. The error detection and recovery system monitors the recovery attempt and will re-try with another routine if the first one fails, or continue with the assembly in the case of success. Failure or success attempts are recorded according to their respective routines.

Note that the rescheduling principle exhibits many similarities to the "Automatic Planning and Scheduling of Assemblies", which is a technique born directly from the CAD world of research. Its goal is to produce the optimal assembly sequence automatically, based directly on the CAD "drawings" of the product's assembly. Work in this field has been carried out, among others, by: Delchambre et al. [5], Moed and Kelly [6], Selke et al. [7] and Laperrière and El Maraghy [4].

Concerning the "Active Recovery" technique, among the most prominent authors are: Gini and Gini [8], Lee et al. [9], Abu-Hamdan and El-Gizawy [10] and Arnström et al. [11].

7.3   General Description of the Error Types

7.3.1   What is an Error?

It needs to be emphasized that an "error" in FAA terminology does not have the same meaning as in control theory. One cannot use the term "failure" either, as some errors in FAA do not directly imply an unsuccessful operation. Once again there are two main schools of thought regarding the definition of error within FAA:

1. Any occurrence that deviates from a "correct" operation – "that which is not correct";
2. A fault in the operation in progress.

The differences are subtle. Either of the principles can be applied to detect an "error".

Principle 1 is based on the comparison of a given sensor signal to its pre-defined correct sensor signal. When the two values do not coincide, the system defines the state as an error (whatever the reason or cause). Note that this does not necessarily imply that an error in the actual assembly operation has occurred. The sensor itself might be malfunctioning, being read a few milliseconds too soon, etc.

The second principle implies that the sensor signal should be compared with the sensor values for a known error. If there are say, two possible causes of error, one can then easily distinguish which of the two has occurred and hence initiate a recovery routine. If neither is recognised, then either we have a new form of error or a sensor malfunction.

As in the previous case, it is best to strike a balance between both principles. Normally, errors that have few specific causes are dealt with using the second principle, while the first is used for more general errors.

7.3.2   FAA as an Open System – External and Internal Errors

It is important to make clear that Flexible Automatic Assembly systems are to be

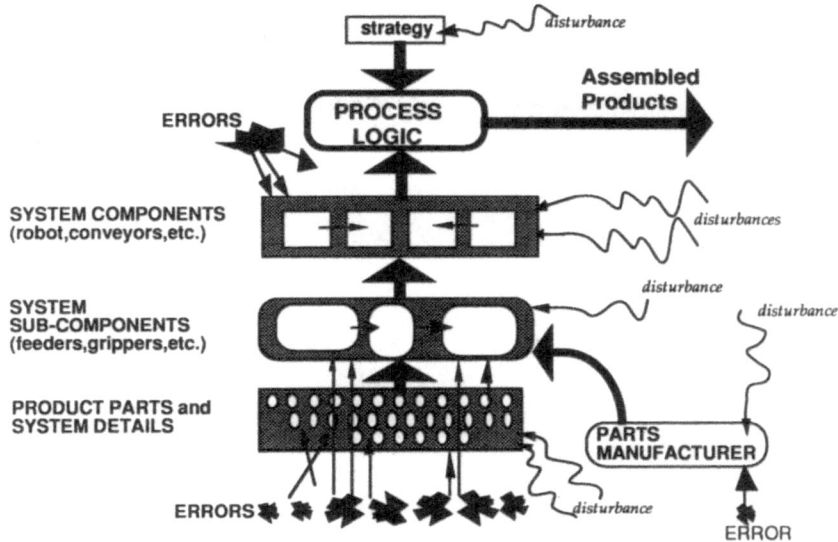

Fig. 7.1. Flexible Automatic Assembly (simplified).

considered as open systems. That is to say, the assembly process suffers disturbances not just from internal sources but also from external ones.

Figure 7.1 illustrates the difficulty in trying to simplify a complex open system. Let us first of all subdivide the types of errors into two categories:

1. *External errors** are all the errors that occur outside the assembly process, e.g. a break in the air pressure link, wrong parts at the assembly station, parts wrongly placed in pallets, programming errors, etc.
2. *Internal errors* are all the errors linked to the assembly process itself, e.g. the robot fails to pick up a part, parts jam during assembly, defective transfer of fixtures, etc.

Obviously these two categories of error are interrelated: external errors cause internal errors. Most external errors are usually attributed to human errors and failures within the supply network (electricity, air etc.). However, this is over-simplistic. The assembly process itself might set high demands on the infra-structures, and small, repetitive internal errors can, in the course of time, generate external errors.

External errors often occur stochastically and are not as common as internal ones. They usually cause fatal errors that can only be recovered manually. Error recovery research has concentrated mainly on internal errors, the exception being for programming errors.

It is a worrying fact that certain disturbances can give rise to temporary external errors without giving clues as to the cause. Lighting disturbances can give major problems to vision systems, but vary so quickly that the user might not detect such

---

* "External errors" in this chapter are treated as "disturbances" in the rest of the book. [Editor]

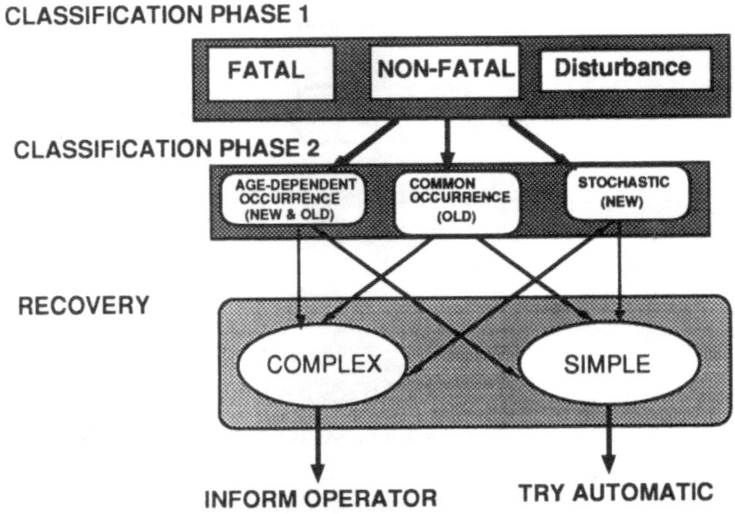

CLASSIFICATION PHASE 1

Fig. 7.2. Simple error clasification scheme.

a cause. The error will often show itself in the robot failing to pick a given part. To trace the error back to a lighting disturbance is far from easy. Other types of disturbances might include strategic disturbances in which the assembly sequence is ordered to change immediately to satisfy urgent customer demands. These stochastic changes of product assembly priority must be planned into the automatic assembly system controller so that no large amounts of re-scheduling and programming are required. Changeover time is in this case the major factor for an efficient implementation of changes of strategy.

Disturbances can hardly be described as internal or external error sources since they are vague stochastic events that require a category of their own, however the literature rarely makes this distinction.

To summarize, there are some causes of error that depend on the assembly process (internal) and others that depend on the immediate world in which the assembly process exists (external). This means that a flexible automatic assembly must be considered to be an open system. Odd stochastic events with hardly traceable causes, which do not cause lasting errors, can be considered as disturbances, and should be given some consideration.

7.3.3   Error Classification

The most common classification of an error is the separation of *fatal* and *non-fatal* errors. In common terms this means that one must make clear for the recovery system what type of error is going to cause a system shutdown and what can be corrected on-line. This is the highest hierarchical subdivision of errors. Compare the very simple sequences shown in Fig. 7.2 and Fig. 7.3.

A certain amount of research has been carried out on how one should define an event as being faulty, i.e. on how to define an incidence of error. One school of thought describes an error as "any situation which is not correct." Philosophically

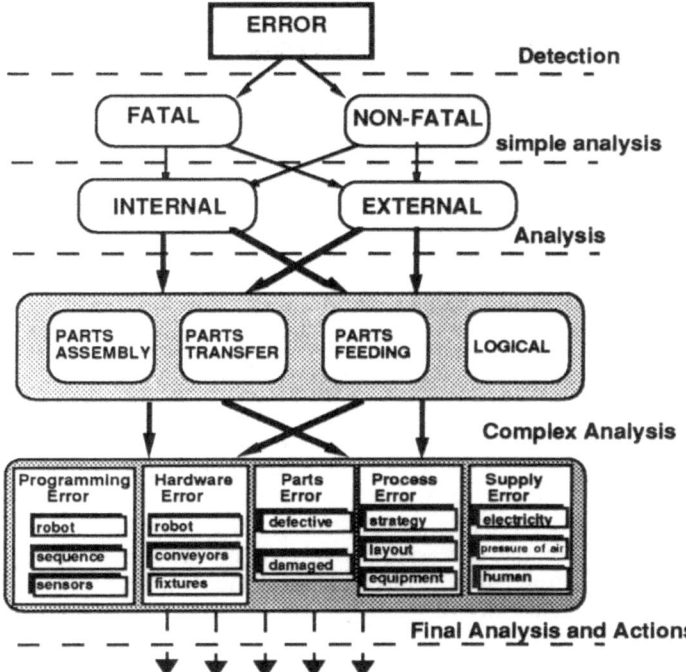

Fig. 7.3. Example of error classification sequence.

speaking this is highly applicable, but it does not solve all the ambiguities that arise in flexible automatic assembly. As in all practical implementations, the easiest route to take is that of formalizing the problem prior to forming a solving strategy. Experience is invaluable. The assembly system must be designed to be flexible at all levels of hardware and software, in such a way that errors are not created by too strict operational constraints. For example, in the case of vision systems, design the grippers in a way that they can compensate for minimal concentricity problems.

The recovery system must also have some form of *historical backing*. Such a facility enables the system to log new types of error while being able to recall previously occurring errors and their causes. Hence the classification of *new* or *old*.

Note that in practice the error recovery system will always have to consult a database to check whether the error in question is known (old) or whether it is a new occurrence (new). Fatal errors are almost always corrected manually. Remember that fatal errors require a system shutdown.

7.4 The Process

In flexible automatic assembly, the programming tasks are mainly made up of data handling : parts scheduling, product priorities, job completion schemes, assembly counts, etc. Up to 95% of the programming requirements consist of data handling

Table 7.1. Details of the FAA system at ATLAS COPCO, Tierp, Sweden

No. of air motor variants	= 43
No. of different parts/variant	= 6–8
No. of parts/variants	= 10–12
Unique parts, total	= 56
Typical batch size range	= 1–200

Operations: Insertion
                Pressing
                Lubrication
                Sub-assembly
                Quality tests
                Orientation with vision
                Feeding/materials handling

and manipulation tasks [3, 12]. Only between 5% and 10% of the programming is dedicated to robot motion. This is a very important aspect of automatic assembly and drastically affects the way in which the user might correct for robot motion inaccuracies.

One can therefore separate the assembly process into two main areas: process control and motion control. An assembly process is simple in its logic but requires the handling of large amounts of complex data. Consider a flexible automatic assembly system which assembles two types of motors, of which there are 43 variants – see Table 7.1. We will have to monitor the assembly of 86 different products which can be assembled at varying time intervals, quantities and product sequences. Each product variant will have its own fixtures, delivery times, quality levels, parts list etc.

7.4.1   Process Control

Everything that has to be externally controlled within the assembly process should be carefully studied and planned prior to programming. By process control we mean task scheduling, materials flow, product priorities, job completions, etc.

A very detailed look must be taken at the product to be assembled. Do not hesitate to adapt the product to fit the process. Here are a few examples of what process control might have to deal with:

Product priority: a time-based scale that informs the system which product to start to assemble. A current product assembly can be halted in favour of another product assembly if an emergency delivery is required.

Task scheduling: works hand-in-hand with the above. A more long-term plan of the various products to be assembled, materials flow, which tools they will require, which programs they will need access to, etc.

Robot scheduling: program sequencing plus possible tasks for the robot during manual assemblies or other non-robot-intensive events (strategies for maximizing robot utilization).

Parts databasing: databases which keep count of parts, to which products they belong, which robot programs they are linked to, which errors they are linked to, etc.

Conveyors: programs which drive the conveyors, stop them, etc.

Fixtures: keeps control of which fixtures are coming in and out of the system, to which products they are associated, etc.

Error detection: routines which the operator can activate/deactivate. They are associated with the process, specific equipment or specific products.

Error recovery: routines which attempt to correct an error. Can be, as above, associated to the process, equipment or product parts (picking, transfer, assembly).

The above process control tasks are only selected examples. Note that an operator must ideally be able to update and add new items to these control structures at any time. This type of flexibility is not often found in industry, but is now beginning to reach the market [3, 10]. A rule-based system linked to several databases seems, for the time being, to be the best suited method for such applications. An Expert System would not make the updating very simple and since these tasks do not present a complex logic, we believe it unnecessary to apply them at this level.

Process errors usually manifest themselves very frequently. The apparent cause of the error might seem to be elsewhere, but if an error is extremely frequent then a mistake must have occurred at the process layout level, mechanical construction level, assembly sequencing etc. This is yet another reason why one must keep records of errors.

Furthermore, it is very important to make the process control mechanism very stable. In other words, disturbances must not complicate the smooth execution of the process. Strategic disturbances, such as those mentioned earlier (sudden priority changes), are easily incorporated into a software control system, but others, such as power failures and fluctuations, are best dealt with by hardware back-up systems. What is important is that fatal errors are avoided, as system shutdowns are costly and also entail start-up procedures from where the operation was interrupted.

7.4.2 Task Execution

It is often at process control level that one applies the assembly rules which drive the robot, conveyors and system as a whole. Hence it is important to insert at this level the means to check on whether the tasks in question are being carried out satisfactorily.

Usually this is done by checking on whether certain digital inputs and outputs have been set. This is, in fact, a very robust way of finding out whether a fixture has arrived, a part is in the gripper etc., but it does not carry any information in itself. We do not know if the *correct* fixture has arrived, if the *right* part is in the gripper, etc.

The important thing, is to decide from the outset what one really needs to know and when. A vision system can be a very flexible tool: it can give the robot precise positional information, it will respond to the process control demands only if the specific part required is found, it can carry out inspection tasks, etc. But it is an expensive choice and must be applied by experienced people – however it is a powerful tool.

Simple sensors are usually used to detect where the task interrupt has occurred. Often these sensors supply enough information for the system to draw its own conclusions. Remember that a particular sequence of simple sensor values can indicate a specific error if we have a process control system with some form of

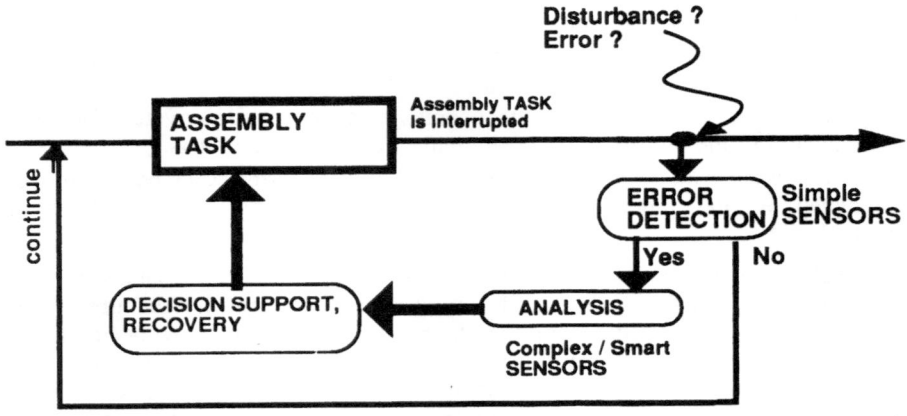

Fig. 7.4. Basic principles of task execution supervision.

historical backing. Hence we can eliminate the need for several smart sensors by planning: many small binary sensors are a more *reliable* source of information than a few very complex sensors.

However, this is not enough in itself because new errors do occur and it is always best to have some form of analysis available. Consider Fig. 7.4: (1) in practice one does not always resort to a secondary analysis with smart sensors; (2) classification of the error begins at the simple sensor level; (3) try to create routines which modify existing robot programs to new conditions, instead of writing several routines with micro-movements, etc. See Section 7.6.2.

Take as an example the situation where there is no part in the gripper. The part in question is metallic. The error is signalled by the absence of 24 V across the gripper fingers. A possible error is the vision system, so first check the vision supply and then try vision picking again. If the error is still present, the robot is possibly picking the part at the wrong height owing to a small camera misalignment. Add 2 mm to part picking program's Z-axis co-ordinates. Try the picking operation again. If successful, record the new program and the error parameters. Note that first we check the vision system by actively following a new attempt. If it works we can also continue the assembly. The analysis is also an attempt to recover from a disturbance – see the broken line in Fig. 7.5. Note also, that no recovery routines are invoked. The system creates a new program by adapting an existing program to the new conditions. This is an example of program adaptability [3].

Finally, do not forget that many error recoveries still require manual intervention. It is not always easy to define the parameters which may guide one to choose manual rather than automatic recovery. Time is a major factor. If an operator is present when an error occurs, he or she might see what the problem is before the process control system has evolved a successful recovery for it. A typical example is an empty parts pallet. The assembly system will try to pick parts from it, fail, and try to reach a conclusion. An operator can act before letting the system try to pick a part. So, it really is not as straightforward as one would like it to be. Operators will rarely be present in the system, but may be called in if needed. Such choices must

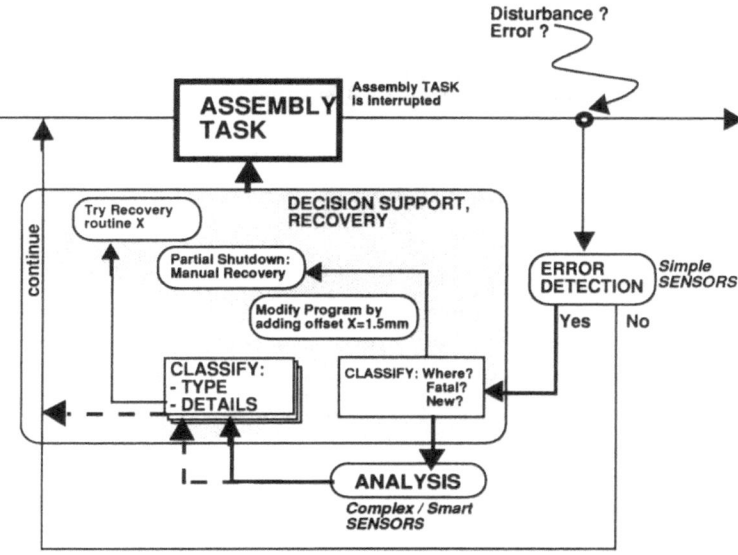

Fig. 7.5. Example of stepwise analysis.

be made at the beginning, since they involve the basic performance requirements of the assembly system and operator. But, on the other hand, most process control systems today are not able to ensure cost-effective automatic error recovery for all the types of errors that could occur.

7.4.3   Motion Control

Let us now consider some of the major areas covered by motion control:

Parts picking: the manner in which we instruct the robot(s) to pick the product parts. Possible choices are pattern picking and vision. This operation requires high robot repeatability accuracy.

Assembly: programming strategies for the final assembly of parts. This operation should be carefully studied, as slight modifications to a single robot program can satisfy the assembly of several parts – see Fig. 7.6. This operation requires very high repeatability levels.

Transfer: programs which move the robot(s) from one operation to another, e.g. move robot from part picking position to assembly position: this series of movements is to be carried out as quickly as possible, without collisions, but poses no high accuracy demands.

Tooling: programs which drive the robot to the grippers and special tools (screwdrivers, feeders etc.). The final position of the program has to be accurate, not how the robot moves to it – compare above.

Recovery routines: special robot motions which are known to solve given problems. They include small micro-movements to adjust the position of a part in a fixture, programs which create slight offsets from the robot's current position, etc. These routines require high accuracy motions.

The following 6 objects are assembled one on top of the other. The Assembly Position varies only in the Z-Axis.

One Robot Assembly Program with 6 Z-axis offset values is sufficient.

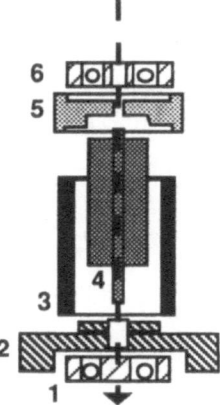

Fig. 7.6. The "hamburger principle".

Cell calibration: automatic calibration routine which corrects for small robot positional errors. These errors are due to robot positional accuracy problems or small hardware positional changes.

As mentioned earlier, it is important to study the product before one begins to plan the cell layout and details. It is often very profitable to adapt the product to the particular assembly process. Figure 7.6 shows a case in point. Six of the product parts to be assembled are stacked one on top of the other. This not only creates a very simple and flexible robot assembly program situation, but simplifies the construction of the product assembly fixtures and minimizes the parts transfer operations.

In automatic assembly applications, the robot(s) perform very repetitive tasks of which only some require extremely high repeatability. High repeatability accuracy is required primarily during the picking of parts and their assembly. The paths between the picking point and the assembly site need not be more accurate than ±1.0 mm, a value that almost all robot types can conform to. The transfer motions described previously are the least likely to give rise to errors.

The most common robot motion problem is for the robot to have difficulty in reaching the desired position. Possible causes of this effect in the robot program are:

1. Tool centre point (TCP) error.
2. Programming error.
3. Error due to mechanical discrepancies (robot, fixtures, etc.).
4. Non-linearities in robot – inaccurate knowledge of such kinematic parameters as steady-state errors in servos, compliance in links or joints, gear backlash or harmonics, and temperature effects.
5. Robot resolvers are uncalibrated - manual recovery.

Error cause 1 is taken to be a programming error. It should not really occur if the FAA system has been programmed correctly; however wear and tear can create small shifts in the grippers so that the TCPs behave in an unwanted way.

These errors are usually corrected in the same way as programming errors. Error cause 2 is easily corrected either by recovery routines, or, even better, by adaptively

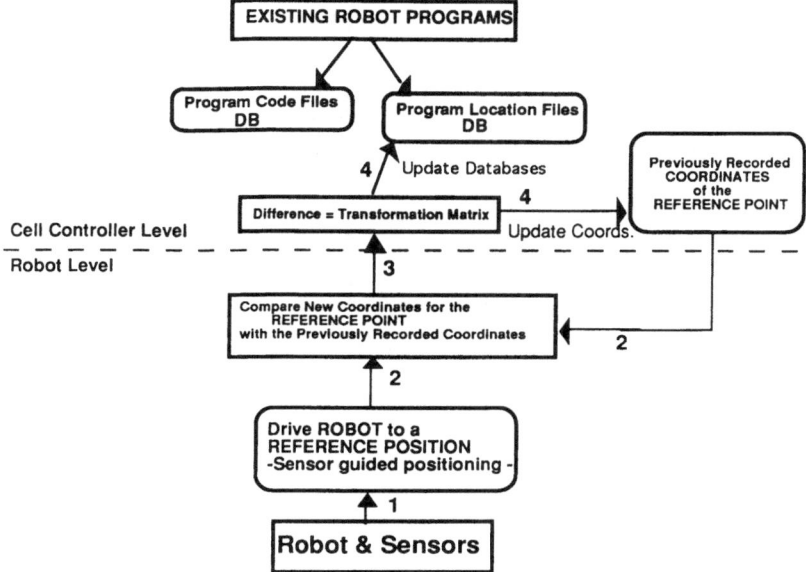

Fig. 7.7. Cell calibration events (simplified).

updating existing programs with the required offsets. Error 4 is usually recovered by robot calibration methods [13]. This error is a real problem in welding or grinding applications, where the robot path as a whole must maintain very high repeatability and absolute accuracy levels. It does not affect assembly as much, since in such processes it is only the final positions which are of interest. Hence we can correct this error by cell calibration methods, also known as pose-error compensation – see Fig. 7.7.

Error cause 3 can also be corrected by cell calibrations if one utilizes an object oriented programming environment. The assembly cell is subdivided into zones – say assembly zone, parts-picking zone and tooling zone [14]. Each zone is assigned a calibration station. All the programs concerning the particular zone are "owned" by the calibration station, which forms a reference point. When the new co-ordinates for the calibration point are found, one can immediately correct all the programs within the zone. By subdividing the cell into "zones" we accomplish two things: (1) minimization of the linearization effects of this linear compensation routine; and (2) creation of groups of programs which can be adapted, to a certain degree, to changes in the physical layout of the system.

This type of error recovery requires robust six-degree of freedom sensors (force/torque, vision etc.) and object oriented programming environments [15, 16].

7.5   Error Detection and Analysis

7.5.1   Sensors

Sensors are used to detect disturbances or to determine conditions that are not pre-defined, within the robots' and the FAA system's environments. They are

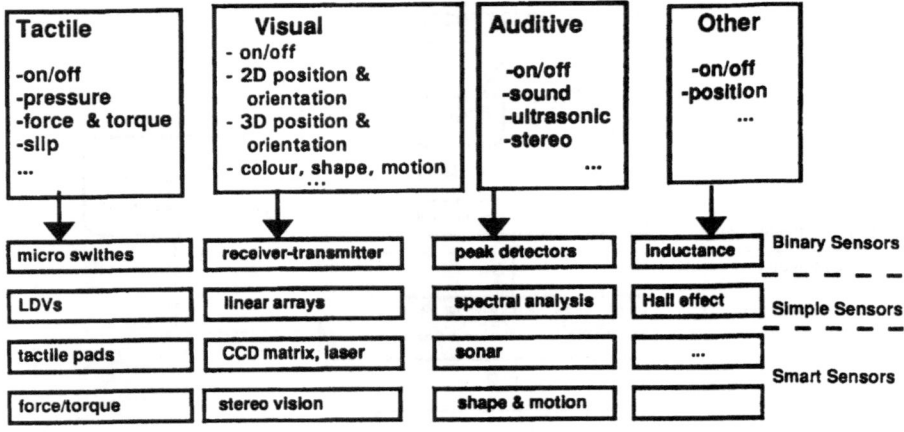

Fig. 7.8. Simple description of sensor types.

used to optimize the FAA system's flexibility. Sensors range from very simple devices to complex systems capable of on-site signal analysis. Sensors can be subdivided into categories in order to clarify the type of physical entity one is trying to detect. Visual sensors deal obviously with light, but in 1, 2 or all 3 dimensions. See Fig. 7.8. The principal elements of sensors are shown in Fig. 7.9.

Binary sensors are very common, very robust, reliable and very cheap. They usually constitute the backbone of most FAA system error detection systems. Secondly there is a small range of sensors that are neither binary nor smart, and thus called "simple" by the authors. These produce analogue signals rather than simple binary levels, and these must be analysed at a later stage.

Smart sensors involve a given level of signal data processing at the sensor level. As much transduction as in data processing is carried out by such sensors. Therefore, a certain amount of the monitoring task is also carried out at the sensor level. These sensors are expensive and not always suited to heavy industrial environments. Care must be taken in applying such technology.

The general requirements of a sensor:

Linearity of actual transducer (minimal hysteresis);

Robustness to industrial environments;

Flexibility of use;

Compliance;

Minimal spatial intrusion;

Real-time data acquisition and processing;

High resolution;

Low cost where possible – but always remembering that performance is more important than price.

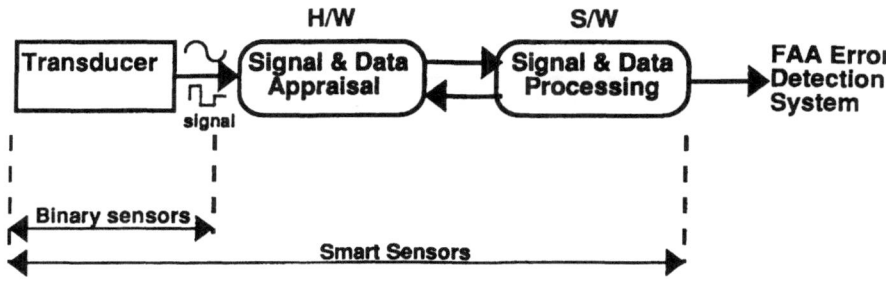

Fig. 7.9. Basic elements in a sensor.

7.5.2 Choosing a Sensor

It is of vital importance to analyse the operation one is going to monitor with a sensor in depth before doing anything else. Certain questions must be answered satisfactorily. The following points should be carefully considered:

Make sure that the previous operation does not influence the correct execution of the current one (e.g. avoid hysteresis);

Analyse all the factors which can create the error in question;

Analyse which of these factors are common, which are stochastic and which occur very rarely;

Lay out the details of how much time, money and space you have to supervise the operation;

Carefully plan the way in which this operation is included within the automatic supervision system.

For example, define what one wants to monitor, how one wants to intervene, and where in the software the sensor feedback should be included. In other words, one must plan the supervision operation prior to choosing a sensor, as shown in Fig. 7.10. Many sensors are flexible enough to be used in several different supervision tasks, so there may be no need to add a new sensor, but only to reorganize the sensor's job sequence.

Once the operation has been analysed, a careful appraisal of the sensor characteristics available must be made. For instance, a sensor might produce information in real-time but at the expense of resolution. Speed does not mean accuracy. Compromises often have to be made. It is advisable to draw up a list of performance constraints for the particular application.

7.5.3 Action Alternatives and Historical Backing

As described earlier, it is advisable to use simple, robust sensors for the detection phase. This decreases costs, software execution times, and the overall reliability of the supervision system is improved. Keep in mind that digital sensors are less sensitive to noise than are analogue sensors. Smart, or more enhanced, sensors can

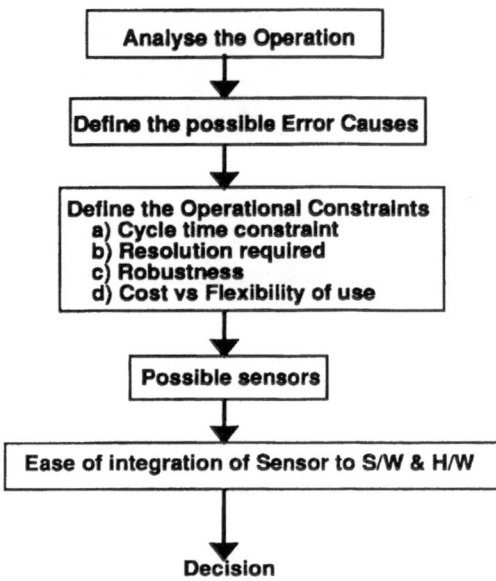

Fig. 7.10. Steps to take when choosing a sensor.

be brought into the scheme after an error or disturbance has been detected, in order to aid the software in the classification phase. See Fig. 7.11.

One of the best pieces of advice is to minimize the use of advanced sensors at the error detection level. However, the authors are well aware of the fact that this is not always possible.

Very much depends on how one structures the software supervision system. It is not always necessary to bring smart sensors into action. A historical back-up of the errors is the vital ingredient here. If a simple sensor gives a signal at a very specific time and system condition, we can trace the error source from a historical specification of these existing conditions and use a known error recovery routine without wasting more time.

Of course this is not an absolute guarantee that error recovery will always be successful. Hence we also need to associate with the error recovery routine a data field representing the routine's success rate. Since an error cause may be recovered by a choice of known routines, the software system simply chooses the one with the highest historical success rate. See Fig. 7.12.

Although this figure is rather complex, it is essential if the chain of events for plausible error detection alternatives are to be understood. Compare Fig. 7.11 with Fig. 7.13.

A possible alternative course of action is to call for manual assistance. This is contrary to the goal of an automatic supervision and error recovery system, but experience shows that an experienced operator can decide the best alternative to take when the error cause is difficult to locate and there is a time constraint. One must not forget that there are countless parameters to consider. For example, if such an error occurs when the last product in a batch is being assembled, and the next

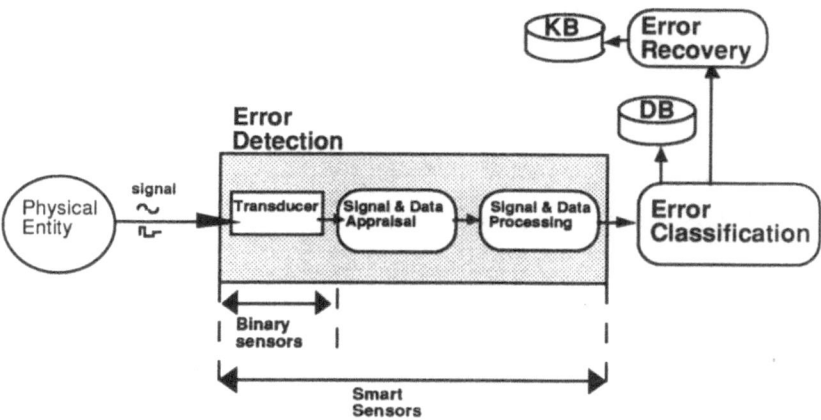

Fig. 7.11. Simplified error detection and classification sequence.

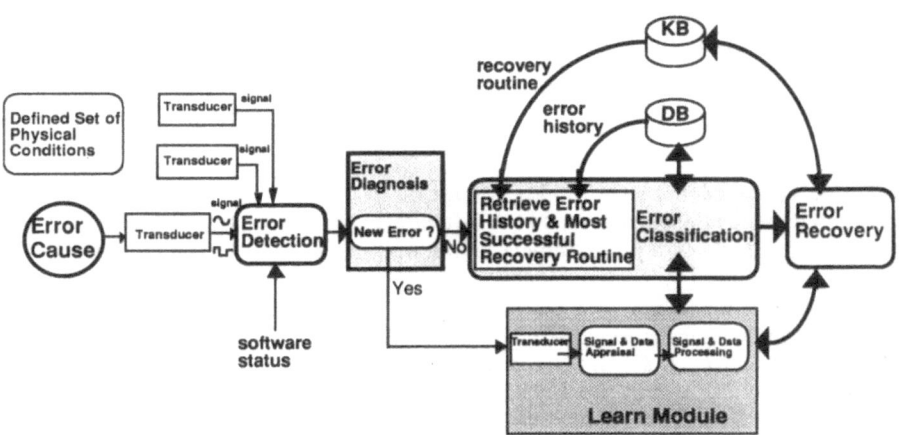

Fig. 7.12. Detailed error detection and classification sequence.

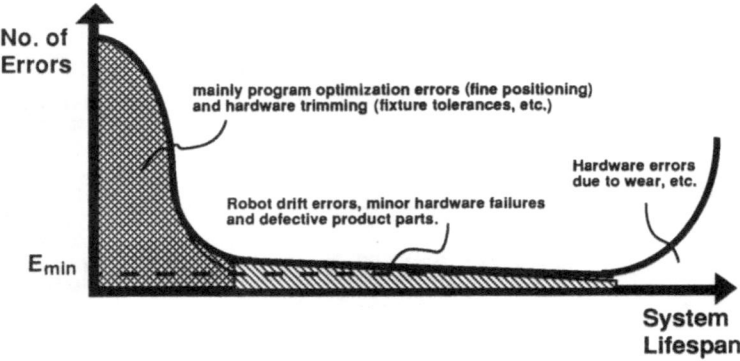

Fig. 7.13. Frequency and type of errors in an FAA system lifespan.

product to be assembled is urgently required, then the operator may decide to jump directly to the next product and assemble the last one manually to save time. This is of course a very difficult decision to take, since it does not take into account the fact that the cause of error may be due to the hardware and so the same error will affect the next product. It is a question of how complex one should make the supervision and recovery system without increasing the overall occurrence of errors caused by the complexity of such a system.

7.5.4   Error Classification Schemes

When designing an automatic supervision and error recovery system, there are two factors to be carefully considered:

1. The software must be easy to upgrade, debug and expand;
2. A good knowledge of the types of errors that will be incurred must be available, as well as the sensors to be used.

Obviously, one cannot have full knowledge of all the types of errors that will be encountered; hence the need to make the software as generic as possible, in order to enable an operator (not a programmer) to add new error recovery routines as they are required.

At the start of the project one concentrates on the known or foreseeable errors. Figure 7.13 gives an idea of what types of errors occur most frequently at given stages of a normal FAA system. As one can see, it is quite important to construct a software system in which robot and machine programs can be easily optimized, and, ideally, without creating any system downtime.

We therefore advise the reader to consider carefully which software approach to use. Hard programming principles (see Section 7.2) are to be avoided. Procedural languages can be used satisfactorily in combination with database systems and some form of user-friendly interface. The best results are usually obtained with rule-based systems or Expert Systems. Consider, however, that an Expert System's inference engine does not make debugging an easy task for a non-programmer, and that new

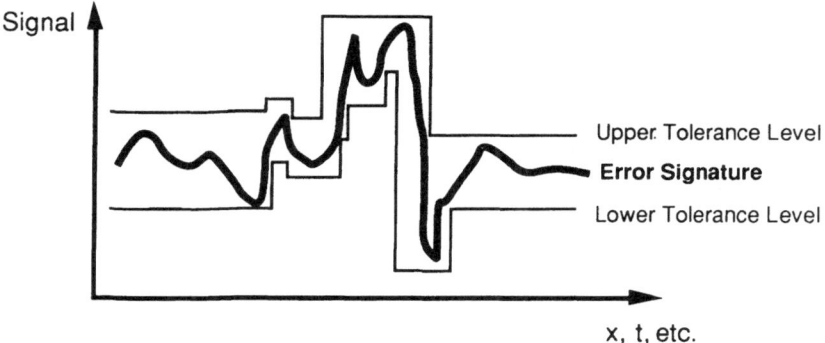

Fig. 7.14. Error signature with set tolerance levels.

error recovery routines are also not so easy to add to a backward/forward chaining system. Object oriented environments are advisable.

Error types and classes were discussed in some detail in Section 7.3. When classifying an error, one usually utilizes a rule base. The software steps hierarchically through the classes and matches the transducer signals to known error signal sets. In the case of Expert Systems, the software will search for a match with classical strategies such as forward and/or backward chaining – the inference engine may be of a deductive, inductive, fuzzy-logic or other nature. As long as known errors occur, the system should perform satisfactorily, proposing recovery strategies and updating the error history. When an unknown error occurs, one must classify this new error within the existing software structure.

New Errors. In Fig. 7.12 we see that the supervision system questions itself on whether the error is new or not. If the answer is "yes", it means that the sensor/transducer in question has reported a signal which does not relate to some known software and physical-world states of the FAA system (see below). One must therefore record this new error signal and its relative software and physical conditions (world state). We know where in the software the error occurred (where and when). We know what state the other sensors showed at the time (world state). All we need to record now is the transducer signal. When using simple sensors, this is relatively easy as the signal is usually on or off. In this case it is usually the world state, software position plus the signal itself which, together, make up the new error signature. When using smart sensors, it is very important to set an upper and a lower tolerance margin for the error signal.

Note that, because robotic assembly exhibits high flexibility, a previously unknown world state can also occur.

Creating a new signature for the error type is not problematic. One must read this signal and store the value with the other world-state signal levels. Then tolerance levels must be set for this signal if complex sensors (digital or analogue) are being used – see Fig. 7.14.

This is complicated as it can create clashes with other error types, so great care is necessary. A mathematical model of the error signature is another way to compare the error signal to tolerance levels. It can be introduced as a confirmation; however, mathematical models cannot be developed automatically. There are no known faultless automatic error learning modules yet available.

Once the error signal has been classified with its signature, tolerance levels and history, the system should ideally propose an error recovery routine. In some cases, depending on the type of error, it can be generated automatically. Unfortunately, this is a very difficult part of error recovery in which a great deal of research is at present being carried out, and hence it is usually done manually (operator-assisted).

Of course, one tactic is to compare the new error with the most similar known error, and use the latter's recovery routines. This can sometimes be a very time-saving move. If it fails, one resorts to learning-in, whether this is operator-assisted or automated (e.g. an Expert System).

7.6 Error Recovery

Let us say that we are at the point where we have detected and classified an error. All that we require now is to start the error recovery routine. As we saw earlier in Section 7.2, the main question to ask oneself is: how do we do this? Recall the two main schools of thought: rescheduling, and active recovery. Rescheduling tries to avoid the negative effects of an error by rearranging the system's sequence of actions. It temporarily or permanently disregards the error. This is acceptable to some degree if the error was of a stochastic nature, but this tactic does not improve the overall performance of the system, it does not add to the knowledge of the process and it does not consider the effects of recurring errors. In our opinion it cannot be considered as representative of the content of "error recovery". How then, does "active recovery" work?

7.6.1 Philosophy

As the word implies, active recovery signifies that the supervision system attempts to correct the error on-line, with either manual or automatic salvage operations.

Levels of application of this school of thought are different because of the practical constraints posed by the various FAA systems. The extreme case is when all error recovery is carried out manually, regardless of whether the error is of fatal nature or simple. The supervisory system simply detects and classifies the error for an operator. An example of such a system exists in another application, that of motor engine problem diagnostics.

Automatic error recovery is today possible to do at both the simple and the complex error levels, but becomes impossible for fatal errors. Consider Fig. 7.13 and Fig. 7.15.

It is lucky that simple errors occur most frequently, such as robot program optimizations, slight fixture position adjustments, a poorly gripped part, etc. This implies that if we can successfully recover from these simpler errors (and some of the more complex ones), we are raising the overall functionality of the FAA system. In other words, we are achieving one of the goals expected of an FAA system – higher reliability.

Philosophically speaking, there are a number of further questions that need to be raised in relation to error recovery. For example: how much time and equipment should one spend on trying to recover an error? One way to answer the question is simple: that amount which corresponds to the implied cost of the error for the

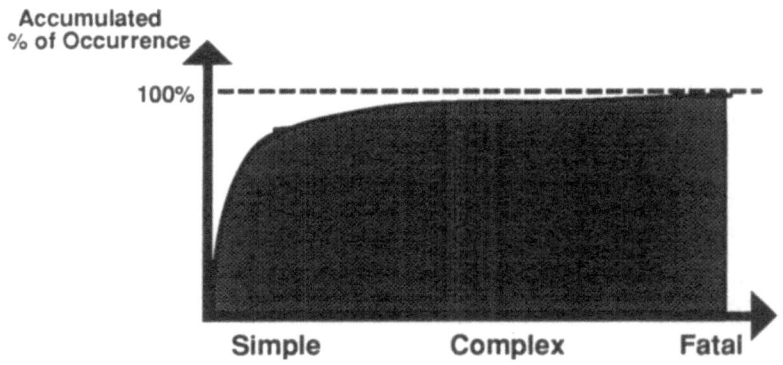

Fig. 7.15. Rate of occurrence of errors in FAA systems.

system during the product's lifetime. However, one should try to make the error recovery system as adaptive as possible. It should, with well-grounded software and hardware, become cheaper to recover from every new error. One should also aim to have self-generating recovery routines where possible – see Section 7.6.2 on adaptability.

The answer to the previous question can also be that in reality a more heuristic approach is common: one starts the system with no error recovery routines. All recovery is carried out manually, and historic and physical data about the errors are collected. In a stepwise manner, one chooses to deal with the simplest and most frequent errors. Routines for these are created. This process is usually on-going until it is estimated that there is no further gain to be obtained from the sequence. See Fig. 7.13. Note, however, that it is of vital importance to work within an optimal software environment. The above implies databasing, object oriented programming and rule-based work for efficient updating.

Since errors usually stem from a large range of causes, they obviously require a correspondingly large set of recovery routines.

Consider the following example: a detected error is classified as being an incorrectly assembled ballbearing. The possible causes may be: (1) the bearing was incorrectly gripped (program or gripper error); (2) the bearing's diameter is outside tolerance values (parts error); (3) the gripper released the bearing too early (program/air supply). The possible recovery routines for cause (1) could be: (a) remove bearing, re-grip on a flat surface and re-assemble; (b) remove bearing, scrap it and pick up a new one; (c) if the bearing is jammed, scrap the whole assembly, update batch size and modify the program.

For cause (2): (a) scrap assembly and start anew – contact the parts manufacturer; (b) try removing bearing and pick a new one. And finally, for cause (3): (a) check air supply – manual service is required if faulty; (b) modify the program by adding the required offset to final position.

Obviously some routines can be used for several error causes. It is also true that certain routines will exhibit a higher rate of success than others. Hence the need to associate a "rate of success" value to each routine in order to aid the software systems decision-making process – No. of successful attempts/No. of calls. This type of historical backing is often used together with an identity parameter which indicates which error cause it was used to correct. After a given time, the automatic error recovery system will have a clear indication of which routines are to be kept and which could be omitted – a sort of survival-of-the-fittest algorithm for the routines [3]. An example of a function used to carry out such work is shown

$$E = t \times (1-P(A))$$

where t = time taken to execute the routine, and
$P(A)$ = probability of success.
Therefore, the lower E is, the better the routine. Note that in this case the time taken to execute the routine is taken into consideration, since the user wants to minimize the cycle times.

7.6.2   Adaptability

Adaptive error recovery is another form of active recovery, theoretically more true to its name than any other form of recovery. Let us consider two examples.

The most common way to use adaptive error recovery is to have smart sensors linked directly to the software. Imagine an active force/torque sensor used in the assembly of parts. If the threshold signal patterns are outside the signature tolerance levels, it immediately goes into an active closed loop to drive the robot to the correct position.

A second example is when a rule-based system can actually adapt existing programs to new situations. If the cause of error is built into a program, the software simply adds an offset to the robot program position and re-tries the assembly. The offset may be chosen on the grounds of historical data, e.g. this particular gripper has a tendency to deviate from x by 2 mm owing to a concentricity problem

The main difference between conventional error recovery and adaptability is that the automatic error recovery system works in a closed loop with either smart sensors or existing programs to arrive at a successful conclusion. The classification scheme is simplified and the execution times can, at least in theory, be greatly reduced.

References

1. Redford A. Is there hope for robots in assembly? Assembly Automation 1991; 11(1)
2. Gini M, Doshi R, Gluch M, Smith R, Zualkernan I. The role of knowledge in the architecture of a robust robot control. In: Proc IEEE Conf Robotics and Automation, 1985
3. Nyström M. Applikationspaket för flexibla automatiska monteringssystem. Licentiate thesis, TRITA-LVE 92-108, ISRN: KTH/TSMS/LA-92/1-SE
4. Laperrière L, El Maraghy. Planning of products assembly and disassembly. Annals of the CIRP 1992; 41(1)
5. Delchambre A, Coupez D. A knowledge based error recovery in robotised assembly. In: Proc 9th Intl Conf. Assembly Automation (ICAA), London, 1988
6. Moed MC, Kelly RB. An expert supervisor for a robotic work cell. In: Proc – SPIE Intl Soc Opt Eng, vol 848, Intelligent robots and computer vision: 6th in a series. SPIE, Cambridge, Massachusetts, 1987
7. Selke K, Swift KG, Taylor GE, Pugh A. A knowledge based approach to robotic assembly. In: Proc Conf UK Research in Advanced Manufacturing, London, 1986
8. Gini M, Gini G. Recovering from failures: a new challenge for industrial robots. In: Proc COMPCON 83 Fall, 27th IEEE Computer Society Intl Conf, Arlington, Virginia, 1983
9. Lee MH, Hardy NW, Barnes DP. Research into automatic error recovery. In: Proc Conf UK Robotics Research, Institute of Mechanical Engineers London, 1984

10. Abu-Hamdan MG, El-Gizawy AS. An error diagnosis expert system for flexible assembly systems. In: Proc IFAC/INCOM '92 Conf Information Control Problems in Manufacturing Technology, Toronto, 1992
11. Arnström A, Onori M, Nyström M. Adaptivity for flexible automatic assembly. In: Proc 23rd Intl Symp Industrial Robots (ISIR), Barcelona, 1992
12. Carlisle B (presented by ADEPT Technologies Ltd). Roundtable discussion overheads. In: 22nd ISIR, Detroit, Michigan, 1991
13. Mooring BW, Pack TJ. Robot calibration in an industrial environment. Intl J Manufacturing Technology 1988; 3(5): 3–6, IFS
14. Onori M, Nyström M. Adaptive techniques for the MkII flexible automatic assembly system. In: Proc IFAC/INCOM '92 Conf Information Control Problems in Manufacturing Technology, Toronto, 1992
15. Pathre US, Driels MR. Simulation experiments in parameter identification for robot calibration. Intl J Advanced Manufacturing Techniqies 1990; 5: 13–33, Springer-Verlag, London
16. Aalto H. Programming of robots using a 3-D model object. Technical Research Centre of Finland (VTT), ESPRIT Project Proposal, 1990

Further reading

Boothroyd G. Automatic assembly. Marcel Dekker Inc, New York, 1982

Hardy NW, Barnes DP, Lee MH. Declarative sensor knowledge in a robot monitoring system. In: Rembold, Ulrich and Hörmann eds, Languages for sensor-based control in robotics, Springer, Berlin, 1987

Kauhaniemi I, McLeod I. Local calibration of robot cell for off-line programmed tasks using touch probes, Napier Polytechnic of Edinburgh & VTT Research Centre of Finland, ESPRIT Project result, 1988

Lee MH, Hardy NW, Barnes DP. Error recovery in robot applications. In: Proc 6th British Robot Association Annual Conf, Birmingham, 1983

8 Automatic Supervision of Machine Tools

S.A. Spiewak

8.1 Introduction

Mechanical failures are by far the most significant non-controllable cause of lost production time in manufacturing systems. De Barr [1] estimates that these failures in conventional machine tools account for four times as much down-time as electrical failures. Kegg [2], Milacic and Majstorovic [3] suggest that in NC machines this ratio is at least 7 to 1. Detailed analysis of the component failure distribution in manufacturing systems can be found in [2–5]. According to these studies, the tool and workpiece changing systems may account for as much as 40% of down-time. Other basic components of machine tools, such as spindles, slides, bearings, gears and lubrication, are responsible for slightly over 10% of the lost production time. A thorough investigation of 44 lathes reported in [5] breaks down all components of these machines into 25 classes, including: control unit (NC, CNC, PLC), electric motor, bearing and spindle. For each class the frequency of failures and the average down-time per failure are estimated. According to these results, malfunctions of bearings and spindles alone account for about the same down-time as the control unit failures.

Present experience indicates that the failure diagnosis accounts for the main component (80%–95%) of the down-time [2, 3]. Once the faulty elements are properly identified, the repairs are performed quickly. Thus, there is a general agreement that effective monitoring and diagnosis (M&D) holds a promise of great reduction of the lost production time. Milacic and Majstorovic suggest that for machining centres involved in unit quantity or small batch production, the effective time of work can be increased "... more than 240%" as a result of automatic supervision of the *machine and process* [3]. Furthermore, they expect a decrease of failures by "... up to 80%."

There are very few publications which allow the estimations of losses in manufacturing systems caused by machinery failures and poor quality of parts produced by impaired equipment before the failure detection. Some indication as to the order of magnitude can be found in [6] and [7]. According to the first paper [6] the introduction of the DEFT diagnostic expert system (DEFT – Diagnostics Expert for Final Test – tests the product, not the manufacturing equipment) in one of IBM's plants saves the company $5 million per year. It cuts by half the number of disk drives that do not pass the quality control and therefore require some form of re-work. Based on a survey carried out in the UK in 1975, the second paper [7] estimates potential savings resulting from M&D as nearly £10 million across all shipbuilding and marine industries, or about £75 million across metal manufacture industries. The same paper [7] suggests that efficient maintenance practices would result in savings that are five-fold higher than the maintenance cost. These savings take into account the decreased cost of repairs and maintenance achieved with M&D. In the petrochemical industry, for example, the cost of preventive maintenance (Automatic Supervision of Machine Tools – ASMT – is an advanced form of preventive maintenance) is less than half of the maintenance cost in the "run to failure" operation [8].

ASMT [9, 10] may be implemented in a variety of ways. The ideal system will: (1) monitor and diagnose the condition of various machine components to detect failures (including latent failures [6]) in their early stages; (2) forecast the development of the detected failures; (3) negotiate a suitable repair time with the production scheduler; and (4) suppress the impact of the existing failures on the product quality. The ASMT systems dealt with in this paper perform only the first two tasks. The last task has been addressed in publications such as [11, 12] and can be considered as an example of the Geometrical Adaptive Control.

The optimum form of ASMT is implementation dependent. Sequential binary devices, such as tool changers [13, 14], require different techniques than those used for devices with an infinite number of states, such as bearings, gears or spindles. The devices belonging to the latter class form the subject of this chapter. Many of these have rotating components, to which a large spectrum of methods developed in the field of machinery monitoring [15–19] can be applied. However, there are also devices that are specific to machine tools, such as guideways, chip removal systems, brakes or chucks [20]. For these devices suitable ASMT techniques need to be developed. Other important elements of machine tools are servo mechanisms. However, the signals most suitable for their monitoring and diagnosis are readily available in CNC systems. For this reason, servo mechanisms are considered "control-external" devices [21] and are not dealt with here.

In the following two sections a generic structure of ASMT systems utilizing *prognostic parameters* [22] is proposed. Two distinct approaches to the generation of these parameters, namely system analysis and signal analysis, are reviewed in Section 8.4. This review is followed by a brief discussion of five representative monitoring and diagnosis systems. Finally, two examples in Section 8.6 illustrate the unique demands of signal processing in ASMT.

8.2   Generic Structure of ASMT

In the reliability theory the task of monitoring and diagnosis can be formulated analytically as a problem of finding an optimal control of random processes [22]. This approach is characterized by the following example involving a single machine tool. A "history" of failures of this machine over a certain period of time is shown as a diagram in Fig. 8.1. The vertical axis represents three distinct states of the machine (the number of states in realistic cases is larger): healthy condition (HC), preventive maintenance (PM) activity and emergency repair (ER). The intervals $t_{m,i}$, $i = 1, 2, ..., n$ shown above the diagram indicate the lost production time. The intervals $t_{p,i}$, $i = 1, 2, ..., n$ shown below the diagram correspond to the normal operation.

The diagram represents a semi-Markov random process (SMP), which may be used as an analytical model of the machine failure-related behaviour. The

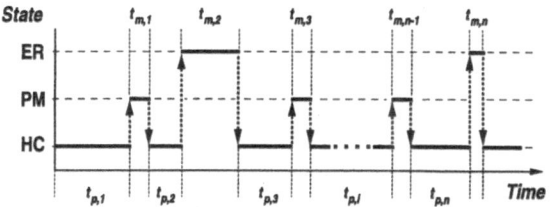

Fig. 8.1. Machine operation record as a three-state random process.

availability of this machine can be defined analytically by the *coefficient of readiness* (for the sake of brevity, all losses are expressed in units of time)

$$g(t) = \frac{t - t_m}{t} \qquad (8.1)$$

where t_m is the total lost production time due to the preventive maintenance and emergency repairs occurring over a time interval t. In the case of ASMT, three important assumptions can be made:

1. The considered time interval, t, is significantly longer than the average period of failure-free operation (e.g. 250 hours [1–3]).
2. The transitions between the states shown in Fig. 8.1 and the instances when these transitions take place are realizations of random variables.
3. There are quantities which can be used as indicators that the whole machine or its individual components approach a "failure state". These quantities, often termed "prognostic parameters", are either directly measurable or obtained as a result of processing of suitable signals from the machine.

The first assumption allows equation (8.1) to be rewritten as

$$g_n = \lim_{t \to \infty} \{g(t)\} = \frac{t_{p,av}}{t_{p,av} + t_{m,av}} \qquad (8.2)$$

where $t_{p,av} = \dfrac{1}{n} \sum\limits_{i=1}^{n} t_{p,i}$ is the average period of healthy operation, and $t_{m,av} = \dfrac{1}{n} \sum\limits_{i=1}^{n} t_{m,i}$

is the average period of maintenance activity, either scheduled or in response to emergencies.

The optimum maintenance strategy is such that it guarantees the highest operational readiness g_n. This strategy can be found by maximizing the expression on the right-hand side of equation (8.2). The optimized variables are the time intervals between the scheduled maintenance activities applied to various components of the machine. To find these optimum values, the probabilities of transitions between the states shown in Fig. 8.1, namely HC, PM and ER, and the distribution of the time intervals between these transitions have to be known. These probabilities and distributions can be defined in the considered case by matrices $\mathbf{P}$ and $\mathbf{T}(\tau)$ respectively. These matrices are

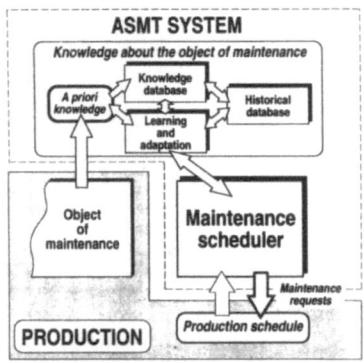

Fig. 8.2. A block diagram of the maintenance system implementing *a priori* information about the failure characteristics of the machine.

$$\mathbf{P} = \begin{bmatrix} p_{\mathrm{HC \to HC}} & p_{\mathrm{HC \to PM}} & p_{\mathrm{HC \to ER}} \\ p_{\mathrm{PM \to HC}} & p_{\mathrm{PM \to PM}} & p_{\mathrm{PM \to ER}} \\ p_{\mathrm{ER \to HC}} & p_{\mathrm{ER \to PM}} & p_{\mathrm{ER \to ER}} \end{bmatrix} \tag{8.3}$$

$$\mathbf{T}(\tau) = \begin{bmatrix} F(\tau)_{\mathrm{HC \to HC}} & F(\tau)_{\mathrm{HC \to PM}} & F(\tau)_{\mathrm{HC \to ER}} \\ F(\tau)_{\mathrm{PM \to HC}} & F(\tau)_{\mathrm{PM \to PM}} & F(\tau)_{\mathrm{PM \to ER}} \\ F(\tau)_{\mathrm{ER \to HC}} & F(\tau)_{\mathrm{ER \to PM}} & F(\tau)_{\mathrm{ER \to ER}} \end{bmatrix} \tag{8.4}$$

where

$p_{x \to y}$ probability of transition from the state x to y;
$F(\tau)_{x \to y}$ probability distribution of the transition time, $t_{x \to y}$, from the state x to y, i.e.
       $F(\tau) = P(t_{x \to y} \le \tau)$,
x, y HC, PM, ER states defined in Fig. 8.1.

A block diagram of the system implementing the above principle is shown in Fig. 8.2. All information about the failure characteristics of the considered machine tool is stored in the *Knowledge database*. This information is condensed into the form of the implemented SMP with three states, and the matrices **P** and **T**(τ). During the initial operation of the maintenance system this is *a priori* information, which may not be accurate. It can be established on the basis of the technical data (reliability) of machine tool components or previous experience with similar machines. However, various algorithms represented by the block *Learning and adaptation* may be used to perform fine-tuning of the system as time goes on and the actual behaviour of the machine is observed.

There is no direct information flow between the *Object of maintenance* and the *Maintenance scheduler* shown in Fig. 8.2. A certain amount of signal processing is usually required to facilitate this flow. Thus, automatic supervision systems utilizing prognostic parameters are featured by a *Signal processing unit* shown in Fig. 8.3, which represents various possible signal processing functions. The addition of this block means also that there are then suitable extensions of the upper layer of the monitoring system (namely *A priori knowledge, Knowledge database, Historical database* and *Learning and adaptation*) which are required to include and maintain all information relevant to sensing and signal processing.

In the general field of monitoring and diagnosis, the complexity of the various functions represented in Fig. 8.2 and Fig. 8.3 depends upon specific applications. In particular, two cases are worth consideration. The first case occurs rarely, when the prognostic parameters provide reliable information about the present and future

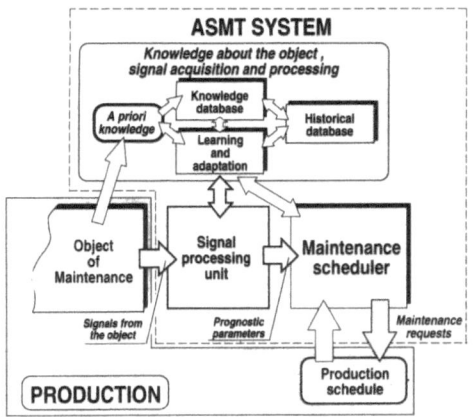

Fig. 8.3. A block diagram of the ASMT system implementing prognostic parameters.

condition of the object. In this case the *Maintenance scheduler* does not need any information regarding the statistical properties of the object reliability and generates an optimum maintenance strategy based exclusively on the prognostic parameters and the production schedule. This represents an ideal *predictive* M&D system.

The other case occurs when the probability of failure has to be unconditionally minimized, for example in commercial aviation, space flights, some military applications or critical components of manufacturing systems. In this case the basic maintenance schedule is derived from the reliability data. In addition, whenever prognostic parameters indicate (although the indications may be false) a possibility of failure, the object is inspected and, if necessary, repaired. The importance of prognostic parameters in this latter type of maintenance has led to the development of a branch of engineering known as monitoring and diagnosis of machinery [15–19, 23].

The typical situation in ASMT is quite different from the extreme cases discussed above. The available prognostic parameters are, as a rule, unreliable. Therefore, maintenance activities in response to *every* indication of a potential failure may lead to higher losses than caused by these failures. It is intuitively understood that efficient ASMT systems have to optimize some "quality index" (cost function) including: (1) the cost of preventive maintenance; (2) the losses due to potentially avoidable failures; and (3) the losses resulting from unnecessary maintenance activities caused by misalarms (Type I errors of monitoring and diagnosis systems [23, 24]). It is also expected that reliable prognostic parameters are instrumental in facilitating a significant reduction of the latter two types of losses.

8.3   Generation of Prognostic Parameters

Measurable quantities in manufacturing systems are rarely suitable as prognostic parameters. Most often these parameters are obtained as a result of extensive processing of one or more signals from the monitored object. It is possible to define the structure of a generic *Signal processing unit* shown in Fig. 8.3 [24–27]. This structure represents various algorithms, which have been implemented or are potentially applicable to monitoring and diagnosis. As is shown in Fig. 8.4, the

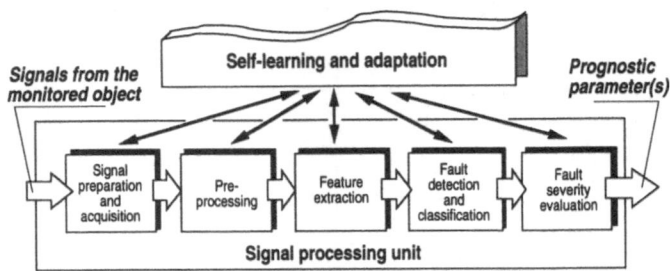

Fig. 8.4. A block diagram of the *Signal processing unit*.

Signal processing unit may comprise up to five cascaded blocks, whose functions are discussed below:

1. *Signal preparation and data acquisition* functions can be as simple as analogue-to-digital conversion and signal storage. In advanced systems the characteristics of signal acquisition channels, such as gain and cut-off frequency, self-tune to maximize the signal-to-noise ratio, minimize quantization error or obtain maximum spectral resolution [28, 29]. The validity of signals, which may be impaired by sensor malfunction or overload, is also checked [30, 31]. Research is under way on self-tuning sensors that compensate for the distortions to the measured signals caused by the working environment [32].

2. The *Pre-processing* block represents all signal processing techniques implemented in software or hardware to emphasize important features of the monitored systems, which are contained in the signals. The functions represented by this block are extremely diverse. Their tasks range from the suppression of noise to time-series-based parametric identification of structures and parameters of dynamic systems. These functions are required to process huge quantities of data in a short time. Although the signal pre-processing techniques are universal across several disciplines of science and engineering, such as communication, radar, speech recognition or control, relatively few "standard" techniques have been widely adopted in monitoring and diagnosis systems for machinery. Some representative techniques are reviewed in Section 8.4.

3. *Feature extraction* represents functions which can be the most intriguing to the maintenance engineers using commercial M&D systems. This block typically transforms a large quantity of its input data into a few numbers which are essential to detect and characterize physical phenomena (e.g. defects in one of the many bearings in the monitored system). Thus, for a practitioner who understands the mechanical design of machines but not necessarily the intricacies of the Fast Fourier Transformation (FFT) algorithm, the *Feature extraction* block can be viewed as a "converter" between more or less abstract data obtained from *Pre-processing* and the physical properties or phenomena in the monitored object. The most fascinating aspect of this conversion is that the important features are usually well camouflaged, and finding them is like solving "Spot the difference" problems. This comparison suggests that in ASMT the extraction of features representing various failures may be extremely difficult. As a rule, there are no "generic" feature extraction algorithms which can be readily used. Instead, there is a broad spectrum of techniques which need to be adopted for each particular type of monitored objects. Some of the most important are the

techniques developed in the areas of: (1) knowledge-based (artificial intelligence) systems including Expert Systems (at present the most frequently used class in commercial systems) and distributed artificial intelligence; (2) pattern recognition; (3) neural networks; and (4) fuzzy sets.

4. *Fault detection and classification* utilizes the set of features extracted by the preceding block to determine whether any abnormal condition exists. If the answer is positive, the locations and types of faults are determined. It is important to acknowledge that, in the majority of cases, the output from *Fault detection and classification* has a certain probability of error. However, this probability may not be known. The sources of potential errors are not only in the algorithms implemented within this block but, as a rule, also in all operations performed by the preceding stages of signal processing. To complicate the situation, errors can be generated during the conversion* of measured mechanical quantities from the monitored objects into electrical signals suitable as inputs to the *Signal processing unit*. A properly designed *Fault detection and classification* block should be capable of minimizing the errors of its output signals caused by the uncertainty, which features the information from the previous stages. This is known as the minimization of Type I (misalarm) and Type II (insensitivity) errors [24]. Methods developed in the fields of hypothesis testing, decision-making and games theory are most suitable for this purpose.

5. The output from the *Fault detection and classification* block can be used as binary (i.e. FAULT or NO FAULT) prognostic parameters. However, it is often desirable to assess the severity of the detected fault(s). This task is performed by the last block in the signal processing sequence, *Fault severity evaluation*.

The flow of information within the generic structure discussed is coarsely indicated by the arrows between the blocks shown in Fig. 8.4. The main stream flows sequentially through the five blocks of the *Signal processing unit*. It begins as the data arriving from sensor(s) and ends as the prognostic parameter(s). For the sake of brevity, the following discussion addresses only one important issue, namely multiple feedback loops which can exist in the information flow. To illustrate the problem a brief example of a feedback loop involving human expertise is considered. In this example a skilled maintenance technician uses the frequency domain analysis to detect possible failures of any of the numerous bearings of the monitored machine. After collecting suitable vibration signals, an FFT analysis is performed and its results are displayed as a waterfall diagram, such as that shown in Fig. 8.5. This diagram is then examined for the presence of "significant" spectral components at, and close to, the characteristic defect frequencies associated with the defect modes of the monitored bearings [37, 38]. The technician *knows* where to look for these frequencies. If the spectral resolution is found to be insufficient, the whole signal acquisition and spectral processing is repeated for a higher sampling frequency; or, if it is suspected that the resolution is correct but the variance of the spectral components is too high, more averaging is applied during the calculation of spectra.

In this example three components of the *Signal processing unit* form the forward path. The first of these components is *Signal preparation and acquisition*, whose characteristics, such as the sampling rate, frequency bandwidth or gain, may be

---

* In sensors, not shown in Fig. 8.3. The location [33, 34], correctness of installation [35, 36] and type [32] of implemented sensors also contribute to these errors.

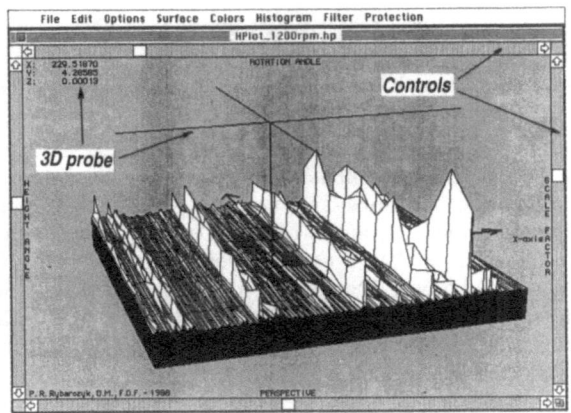

Fig. 8.5. Multiple spectra visualized as a "waterfall diagram" for improved detection of frequencies associated with bearing defects [97, 98]. The axes are as follows: x (longitudinal) – frequency; y (vertical) – spectrum magnitude; z (depth) – time.

adjusted by the operator. Next, *Pre-processing* converts the measured vibration signal from the time domain into the frequency domain using the Fourier transformation. Several parameters of this transformation, such as the shape of the weighting window (the most frequently used shape is known as Hanning [39]), the number and size of the averaged portions of the input signal, the amount of overlap between these portions and zero padding for increased resolution, may be tuned [39]. The third component of the considered forward path is *Feature extraction*. Its function is in this case simple: merely to present several spectra as the three-dimensional "solid" (or surface) in Fig. 8.5 and to facilitate extensive examination of this surface. The operator, who actually detects the features, uses four "controls" shown as scroll bars along the screen edges, and menus shown at the top of the screen* to select the desired portion of the surface, rotate and magnify it and quickly get co-ordinates of selected points with a "3-D probe" controlled via a "joy-stick" or a "mouse". The menus also allow rapid generation of "contour maps" and cross-sections of the surface. All these functions serve one purpose, namely to help the operator to identify and extract features associated with any disorder of the monitored machine. As already mentioned, these features are "buried" in a huge quantity of numbers generated by the Fourier transformation (for example, 128 spectra with 8129 lines resolution each, contain approximately 1 million pieces of data, equivalent to 1000 pages of mathematical tables!).

In the example considered, the adjustments of characteristics of the forward path of signal processing by the operator represents the feedback loop. In the advanced ASMT systems such feedback has to be generated by suitable algorithms, which can be built either into individual blocks of the *Signal processing unit* or implemented in the *Knowledge about the object, signal acquisition and processing* block.

Each block of the ASMT system shown in Fig. 8.3 may be implemented in a variety of ways and utilize a broad spectrum of algorithms. In any particular

---

* This interactive data visualization and analysis has been implemented in the prototype ASMT system developed at the University of Wisconsin–Madison [26].

implementation the internal structures of these blocks, i.e. the algorithms and their interactions, are determined by the knowledge, practical experience and personal preferences of their designers. The feedback loops between the blocks also depend on these factors. The obvious question arises about how good, objectively, these various ASMT systems are. There is no easy answer. The majority of commercial systems are in some sense optimized, mainly by means of heuristic approaches. As this is commonplace with any non-analytical optimization of complex functions, there is no guarantee that the performance of such optimized systems is close to the global maximum. Since this maximum is not known, it is impossible to determine whether the performance of available systems is close to it. Furthermore, the lack of a suitable "yardstick" makes any comparison of different ASMT systems difficult and strongly subjective.

8.4 System and Signal Analysis

Typical faults in a mechanical system include: (1) failures in rolling element bearings, such as cracks and pitting; (2) excessive unbalance of rotating components; (3) failures of gears; (4) loosening of joints; and (5) cracking of structural components. Many of these faults can be thoroughly assessed by visual inspection and suitable instruments, if the investigated elements are removed from the machine. This solution is rarely practical. Instead, traditional preventive maintenance practices [4] involve periodical inspection of the accessible components without removing them from their locations.

An alternative and indirect way of obtaining prognostic parameters involves various signals generated by machines during their operation. Two approaches to indirect M&D are possible. The first approach, termed "off-line M&D", requires specific sequences of machine states, such as certain feeds and rotational speeds occurring and lasting according to pre-determined patterns. These sequences are generated by suitable test programs. In addition, some testing signals, such as external forces from impact hammers or exciters, may be applied. The second approach, termed "on-line M&D", is performed continuously during normal operation of monitored machines. This latter approach is transparent to the user and embodies the desirable features of truly automatic supervision of machine tools. Kegg [2] wrote about such M&D as "... the dream of maintenance managers in large factories to be able to diagnose mechanical machinery conditions through sound and vibration ... and be able to anticipate troubles before they happen." On-line M&D will now be considered.

The entire machine tool and almost all of its components are dynamic systems. Indirect assessment of their features for the purpose of ASMT necessitates the use of the equivalent mathematical models. A broad spectrum of methods developed mainly in control theory [30, 40–47] and statistics [48–51] is available to identify the structures (but only in the case of empirical models) and coefficients of these models. Once the mathematical models are identified, the required physical parameters of the systems investigated can be estimated. This process is illustrated by the block diagram shown in Fig. 8.6. Further information on this subject can be found in [27, 52–54].

The majority of methods developed in control theory assume the availability of the input signals to the *System*. Since in ASMT these signals are rarely available, the identification methods developed in statistics for the analysis of time series are

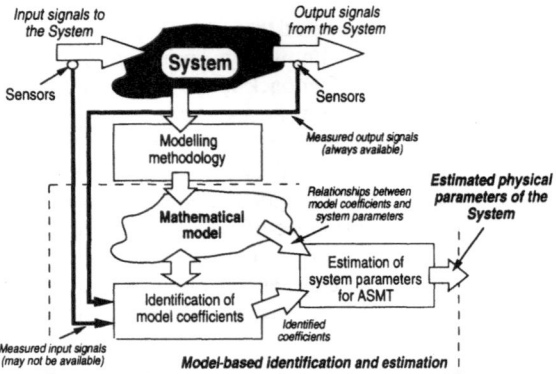

Fig. 8.6. A block diagram of the generic algorithm for parameter estimation.

particularly useful. These methods utilize so called Auto-Regressive (AR), Moving Average (MA) and Auto-Regressive Moving Average (ARMA) models [19, 48, 49]. The most comprehensive of these models, ARMA, has the form*

$$y_k = -\sum_{i=1}^{m} a_i y_{k-i} + \sum_{j=0}^{n} b_j e_{k-j} \tag{8.5}$$

where $y_{k-i} = (k-i)$th element of the considered time series, namely output signal from the *System*; $i = 0, 1, ..., n$;

$e_{k-j} = (k-j)$th element of a hypothetical time series representing the signal (input or disturbance) exciting the *System*; $j = 0, 1, ..., m$; the most frequently used is the uncorrelated times series with the Gaussian amplitude distribution (white noise);

a_i, b_j = autoregressive and moving-average coefficients, respectively.

The development of time-series modelling techniques was motivated by phenomena whose physical interpretation and description were either impossible or extremely difficult, such as the stock market or sunspot activity. Since model structures and the coefficients were needed for the prediction of the future behaviour of these phenomena, suitable methodologies have been developed for the empirical identification of these structures and coefficients. On the other hand, no methodologies that would allow one to define the analytical relationships between the obtained empirical models and the physical nature of the underlying phenomena were sought.

This missing link, very important for the monitoring and diagnosis of mechanical devices, has been developed in control theory for the analytical description of discrete-time systems [45]. In particular, the ARMA model (Equation (8.5)) is equivalent to the difference equation describing linear time-invariant dynamic systems [48]. Thus, using suitable methods, such as the Z transformation formalism, ARMA models equivalent to the investigated physical systems can be developed. These mathematical models have three important features:

1. Their structures determined by the coefficients n and m are known. Consequently, the coefficients a_i and b_j can be estimated more accurately than in the "classical" time-series analysis.

---

* For clarity a single output system is considered. All signals acting on this system, both inputs and disturbances, are represented by one noise signal. The extension to multi-input, multi-output (MIMO) systems can be found in [49, 50, 55].

2. The identification of the coefficients a_i and b_j does not require the knowledge of the actual signals exciting the systems.
3. The coefficients a_i and b_j are known analytical functions of the physical parameters of the identified system. Once these coefficients are identified, the actual values of the physical parameters can be evaluated.

A distinctive feature of the above methods is the use of *a priori knowledge about the investigated systems*. This knowledge is condensed into the form of suitable mathematical models. The signals recorded from the real systems are analysed in the context of these models to obtain the desired quantities, namely estimates of the physical parameters of the actual systems. A class of methods which utilize *a priori* knowledge about the system will be henceforth referred to as *System Analysis*. In addition to the discussed parametric identification developed in control theory, this class contains Modal Analysis and Operational Deflection Shape (ODS) Analysis [56–58]. Since *System Analysis* methods, represented in Fig. 8.4 by the *Preprocessing* block, are capable of estimating the values of physical parameters, they usually eliminate the need for *Feature extraction*. Moreover, they radically simplify *Fault detection and classification*, and *Fault severity evaluation*. The identified values represent the most convenient prognostic parameters needed by the *Maintenance scheduler*.

The applicability of system analysis methods is often limited by the following factors:

1. Insufficient *a priori* information about the monitored system (e.g. no mathematical model of acceptable quality is available);
2. Complex nature of the system (e.g. non-linearities, non-stationarity), which renders useless the identification algorithms known at present;
3. Long analysis time caused by the numerical complexity of identification algorithms;
4. Lack of unique mapping between the identified model coefficients and the required parameters of physical systems.

Since the long analysis time and the complex nature of monitored systems were in the past insurmountable obstacles, a broad spectrum of *Signal Analysis* methods, often collectively referred to as Signature Analysis, has been developed in the field of machinery monitoring [15, 17–19]. The best known representative of this class is Fast Fourier Transform (FFT). Other examples, and their acceptance in the commercial systems, are illustrated in Fig. 8.7 [59]. The most common devices and types of failures that can be monitored by these methods include the following:

Bearing defects (inner and outer raceways, rolling elements, cages);
Gear defects (spalling, broken teeth);
Unbalance (single and multiple plane);
Misalignment (parallel, angular, shaft-bearing);
Resonance (structural and shaft-bearing);
Looseness (loose assembly, poor tolerancing).

The signal analysis algorithms are, in general, unable to estimate the values of the physical parameters of the monitored systems. This makes the tasks of *Feature extraction, Fault detection and classification* and *Fault severity estimation* extremely important. To illustrate the problem, an example of bearing monitoring

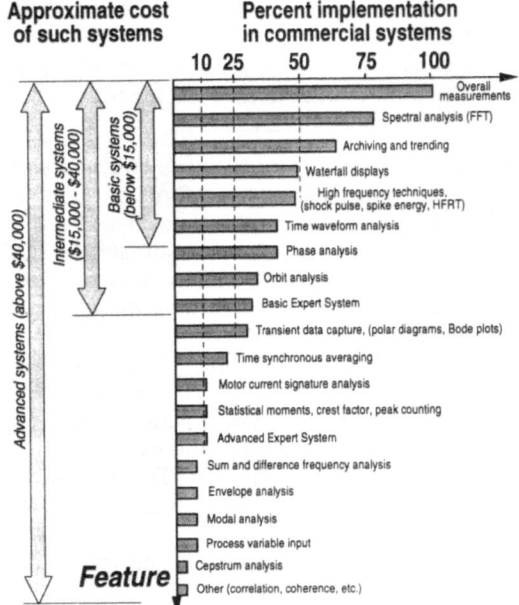

Fig. 8.7. Most frequently used signal analysis methods and their acceptance in commercial M&D systems [59].

involving the High Frequency Resonance Technique (HFRT) is considered [60]. The task of each block of the *Signal processing unit* shown in Fig. 8.4 is briefly discussed.

First, the signal from an accelerometer mounted in the vicinity of the monitored bearing is amplified to a suitable level, passed through a band-pass filter (5–25 kHz) and recorded at a high sampling rate (50 kHz). These operations are represented by the *Signal preparation and acquisition* block. Next, short-time signal processing techniques are employed to obtain signal features such as energy, zero crossing rate and the median. Other features, such as the envelope obtained by means of hardware detectors or Hilbert transformation, can also be used [61]. This step, which produces intermediate compressed data, is represented by the *Pre-processing* block. The same block performs a correlation analysis (an efficient method of calculating the correlation function involves the FFT algorithm [62]) of the intermediate data. This latter analysis results in further data compression, but first of all it emphasizes the presence of so-called "characteristic defect frequencies" [37, 38] related to various forms of bearing defects. The detection and extraction of these frequencies is performed by *Feature extraction*. The obtained frequencies are compared next in *Fault detection and classification* block with the expected characteristic defect frequencies of the monitored bearing. If a match is detected, the corresponding failure type is flagged. The probability of misalarm is also evaluated. A broad range of methods can be applied, including discriminant functions [60, 63–68], neural networks [69], fuzzy set [64, 70, 71] and grey system [72] techniques or knowledge-based (expert) systems [3, 6, 73–75]. These latter systems contain huge database of rules (e.g. 700) and bearing data, which are needed to calculate the expected characteristic defect frequencies, and are most widely used in advanced commercial M&D systems.

Once the failure is detected and diagnosed, its severity is assessed by the last block of the *Signal processing unit*. The assessment may involve features used at the previous stage. Often, other features need to be utilized, such as the energy level at the characteristic frequency of the detected failure or the overall energy of the measured signal. It is intuitively obvious that accurate relationships between the available features and the failure severity are in general not known. As a consequence, prognostic parameters generated by the *Fault severity evaluation* block are prone to significant and usually unknown errors.

System Analysis and Signal Analysis approaches may be combined. For example, Chen [76] proposes a method according to which the unbalance of a spindle assembly is continuously monitored by analysing the prediction error of an AR(5) model identified during the failure-free operation of the spindle. Three features of the prediction error are considered: normalized variance, kurtosis and autocorrelation. A multiple voting scheme is used to detect the unbalance and assess its severity. Coefficients of the implemented AR(5) model are identified according to the "classical" approach in statistics, that is, without taking into account the physical structure of the spindle.

8.5   Example ASMT Systems

The number and diversity of commercial and experimental systems suitable for ASMT is too large to allow their evaluation and comparison here. To provide an overview of structures and signal processing methods employed in these systems, five examples are considered. They are as follows:

1. CHARLEY, an Expert System based software package for off-line monitoring and diagnosis of manufacturing systems [73, 74].
2. Maintenance Machinery Monitoring System (MMMS) developed for ASMT in the automotive industry.
3. A modular multi-processor system developed in Aachen [77, 78].
4. Computer Assisted Fault Diagnosis (CAFD) system developed in Stuttgart [27].
5. Monitoring and Diagnosis Shell (MDS) developed in Madison [26].

Written in C, CHARLEY diagnostic system is easily transportable to many hardware platforms. At present, its inference engine runs on Sun 3/60. This central computer is connected over the local area network with personal computers used for loading of vibration signals and handling user dialogues with the Expert System. CHARLEY utilizes three sources of information to diagnose 72 different types of machinery failure in car manufacturing plants. These sources are:

Vibration Signals (VS) recorded by means of manual data loggers.

Machine Information Data (MID) which contains descriptions of individual machine components and interconnections between them.

Knowledge Base (KB) of over 1000 heuristic vibration analysis rules. This database encapsulates the experience of a maintenance wizard, Mr. Charley Amble, who was employed at one of GM's plants for over twenty years.

CHARLEY is an excellent tool for reactive maintenance. It has proven to be 95% successful with the diagnosis of machine failures – it surpassed its human mentor. Used by dedicated and skilful maintenance personnel it also facilitates some preventive and predictive maintenance. On the other hand it cannot be considered a truly ASMT system, since it requires manual data collection and input, and it communicates only with its operator. To expand the scope of plant maintenance by including on-line and unmanned monitoring and diagnosis, the development of another system was sponsored by the General Motors Corporation. At present, there is no information available in the literature about this new system, termed Maintenance Machinery Monitoring System (MMMS). However, there is a commercial product, Machine View [79], which closely resembles MMMS.

Machine View runs on a network of IBM PC computers. One of these computers, designated as the Management and Diagnostic Station, collects, archives and processes information supplied from the remaining computers, termed Data Collection Stations (DCS). DCSs continuously collect data from the monitored object(s), and process these data and flag alarms when threshold levels set by the user are reached. Signal processing algorithms employed in both stations belong to the "signal analysis" class. The most sophisticated of them is the Fourier transformation. Since this method of signal processing provides poor data compression and generates many "convolved" features related to various forms of machine malfunction, newer products (MasterTrend and Nspectr II [80, 81]) from the developer of Machine View utilize "automated expert system" with a database of heuristic rules applied to analyse approximately 50 features from the spectra generated by the DCSs [80].

The Machine View with its latest extensions, cast into the generic architecture of ASMT shown in Fig. 8.3 and Fig. 8.4, is featured by simple *Signal preparation and acquisition* and *Pre-processing* blocks. By contrast *Feature extraction* and *Fault detection and classification* are sophisticated, to compensate for the deficiencies of the signal analysis approach employed. At the level of *Knowledge about object, signal acquisition and processing* (Fig. 8.3), there is a *Knowledge database* with heuristic rules of the spectral analysis applicable to vibration signals from the rotating machinery. There is also a *Historical database* with archived spectra (other features of signals, such as "time histories" of rms, peak-to-peak values etc, can also be stored). Most noticeable is the absence of *Failure severity evaluation* and *Maintenance scheduler* in the system considered. All alarms and diagnoses are displayed to the user, who is expected to take relevant actions. Thus, in the context of this chapter, the essential difference between CHARLEY and the extended Machine View is the ability of the latter to acquire the vibration data automatically and display alarms to the maintenance personnel.

Two important factors which inhibit the application of systems similar to Machine View for ASMT are: (1) high cost resulting from the application of several PC-like computers integrated into a network; and (2) limited interactions with CNC controllers (such interactions are necessary for the integration of the machine and process supervision [26]). A modular multi-processor system developed at RWTH Aachen alleviates these problems by employing robust industrial computers contained in one enclosure and communicating over a common bus. The system supports four classes of signal processing functions:

1. Pattern calculation, such as trending or filtering;
2. Arithmetic and logical operations, such as addition or multiplication;
3. Signal analysis, such as FFT or cepstrum; and
4. Signal comparison.

Two distinctive features of the system considered, which has been developed for automated supervision of machine tools and processes, are: (1) programmability; and (2) its event-driven nature. The first feature allows easy tuning of the system to specific characteristics of monitored and diagnosed machines and processes. The other feature is an outcome of recognizing the root cause of poor performance of the signal analysis methods in ASMT, i.e. strong fluctuations of vibration signal signatures due to the broad spectrum of operating conditions. By allowing the system to take "snapshots" of data in response to events determined by suitable combinations of signals from the monitored object, these signatures can be analysed in the context of their relevant machine or process states. Neither CHARLEY nor Machine View support truly event-driven data acquisition.

A common feature of the three systems considered thus far is the application of relatively simple signal analysis techniques to vibration signals. In fact, this feature is shared by a large number of machinery monitoring and diagnosis packages developed in other fields of engineering [15, 59]. A qualitatively different approach, which employs system analysis, is presented in [21, 25–27, 30, 52, 53, 82–85]. The Computer Assisted Fault Diagnosis (CAFD) system described in [27, 52] relies heavily on the identification of the physical parameters of the monitored objects. On-line identified values of these parameters are statistically tested for the presence of symptoms associated with possible failure modes. Three classes of symptoms are considered: (1) analytic; (2) heuristic; and (3) "process history and fault statistics related". A knowledge-based (Expert) system is employed to determine root causes of the observed symptoms and estimate the severity of detected failures. The software implementation of CAFD is modular, with various components written in one of the following languages: Fortran, Pascal, C or Prolog. The entire system can be installed either on hierarchical computer networks used for process control or on individual computers, such as DEC LSI or PC/AT.

Three out of four of the systems considered employ one of the best known artificial intelligence techniques – the Expert System. CHARLEY and the extended Machine View apply heuristic rules to *recognize all possible* failure symptoms in the spectra of processed vibration signals and to determine types of failures according to the observed symptoms. CAFD, on the other hand, *synthesizes* failure symptoms from physical parameters of monitored devices. There is no doubt that the latter approach simplifies the detection and classification of failures and improves the reliability, sensitivity and accuracy of the prognostic parameters. Further improvement can be accomplished by: (1) the implementation of syndromes (characteristic patterns of symptoms) generated by various signal and system analysis methods; and (2) the integration of information from several sensors [63, 65, 86] and independent techniques [15, 87, 88], such as sound [23, 67, 89–91], debris [92], temperature, or power consumption [93] analyses.

There is a large variety of system and signal analysis methods documented in the literature. Only a small portion of these methods has been implemented in commercial systems. As a rule, these systems have rigid internal structures that their users must adapt to. A diametric approach is to allow the users to tailor ASMT systems according to their needs. An example of this approach is the *LabView* package [94], in which various signal processing and pattern recognition algorithms are represented by "icons" available from the "toolbox". The user picks up suitable icons and connects them on the computer screen by means of "wire" to construct various data acquisition architectures and signal processing systems. Customizing concepts can be carried to the extreme, in which the users can easily write their own code to perform different types of analysis. Examples of this type of software include *Mathematica* [95] and *SIMULAB* [96]. The major disadvantages with these latter programs are the high knowledge level required to use them effectively, and poor suitability for automation.

A prototype ASMT system developed at the University of Wisconsin–Madison [26, 97, 98] is an attempt to strike a balance between the flexibility and simplicity of development. This system, termed Monitoring and Diagnosis Shell (MDS), consists of two parts. The first part, providing front-end functions, incorporates multiple programs (signal and system analysis, pattern recognition, artificial intelligence etc.), which run simultaneously under a multi-tasking operating system OS/2 on IBM PC. The number of routines is limited only by the disk space. These routines may be thought of as a series of "canned" programs that may be put together according to the user's needs. Because of the modular software architecture, the addition of new programs for continuing improvement and fine tuning is straightforward. The second part of MDS consists of a modular, multiple transputer based, distributed system for data acquisition and parallel processing. This system can be accessed only via the "front-end" routines.

Algorithms of signal processing and decision-making are defined by means of an English-like Monitoring and Diagnosis Language (MDL). Its smallest units, *Instructions*, correspond to individual programs executed under OS/2. A collection of instructions can be performed sequentially or in parallel. Sets of logically related instructions can be enclosed into higher-level constructs – *Commands. Commands* can be nested to facilitate writing concise MDL programs. For example, to read data from a remote acquisition and processing unit, perform spectral analysis of these data and plot the result, the command shown in Fig. 8.8a may be written.

The body of this command is enclosed between the statements "SEQ" and "END". The first two lines are instructions. They reserve memory segments needed for the raw data and spectrum. The next line is a command to read data from the required channel. This command comprises instructions (not shown) which set the gain, bandwidth and the trigger event code (in the "event-driven" mode) for the data acquisition, and transfer the data into the "rawData" buffer. Next, the instruction FFTProcessing (...) is utilized to generate a power spectrum of the data acquired. This command has several formal parameters, whose meaning is explained in Fig. 8.8b. The spectrum is displayed on the screen by the Plot (...) instruction. Finally, the last two lines release the memory segments utilized inside the considered command.

A short code in MDL to display spectra of *n* signals (in channels 1, 2, ..., *n*) can be written utilizing the command defined above as

```
PAR[ALLEL]
    @ GetAndPlotFFT ( 1, *, *, 16384, 2048)
    @ GetAndPlotFFT ( 2, *, *, 16384, 2048)
    ................
    @ GetAndPlotFFT ( n, *, *, 16384, 2048)
END
```

where asterisks on the "gain" and "bandwidth" positions in the command parameters lists instruct the data acquisition system to optimize these parameters automatically for the highest signal-to-noise ratio and spectral resolution. After the above sequence is executed, *n* windows are displayed on the terminal. Each window is driven by a "copy" of the Plot (...) program running under OS/2. The user can

a !!@ GetAndPlotFFT ("channel", "gain", "bandwidth", "dataSize", "spectrumSize")

    SEQ[ENTIAL]

       CreateMemorySegment (rawData, dataSize)

       CreateMemorySegment (spectrum, spectrumSize)

       @GetData (channel , gain , bandwidth, dataSize, rawData)

           !! the '@' character indicates that the above line is a command

       FFTProcessing (Power, rawData, , spectrum, spectrumSize, dataSize, average, overlap, HANNING)

       Plot (spectrum , spectrumSize)

       CloseMemorySegment (rawData)

       CloseMemorySegment (spectrum)

    END

b

FFTProcessing (func, in1, in2, out, FFTSize, dataSize, av, ov, window)

Parameter	Description
func =>	Function to be executed, e.g., 'Power', 'Magnitude', 'Cross-spectrum', 'Cross-correlation'.
in1 =>	Memory block for the first input signal.
in2 =>	Memory block for the second input signal, if needed.
out =>	Memory block for the result.
FFTSize =>	spectrum size, e.g., 512, 1024, 2048 or 4096.
dataSize =>	Total data size.
av =>	Averaging control, e.g., average, no_average.
ov =>	Overlap control, e.g., overlap, no_overlap.
window =>	Window type, e.g., 'Square', 'Parzen', 'Hanning' ,'Welch'.

Fig. 8.8. a A command to read data, calculate its power spectrum and display this spectrum on the screen. **b** Formal parameters of the FFTProcessing (...) instruction. Note that "!!@" declares a command and "!!" indicates a comment.

arrange the windows on the screen and perform within each of them several operations on the displayed data (e.g. zoom, scale or measure co-ordinates of points with a mouse-driven cross-hair).

The first task of the MDS system, designed to extend the capabilities of MMMS,* was to verify the suitability of traditional methods of monitoring and diagnosis, based on spectral analysis of acceleration signals, to on-line monitoring of machine tools (an example of such analysis is shown in Fig. 8.5, which is a snapshot of the computer screen generated by one of the MDL instructions). The results obtained so far indicate generally poor performance of these methods. This should not be a surprise, for at least two reasons. First, the acceleration signals are affected not only by the failures of various mechanical components but, most of all, by the operating conditions of the machines (feeds and speeds). Second, the Fourier transformation is unable to separate the impact of various phenomena on the overall spectra. The research performed under shopfloor and laboratory conditions indicates that many limitations of the existing systems can be circumvented by: (1) the implementation of system analysis methods; and (2) the utilization of a broad range of signals

---

* Under OS/2, MMMS and MDS run simultaneously, MDS has access to the MMMS databases.

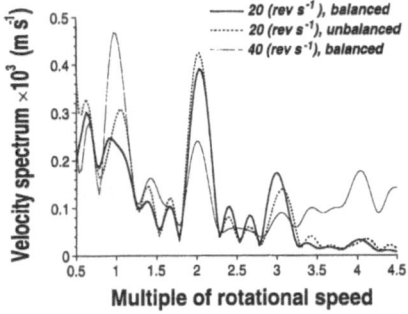

Fig. 8.9. Spectra of acceleration signals obtained at 20 and 40 rev s^{-1} spindle speeds.

including, in addition to accelerations, such quantities as displacements, forces, noise, acoustic emission, temperature and power consumption [56, 99, 100].

8.6   Examples of Signal and System Analysis

Three representative examples of data processing with the MDS system are considered in this section. The data used in the first example are from a small spindle running idle on a laboratory set-up [59]. Such a spindle would be typically used on transfer lines to machine valve seats in car engines. The spindle was instrumented with high-performance sensors. The spectra of signals obtained at two rotational speeds from an accelerometer placed on the spindle housing, approximately 10 mm from the front bearings, are shown in Fig. 8.9. The magnitude of the velocity (integrated acceleration) spectrum is plotted versus the "Multiple of rotational speed" number.

The dominant peak in the spectrum obtained for a well-balanced spindle at 20 rev s^{-1} is at the second multiple. According to the well-established rules of spectral analysis, this peak may indicate: (1) misalignment, or (2) a bent shaft [101, 102]. There is also a smaller peak at the first multiple, which may indicate: (3) unbalance. Finally, a peak at the third multiple, together with the other peaks, may indicate: (4) strain, (5) looseness, or (6) rubbing. When a substantial unbalance is introduced at the same rotational speed, a noticeable increase of the spectrum magnitude at thefirst multiple is observed, accompanied by a smaller increase at the second multiple.

At 40 rev s^{-1} the spectrum pattern of the same well-balanced spindle is significantly different. The dominant peak appears at the first multiple and there is a smaller peak at the second multiple. Based on the results obtained at the lower spindle speed, this can be interpreted as an indication of a strong unbalance. Such a conclusion is, of course, incorrect. This suggests that the spectral analysis of the housing acceleration signals may be reliable at nearly constant rotational speeds and in the absence of strong distortions, which may originate from varying cutting forces.

The second example utilizes data obtained during milling on a modern horizontal spindle CNC machining centre with a 56 kW main drive. A centre like this would be typically used to manufacture stamping dies in the motor car industry. The tool is a 250 mm diameter face milling cutter with four inserts. The spindle housing acceleration measured near the front bearings in the vertical direction over the period of one spindle revolution is shown in Fig. 8.10a. Figure 8.10b shows the cutting force component measured along the same direction. Spectra of these signals

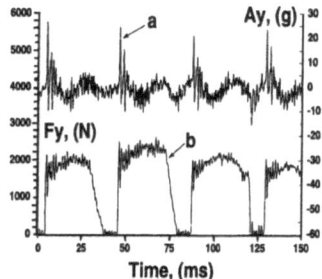

Fig. 8.10. The acceleration (a) and force (b) signals in milling.

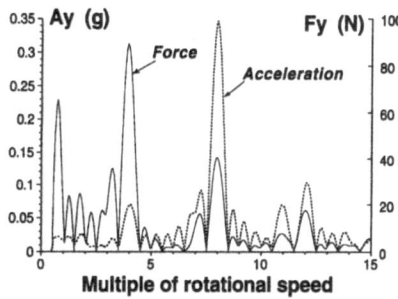

Fig. 8.11. Spectra of the acceleration and force signals shown in Fig. 8.10.

at the spindle speed of 6 rev s^{-1} are plotted in Fig. 8.11 versus the "Multiple of rotational speed" number.

The acceleration spectrum is confusing when analyzed alone. Its very strong peak at the 8th multiple and significantly smaller peaks at the 4th, 7th, 11th and 12th multiples are not a symptom of any typical disorder. The reason is that three of these peaks, namely the 4th, 8th and 12th, represent the acceleration component excited by the cutting force. Indeed, the milling cutter with 4 inserts generates four similar pulsations every revolution, as shown in Fig. 8.10b. The spectrum of this signal has significant peaks at the 4th, 8th, 12th etc. multiples of the rotational speed. If the mechanical structure of the machine behaved as a proportional system, the acceleration spectrum would resemble the force spectrum (and the displacement spectrum would be an exact replica of it).

Even more confusing is the analysis of the acceleration spectra at various spindle speeds. Such spectra obtained at 6 and 15.83 rev s^{-1} are shown in Fig. 8.12. It is worth noting that the spectrum pattern at the higher rotational speed better resembles the force spectrum from Fig. 8.11 than does the spectrum obtained at the lower speed (force spectra are similar in both cases). Nevertheless, the extraction of meaningful patterns signalling specific machine disorders from such acceleration spectra is a futile exercise. It is necessary to analyse these spectra in the context of the cutting forces, i.e., by applying system analysis.

Indeed, such analysis involving input and output signals (e.g. ARMAX models [40, 41, 42, 47] or exclusively output signals (e.g. ARMA models, [48, 49, 51]) comes through this problem with flying colours. It allows one to estimate a transfer function of the spindle assembly (the system) with the force and acceleration signals as its input and output, respectively. The gain of this transfer function, plotted in Fig. 8.13, reveals two resonances in the frequency range 0–100 Hz (the same frequency range as in the preceding spectral analysis). The stronger of these

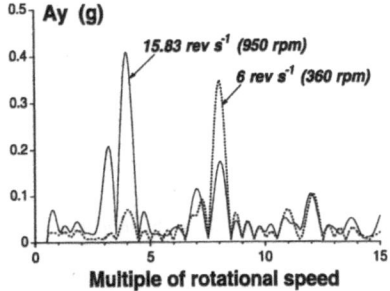

Fig. 8.12. Spectra of acceleration signals in milling at 6 and 15.83 rev s^{-1} spindle speeds.

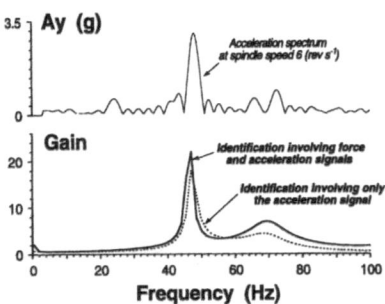

Fig. 8.13. The gain versus frequency plots of estimated transfer functions of the spindle assembly.

resonances is at approximately 47 Hz, and the weaker is at about 69 Hz. As an aside it is interesting to note that these resonances explain the peculiar shape of the acceleration spectrum at the lower spindle speed (this spectrum is repeated for comparison in Fig. 8.13). However, the real importance of the identified transfer function is for the assessment of the physical parameters of the machine,* as reliable features for on-line monitoring and diagnosis.

8.7 Conclusions

Numerous monitoring and diagnosis methods and systems have been developed over the last fifty years for use in capital-intensive industries and in applications where the probability of failure must be minimized regardless of the cost. Although these methods and systems provide a solid foundation for the design of automatic supervision of machine tools, a thorough analysis is necessary to select those algorithm and system structures that are compatible with the nature of machine tools and manufacturing processes.

Perhaps the most distinctive features of machine tools relevant to their automatic supervision are: (1) a broad range of working conditions; and (2) the presence of strong stochastic disturbances. These features render useless many algorithms of monitoring and diagnosis which have been widely accepted and trusted in other applications. The algorithms belonging to the Signal Analysis class, such as the Fourier transformation, are particularly sensitive to the varying environment. The algorithms belonging to the System Analysis class are in general capable of coping with disturbances and varying operating conditions. However, their application is hampered at present by the complexity of machine tools and manufacturing

---

* Such as the stiffness in the vertical direction. In the machining centre considered, the spindle assembly travels vertically driven by a leadscrew. The spindle and leadscrew can be approximated in the simplest way by a mass–spring–dashpot (one degree of freedom) system.

processes and the absence of suitable models. The derivation of such models requires a systematic approach and well-posed decomposition of the modelled devices [26].

There are numerous commercial systems which may seem suitable for ASMT. However, they implement as a rule algorithms belonging to the Signal Analysis class. Consequently, the performance of these systems under shopfloor conditions is expected to be poor, which will manifest itself in frequent misalarms and insensitivity to the legitimate failure conditions. Efforts are being made to improve the robustness of commercial systems by implementing sophisticated methods from the field of artificial intelligence, such as Expert System, neural network or fuzzy set techniques. These efforts are unlikely to succeed unless the "quality" of information obtained from the measurable signals is improved.

The generic structure of ASMT systems presented in this chapter facilitates a systematic investigation of the entire problem. This structure defines distinct phases of signal processing for the purposes of: (1) information compression and highlighting failure sensitive features; (2) failure detection; (3) diagnosis; and (4) severity evaluation. Moreover, the proposed structure accounts for the stochastic nature of results from the detection and diagnosis algorithms, and allows for the minimization of losses due to Type I (misalarm) and Type II (insensitivity) errors.

The proposed generic structure can also be considered as a template facilitating the analysis and comparison of various monitoring and diagnosis systems. A brief example of such an analysis in this chapter reveals some similarities, but also striking differences between the systems already implemented under shopfloor conditions and laboratory prototypes. Both classes employ sophisticated signal processing and artificial intelligence algorithms. However, the systems of the former class use signal analysis methods and Expert Systems with shallow (i.e. containing only heuristic rules [6]) knowledge bases. The purpose of employing artificial intelligence is: (1) the detection of failure related features; (2) failure detection according to these features; (3) failure classification; and (4) severity evaluation. On the other hand, the laboratory prototypes employ system analysis techniques and coherent shallow and deep (containing rules derived from first principles and simulations) knowledge bases. Since they synthesize symptoms and syndromes of failures, the artificial intelligence is utilized to perform only tasks (2)–(4). The result is a significant improvement in the sensitivity and the reliability.

Given a broad range of operating conditions and limited availability of signals, there is no single system or signal analysis algorithm which guarantees reliable results. Instead, it is necessary to integrate within each ASMT system the whole spectrum of algorithms and apply them according to specific situations. The most suitable means of attaining such integration are the artificial intelligence methods employing deep knowledge. Unfortunately, the required knowledge is not available in a convenient form at present. This situation of having sophisticated signal and system analysis algorithms, powerful artificial intelligence techniques and advanced computers, but lacking the knowledge of how to make them work together, is like having a high-performance car with an empty fuel tank. Once this situation is properly recognized, the solution is obvious. With regard to ASMT this means there is a need to concentrate research on: (1) development of models of manufacturing equipment and processes; (2) verification of the reliability and accuracy of the available system and signal analysis methods in various situations; and (3) establishing a deep knowledge base about the applicability and performance of these methods. It is also necessary to acknowledge the inherently stochastic nature of prognostic parameters generated by the indirect condition assessment, and to apply suitable methods to derive the optimum maintenance schedules.

Acknowledgments

The author wishes to thank the General Motors Corporation for providing unique laboratory facilities at the Technical Center and instrumented production equipment at the Vanguard plant. A great deal of technical help from colleagues in Warren and Saginaw is greatly appreciated. Grants from the National Science Foundation (DDM-9009908 and DDM-9202885), which made possible the research of self-tuning Signal Preparation and Acquisition systems, is also greatly appreciated.

References

1. De Barr A. The life and reliability of machine tools. In: Technology of machine tools, vol 3. Lawrence Livermore Laboratory, Livermore, California, 1980, pp 8.16.1–8.16.16
2. Kegg L. On-line machine and process diagnostics. Annals of the CIRP 1984; 33(2), 469–473
3. Milacic V, Majstorovic V. The future of computerized maintenance. In: Milacic VR, McWaters JF (eds) Diagnostic and preventive maintenance strategies in manufacturing systems. North-Holland, Amsterdam, 1988, pp 139-181
4. Weck M. Handbook of machine tools. Wiley, Chichester, 1984
5. Storr A, Hardtner M, Diehl G, Schneider J. New diagnostic methods for faults external to the controller in manufacturing systems. In: Milacic VR, McWaters JF (eds) Diagnostic and preventive maintenance strategies in manufacturing systems. North-Holland, Amsterdam, 1988, pp 43–63
6. Wang SS. Diagnostic expert systems for industry. In: Proc 2nd Intl Machinery Monitoring and Diagnostics Conf, Union College, Schenectady, New York 1990, pp xvii–xxxi
7. Neale MJ, Woodley BJ. Condition monitoring methods and economics. Symposium of the Society of Environmental Engineers, London, 1975
8. Petersen DG. Using computer based instrumentation to automate and simplify machinery vibration analysis. In: Proc 1st Intl Machinery Monitoring and Diagnostics Conf, Union College, Schenectady, New York, 1989, pp 356–362
9. Szafarczyk M, Chisholm AWJ. Automatic supervision in manufacturing systems. Principles, classification and terminology. In: Proc III Intl Conf Automatic Supervision, Monitoring and Adaptive Control in Manufacturing – AC'90, Polish Academy of Sciences, 1990, pp 7–22
10. Dinsdale, J. Automatic supervision of machine tools. In: Proc III Intl Conf Automatic Supervision, Monitoring and Adaptive Control in Manufacturing – AC'90. Polish Academy of Sciences, 1990, pp 101–121
11. Rao SB, Wu SM. Compensatory control of roundness error in cylindrical chuck grinding. ASME Winter Annual Meeting, Paper No 81-WA/PROD-6, 1981
12. Bin HZ, DeVries MF. Microprocessor based compensation of leadscrew drive kinematic errors by a forecasting technique. In: Davies BJ (ed) Proc 24th IMTDR Conf, Macmillan Press, 1983, pp 347–354
13. Sata T, Takata S, Sato M, Suzuki K. Monitoring and diagnosis system of machine tools. In: Rembold U (ed) 2nd IFAC/IFIP Symp Information Control Problems in Manufacturing Technology, 1979, pp 73–82
14. Takata S, Sata T. Model referenced monitoring and diagnosis – application to the manufacturing system. Computers in Industry 1986; 7(1): 31–43
15. Collacott RA. Mechanical fault diagnosis. Chapman and Hall, London, 1977
16. Pau LF. Failure diagnosis and performance monitoring. Marcel Dekker Inc, New York, 1981
17. Collacott RA. Vibration monitoring and diagnosis. Techniques for cost-effective plant maintenance. Wiley, New York, 1979
18. Mitchell JS. An introduction to machinery analysis and monitoring. Penn Well Publishing Company, Tulsa, Oklahoma, 1981
19. Braun S. (ed) Mechanical signature analysis theory and applications. Academic Press, London, 1986

20. Tonshoff HK, Noske H. Machine tool monitoring applied to lathe chucks. Annals of the CIRP 1990; 39(1): 519–523
21. Pritschow G. Automatic supervision of control systems. In: Proc III Intl Conf Automatic Supervision, Monitoring and Adaptive Control in Manufacturing – AC'90, Polish Academy of Sciences, 1990, pp 73–99
22. Gertsbakh IB. Models of preventive maintenance. North-Holland, Amsterdam, 1977
23. Lyon RH. Machinery noise and diagnostics. Butterworth, Boston, Massachusetts, 1987
24. Deutsch R. System analysis techniques. Prentice-Hall, Englewood Cliffs, New Jersey, 1969
25. Iserman R. Process fault detection based on modeling and estimation – a survey. Automatica 1984; 20(4): 387–404
26. Spiewak SA. A predictive monitoring and diagnosis system for manufacturing. Annals of the CIRP 1991; 40(1): 401–404
27. Isermann R, Freyermuth B. Process fault diagnosis based on process model knowledge – Part I: Principles for fault diagnosis with parameter estimation. ASME J of Dynamic Systems, Measurement and Control 1991; 113: 620–626
28. Chen YB. Sensing noise cancellation by optimal filtering. In: Proc 3rd Intl Machinery Monitoring and Diagnostics Conf, Union College, Schenectady, New York, 1991, pp 287–291
29. Wen X, Tang B, Zhu K. The application of adaptive noise cancelling technique on machinery faults diagnosis. In: Proc 1st Intl Machinery Monitoring and Diagnostics Conf, Union College, Schenectady, New York, 1989, pp 869–872
30. Willsky AS. A survey of design methods for failure detection in dynamic systems. Automatica 1976; 12: 601–611
31. Singer RM, King RW, Mott J. Use of a pattern recognition scheme to compensate for critical sensor failures. In: Proc 1st Intl Machinery Monitoring and Diagnostics Conf, Union College, Schenectady, New York, 1989, pp 863–868
32. Spiewak SA, Di Corpo J. Adaptive compensation of dynamic characteristics for in-process sensors. ASME J of Engineering for Industry, 1991;113:198-206
33. Norris GA, Skelton RE. selection of dynamic sensors and actuators in control of linear systems. ASME J of Dynamic Systems, Measurement and Control 1989; 111: 389–397
34. Stein JL, Park Y. Measurement signal selection and simultaneous state and input observer. ASME J of Dynamic Systems, Measurement and Control 1988; 110: 151–159
35. Sweitzer KA. A mechanical impedance correction technique for vibration tests. Sound and Vibration April 1988, pp 30–34
36. Bowers SV, Piety KR, Piety RW. Real-world mounting of accelerometers for machinery monitoring. Sound and Vibration Feb 1991, pp 14–23
37. Schiltz RL. Forcing frequency identification of rolling element bearings. Sound and Vibration May 1990, pp 16–19
38. Kim RH. Analysis of the velocity response of a spherical roller bearing. In: Niskode PM, Doepker PE (eds) Vibration analysis to improve reliability and reduce failure, ASME Design Automation Conference, 1985, pp 39–44
39. Marple SL Jr. Digital signal analysis with applications. Prentice-Hall, Englewood Cliffs, New Jersey, 1987
40. Ljung L. System identification: theory for the user. Prentice-Hall, Englewood Cliffs, New Jersey, 1987
41. Soderstrom T, Stoica P. System identification. Prentice-Hall International, London, 1989
42. Isermann R. Digital control systems. Springer, Berlin, 1981
43. Goodwin GC, Payne RL. Dynamic system identification: experimental design and data analysis. Academic Press, New York, 1977
44. Eykhoff P. System identification: parameter and state estimation. Wiley, New York, 1974
45. Åström KJ, Wittenmark B. Computer controlled systems: theory and design. Prentice-Hall, Englewood Cliffs, New Jersey, 1984
46. Park Y, Stein JL. Steady-state optimal state and input observer for discrete stochastic systems. ASME J of Dynamic Systems, Measurement and Control 1989; 111: 121–127
47. Fassois SD, Lee JE. Suboptimum maximum likelihood identification of ARMAX processes. ASME J of Dynamic Systems, Measurement and Control 1990; 112: 586–595
48. Box GEP, Jenkins GM. Time series analysis, forecasting and control. Holden-Day, Oakland, California, 1970
49. Pandit SM, Wu SM. Time series and system analysis with applications. Wiley, New York, 1983
50. Pandit SM. Modal and spectrum analysis: data dependent systems in state space. Wiley, New York, 1991

51. Fassois SD, Eman KF, Wu SM. A suboptimum maximum likelihood approach to parametric signal analysis. ASME J of Dynamic Systems, Measurement and Control 1989; 111: 153–159

52. Isermann R, Freyermuth B. Process fault diagnosis based on process model knowledge – Part II: Case study experiments. ASME J of Dynamic Systems, Measurement and Control 1991; 113: 627–633

53. Iserman R. Estimation of physical parameters of dynamic processes with application to an industrial robot. In: American Control Conf, American Control Council, San Diego, California, 1990, pp 1396–1401

54. Geiger G. Fault identification using a discrete square root method. Intl J of Modeling and Simulation 1986; 6(1): 26–31

55. Shin YC. System identification of multivariate systems with feedback. ASME J of Dynamic Systems, Measurement and Control 1990; 112: 283–291

56. Powell CD. Machinery troubleshooting using vibration analysis techniques. Sound and Vibration, Jan 1992, pp 42–54

57. Peters J. Dynamic analysis of machine tools using complex modal method. Annals of the CIRP 1976; 25(1): 257–261

58. Tlusty J. Experimental and computational identification of dynamic structural models. Annals of the CIRP 1976; 25(2): 497–503

59. Carlson J, Spiewak SA. Vibration based machine condition monitoring: the state of practice. In: Proc 16th Annual Meeting, Vibration Institute, 1992, pp 87–95

60. Li CJ, Wu SM. On-line detection of localized defects in bearings by pattern recognition analysis. ASME J of Engineering for Industry 1989; 111: 331–336

61. Bell DH. An enveloping technique for early stage detection and diagnosis of faults in rolling element bearings. In: Niskode PM, Doepker PE (eds) Vibration analysis to improve reliability and reduce failure, ASME Design Automation Conf, 1985, pp 65–69

62. Press WH, Flannery BP, Teukolsky SA, Vetterling WT. Numerical recipes – the art of scientific computing. Cambridge University Press, Cambridge, 1986

63. Dornfeld DA. Unconventional sensors and signal conditioning for automatic supervision. In: Proc III Intl Conf Automatic Supervision, Monitoring and Adaptive Control in Manufacturing – AC'90, Polish Academy of Sciences, 1990, pp 197–233

64. Baikiotis C, Raymond J, Rault A. Parameter identification and discriminant analysis for jet engine mechanical state diagnosis. In: IEEE Conf Decision and Control, 1979, pp 648–650

65. Danai K, Chin H. Fault diagnosis with process uncertainty. ASME J of Dynamic Systems, Measurement and Control 1991; 113: 339–343

66. Chin H, Danai K. A method of fault signature extraction for improved diagnosis. ASME J of Dynamic Systems, Measurement and Control 1991; 113: 634–638

67. Sata T, Kimura F, Matsushima K, Takata S, Ootsuka J. Identification of machine and machine states by use of pattern recognition technique. Manufacturing Systems 1984; 14(3): 273–285

68. Monostori L. Pattern recognition based learning and decision making in complex machine tool monitoring systems. In: Milacic VR, McWaters JF (eds) Diagnostic and preventive maintenance strategies in manufacturing systems. North-Holland, Amsterdam, 1988, pp 113–123

69. Kim DS, Shin YS, Carlson DK. Machinery diagnostics for rotating machinery using backpropagation neural network. In: Proc 3rd Intl Machinery Monitoring and Diagnostics Conf, Union College, Schenectady, New York, 1991, pp 309–320

70. Xu J, Zhang S, Liu Y. Fuzzy diagnosis method in machinery condition monitoring and failure diagnosis. In: Proc 2nd Intl Machinery Monitoring and Diagnostics Conf, Union College, Schenectady, New York, 1990, pp 227–233

71. Fang M, Rong Y, Tzou H-S. Diagnostic monitoring of robot operation through fuzzy-logic based multi-assessment. In: Proc 3rd Intl Machinery Monitoring and Diagnostics Conf, Union College, Schenectady, New York, 1991, pp 207–212

72. Kuhnell BT, Luo M, Wang R. Forecasting machine condition using grey system theory. In: Proc 2nd Intl Machinery Monitoring and Diagnostics Conf, Union College, Schenectady, New York, 1990, pp 103–107

73. Emerling L, Winkler BE. Expert system Charley. In: Proc Presstech Conference, SME, 1991, pp 121–125

74. Bajpai A, Marczewski RW. Charley: an expert system for general purpose diagnostics of manufacturing equipment. In: Schorr H, Rappaport A (eds) Proc Conf Innovative Applications of Artificial Intelligence, Stanford, California, 1989, pp 178–182

75. Petersen DG. Development of an integral PC-based predictive maintenance expert system for automated machine diagnostics. In: Proc 2nd Intl Machinery Monitoring and Diagnostics Conf, Union College, Schenectady, New York, 1990, pp 8–12

76. Chen YB. Machine diagnostic monitoring by prediction error analysis and information measures. In: Proc 1st Intl Machinery Monitoring and Diagnostics Conf, Union College, Schenectady, New York, 1989, pp 553–559

77. Weck M, Vorsteher D, Monitoring and diagnosis system for numerically controlled machine tools. In: Davies BJ (ed) Proc 24th IMTDR Conf, Macmillan Press, 1983, pp 229–237

78. Weck M. Development and application of a flexible, modular monitoring and diagnosis system. Computers in Industry 1986; 7: 45–51

79. MachineView. Computational Systems Inc., Knoxville, Tennessee, 1988

80. Petersen DG. Development of an integral PC-based predictive maintenance expert system for automated machine diagnostics. In: Proc 2nd Intl Machinery Monitoring and Diagnostics Conf, Union College, Schenectady, New York, 1990, pp 8–12

81. Nower DL. An expert system for routine machinery analysis. In: Proc 3rd Intl Machinery Monitoring and Diagnostics Conf, Union College, Schenectady, New York, 1991, pp 358–365

82. Robinson SD, Russel RH. Signature analysis extended – a system approach to machinery dynamics. In: Proc Signature Analysis in the 80's Conf, Vibration Institute, 1980, pp 1–13

83. Jackson R, Robinson SD. System analysis approach for rotating equipment. In: Proc Machinery Vibration Monitoring and Analysis Meeting, Vibration Institute, 1980, pp 2–16

84. Rogers P. Genuine modal testing of rotating machinery. Sound and Vibration Jan 1988, pp 36–42

85. Spur G, Kirchheim A, Schule A. Monitoring the cutting process in multi-spindle lathes. Proc 1st Intl Machinery Monitoring and Diagnostics Conf, Union College, Schenectady, New York, 1989, pp 108–111

86. Chryssolouris G, Demroese M, Subramaniam V. Decision making and sensor synthesis for manufacturing processes. In: Proc 1992 NSF Design and Manufacturing Systems Conf, SME, 1992, pp 995–1000

87. Lee LD, Skeirik R. PC-base integration of lubricant, wear particle and vibration analysis data in a predictive maintenance system. In: Proc 2nd Intl Machinery Monitoring and Diagnostics Conf, Union College, Schenectady, New York, 1990, pp 480–484

88. Meher-Homji CB, Meher-Homji FJ, Chinoy RB. Vibration and debris analysis for condition monitoring of gear boxes – an integrated approach. In: Proc 1st Intl Machinery Monitoring and Diagnostics Conf, Union College, Schenectady, New York, 1989, pp 494–501

89. Takata S, Ahn JH, Miki M, Miyao Y, Sata T. A sound monitoring system for fault detection of machine and machining states. Annals of the CIRP 1986; 35(1): 289–292

90. Takata S, Ahn JH. Overall monitoring system by means of sound recognition. In: Milacic VR, McWaters JF (eds) Diagnostic and preventive maintenance strategies in manufacturing systems. North-Holland, Amsterdam, 1988, pp 99–111

91. Tarbet MA, Trevillion WL. Using sound data to support vibration analysis of rotating machinery. In: Proc 2nd Intl Machinery Monitoring and Diagnostics Conf, Union College, Schenectady, New York, 1990, pp 543–548

92. Lukas M, Anderson DP. Developments in instrumentation and automation in techniques for machine condition monitoring through oil analysis. In: Proc 1st Intl Machinery Monitoring and Diagnostics Conf, Union College, Schenectady, New York, 1989, pp 479–487

93. Haynes HD, Kryter RC. Condition monitoring of machinery using motor current signature analysis. In: Proc 1st Intl Machinery Monitoring and Diagnostics Conf, Union College, Schenectady, New York, pp 690–695

94. LabView. National Instruments Corp, Austin, Texas, 1991

95. Mathematica. Wolfram Research Inc, Champaign, Illinois, 1991

96. SIMULAB. The MathWorks Inc, Natick, Massachusetts, 1991

97. Spiewak SA. The enhancement of monitoring and failure prediction capabilities of the MMMS system. Technical Report to General Motors Technical Center, 1990

98. Di Corpo J. Predictive monitoring for metal cutting machine tools. MS thesis, University of Wisconsin-Madison, 1990

99. Birla SK. Sensors for adaptive control and machine diagnostics. In: Technology of machine tools, vol 3. Lawrence Livermore Laboratory, Livermore, California, 1980, pp 7.12.1–7.12.70

100. Tlusty J, Andrews GC. A critical review of sensors for unmanned machining. Annals of the CIRP 1983; 3(2): 563–572

101. Taylor JI. Evaluation of machinery condition by spectral analysis. In: Proc Machinery Vibration Monitoring and Analysis Meeting, Vibration Institute, 1980, pp 1–15

102. Tranter JT. The application of computers to machinery predictive maintenance. Sound and Vibration Dec 1990, pp 14–19

9 Automatic Supervision of Control Systems

G. Pritschow

9.1 Diagnosis

9.1.1 Introduction

In order to increase output as well as to protect systems and operators, it is increasingly becoming standard practice to integrate the monitoring and diagnosis functions of manufacturing into the control systems. The term "diagnosis" is of Greek origin and means "the detection and determination of an illness." For technological processes the "illness" corresponds to "disturbances" which affect the process adversely. The disturbances can be located, as Fig. 9.1 shows, in the machine system, the process itself or the operation.

As soon as the monitoring system has detected a functional disturbance, it is the task of the diagnosis system to determine its location, type and cause. Following diagnosis, an adequate reaction (or "therapy") which nullifies the influence of the disturbance on the process is essential.

The monitoring and the diagnosis can be done either manually or completely automatically. Automatic diagnosis systems are based on an analysis of the system, module-by-module, and knowing their behaviour under normal conditions, automatic location of changes followed by observations and tests to draw up conclusions. This diagnosis knowledge is part of the automatic system.

Since a great number of factors can lead to many undesirable kinds of behaviour, complete automatic monitoring and diagnosis are only possible in exceptional cases, depending on the application. For non-automatic cause detection, specialists are needed to carry out a manual diagnosis. This manual diagnosis can be done in a computer-aided form by so-called "Expert Systems", whereby the type of disturbance is determined via a dialogue of questions and answers with the help of structured knowledge.

Figure 9.2 shows the possible conditions should a disturbance be located during the process. If a fault is detected by the monitoring unit, the system condition changes from the "regular" to the "disturbed" condition.

At this point the diagnosis system comes into operation. It detects the causes of the functional disturbance by means of the following information:

Location of the fault (e.g. output A1 does not work);

Type of fault (dangerous, not dangerous); and

Cause of fault (frequently only possible with the help of the protocol and following

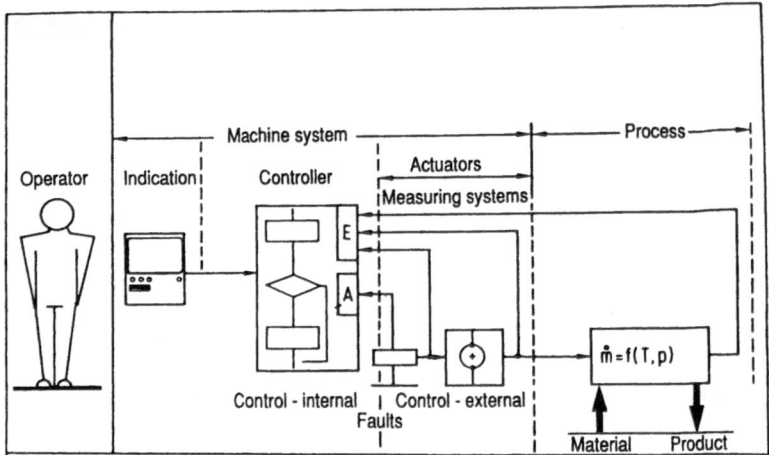

Fig. 9.1. Sources of disturbances in technological processes.

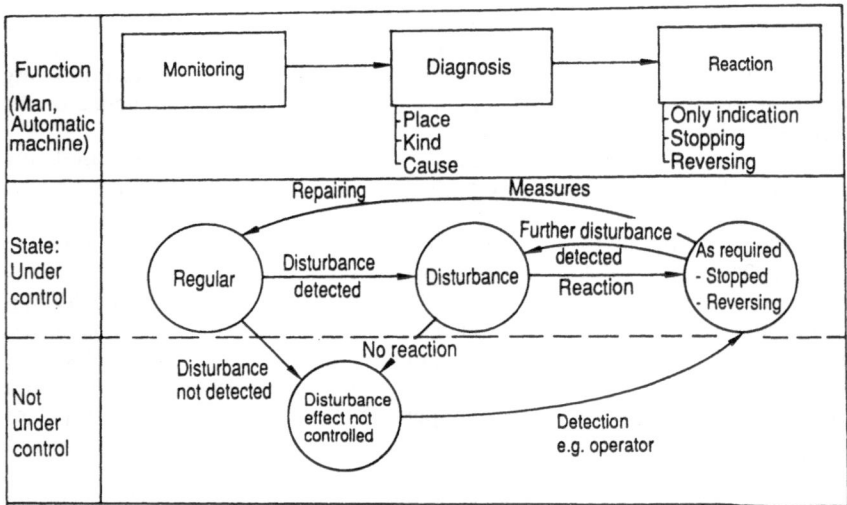

Fig. 9.2. Process condition and effectiveness of the monitoring and diagnosis functions.

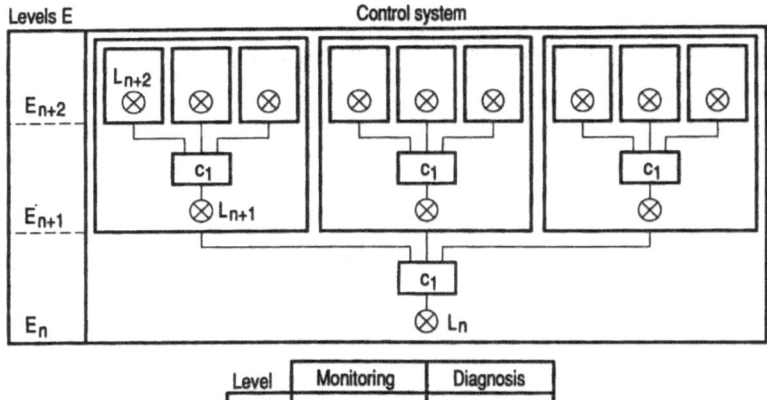

Fig. 9.3. Principles of automatic self-monitoring.

evaluation by the maintenance personnel – in this case, for example, a short circuit in the relay has adversely affected the switching transistor).

According to type and cause of the disturbance, a decision can be made about the best reaction to put into operation. This reaction is graded according to the type of fault, and can range from an indication and a protocol, to automatically switching off the process or switching over to other units.

Since it cannot usually be guaranteed that all possible disturbance conditions can be detected automatically, the system might come to a point where disturbance effects bring it to an uncontrollable condition. This condition might also result if disturbances were detected but no effective reactions were programmed. In such cases, an interaction by human operators is unavoidable. This shows that the "ghost factory" of the future, meaning a completely unmanned factory, is unrealistic for complex systems. A process disturbed into an uncontrollable condition has too great a potential for damage.

Figure 9.3 shows a typical modularly structured control system, and from this it can be shown that monitoring and diagnosis must be two separate processes.

The monitoring L_n of the principal system reacts when a disturbance is indicated by a sub-module L_{n+1}. Thus, the monitoring systems L_{n+1} of sub-system level E_{n+1} are the diagnosis systems for the principal system of the level E_n, for with L_{n+1} the disturbance can be located. Thus, in general, a monitoring indication depends on an information source that has a diagnosis information, i.e. monitoring always includes a certain diagnosis knowledge about the location of the disturbance.

9.1.2   Diagnosis of Control-External and -Internal Faults

The control equipment of a system refers to all the electronic and electrical elements, including the connecting elements of the drive and measuring systems, and as far as hydraulic or pneumatic systems are concerned, also all the equipments

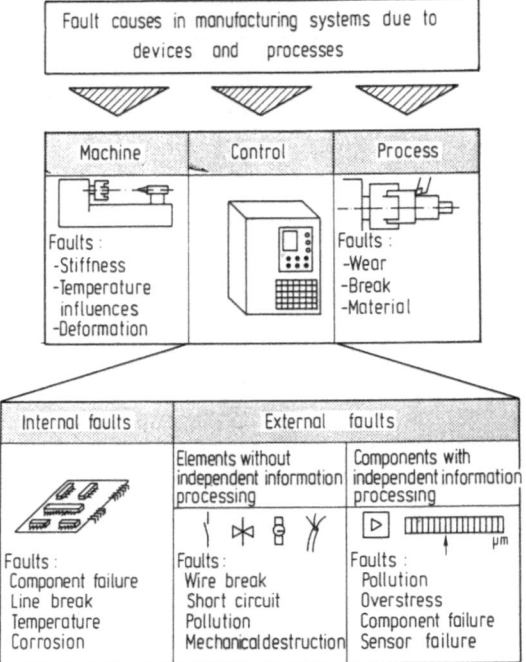

Fig. 9.4. Equipment and process-conditioned causes of faults in manufacturing systems.

used in information processing. The latter, however, will not be considered in any detail in the discussion that follows. Modern diagnosis technology is based on electronics, and such equipment is generally available today for most control systems in the form of microcomputers. Thus, the monitoring and diagnosis component is part of the machine system which in itself has a potential to fail. However, as experience shows, when electronic controllers are used as the central part of the control system (processor, storage, bus, current supply), failure rates are very low in comparison to the system as a whole, so such control electronics are the preferred choice in diagnosis.

For a better classification of the diagnosis strategies, it is reasonable to divide the faults to be observed into internal and external control faults, and thus to relate the faults according to the monitoring and diagnosis functions (Fig. 9.4).

Disturbances are caused by internal faults (e.g. by defective electronic parts, wire breaks or software faults) and by external faults. The latter are differentiated, according to their occurrence, into elements without information processing of their own, such as end-switches and valves, and elements with information processing of their own, such as drive systems, measuring systems and amplifiers.

Such causes of system disturbances, found by disturbance estimation, are shown in Fig. 9.5. It can be seen that disturbances in the control hardware have been reduced considerably [1] by the introduction of storage-programmable control systems (PLC or computers).

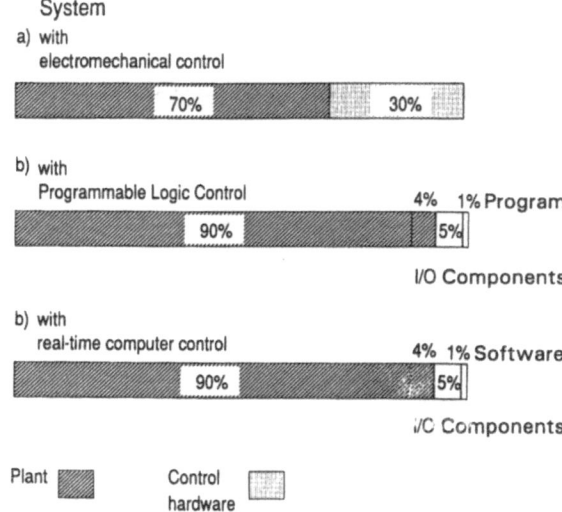

Fig. 9.5. Causes of disturbance with differently controlled production systems.

9.1.3   Methods for Automatic Diagnosis and Supervision

There are two basic principles for automatic monitoring and diagnosis: (1) checking the functionality of a module by means of specified tests; and (2) checking the functionality by observation.

A testing procedure is characterized by the fact that the module to be checked is connected to a testing function, which means that the checking is done actively. Since this generally has a harmful effect on the process, it must be possible to decouple the checked module from the process itself while testing is in progress. Examples of such tests are the identification of a drive by pseudo-binary noise signals, and the checking of a processor by means of a watch-dog timer.

The characteristic of an observation is the passive checking of the normal system behaviour of a module by observing its input and/or output values (Fig. 9.6).

For the testing and observations, there are a great number of procedures available so the choice has to be made carefully. Thus, signature analysis is a well-known procedure for checking storage components, while Fourier analysis is a typical procedure for observing system transfer behaviour by noting the input and output signals during its operation (Fig. 9.7).

Both with testing and observing, the results need to be compared. The comparison has different criteria according to the tasks and diagnosis strategy involved. The result of a signature analysis is compared with a given bit pattern, from which the decision "defective" or "in order" can be made, while the result of a time-length observation is typically subject to a constraint comparison that is obtained from tables. The comparison values can be stored before the operation, transmitted with the help of operators, or by self-learning during the test or in an operation phase built into the system itself. Self-learning procedures are especially valuable for those variables and their combinations of which the constraint values are not known *a priori*. Test and observation routines serve for checking both individual components

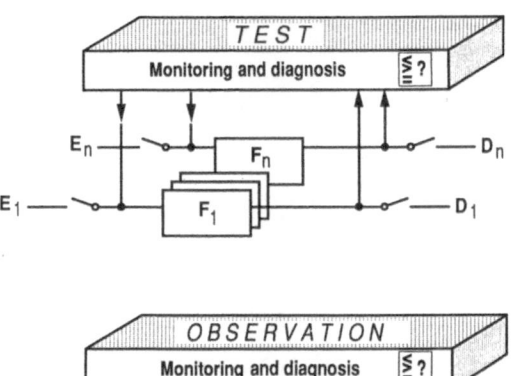

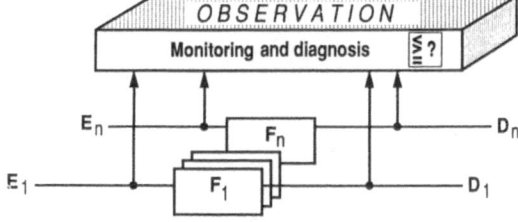

Fig. 9.6. Basic principles of automatic self-monitoring.

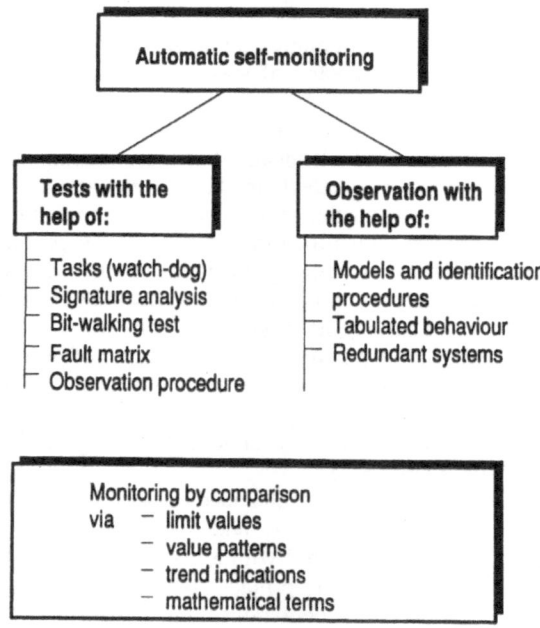

Fig. 9.7. Procedures for automatic self-monitoring.

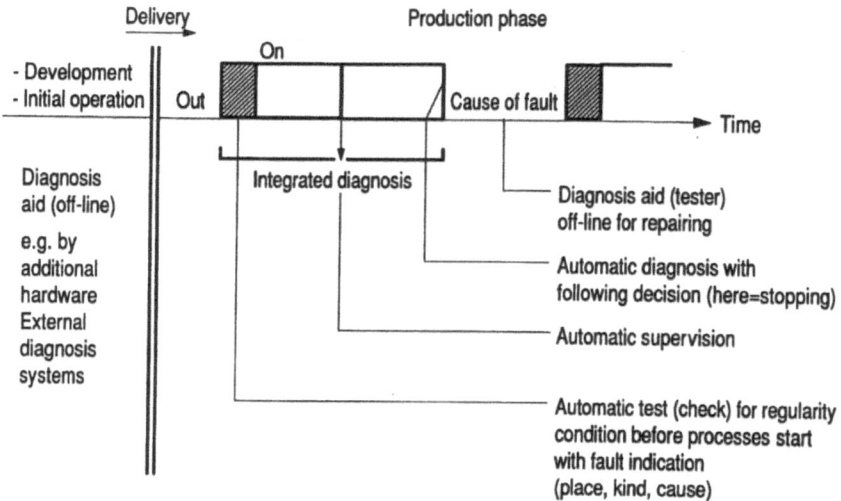

Fig. 9.8. Various applications of diagnosis technology.

and systems. The routines can be used either by external diagnosis systems or additional programs contained in the control system. There is a growing tendency for such aids to be stored in the control system as permanently loaded service routines.

When following a system through from its development phase to its use in production, various diagnosis applications can be differentiated (Fig. 9.8).

During the phases of development and initial operation, the diagnosis routines do not have to be components of the system. Programmable logic controllers, for example, can be extended by using so-called external test systems.

During the production phase, however, permanent automatic monitoring and diagnosis are required. The corresponding routines therefore run on-line, parallel to the process course. If the diagnosis system is part of the control, and such is the usual case, one speaks of "integrated diagnosis".

9.2 Diagnosis Processes

9.2.1 Diagnosis of Control-Internal Faults

Faults which occur in the hardware of NC could, in principle, be tolerated by existing hardware redundancies. For financial reasons, however, this is not done as a rule. Instead, individual hardware components are supervised by small hardware supplements such as watch-dog timers and test bits. Another usual method is to check all the hardware components, if possible, by test programs (e.g. for the initial operations and every time the equipment is switched on).

Tests to supervise and diagnose the hardware of NC have for example been described in [8] and[9].

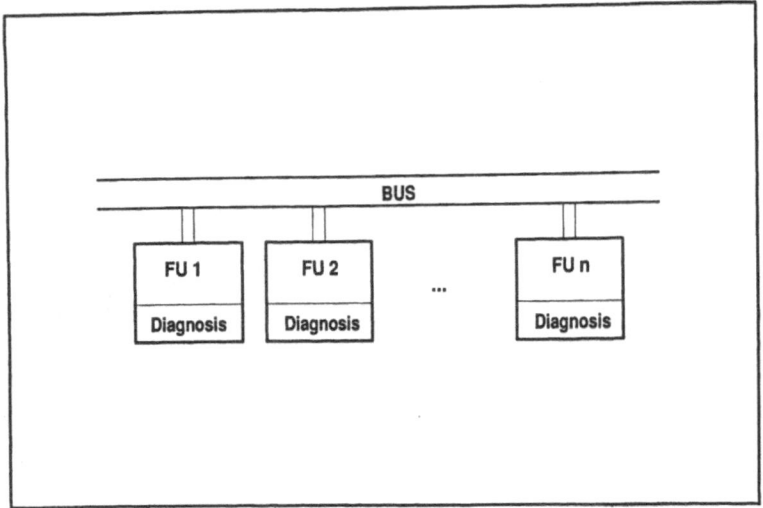

Fig. 9.9. Supervision and diagnosis measures integrated into functional units.

Since large software packages are not free of faults, it is necessary to take supervisory measures within the software. Various algorithms and procedures are available for these. Statical diagnosis procedures use known fault patterns to find the appropriate cause of any faults and to initiate appropriate reactions. In this case, statical diagnosis procedures mean that an already existing fixed arrangement between fault pattern and fault causes is used [8]. Dynamical diagnosis systems are capable of automatically recording correctly operating processes to a certain extent, and then of comparing the actual processes with these reference standards.

9.2.2   Diagnosis Configuration

Today, automation systems are modularly structured, so that diagnosis is based on accessible single components. There are several possibilities of realization. One method is the integration of the supervisory and diagnosis functions into the software already used during the development of the control functional units (Fig. 9.9).

Examples of this are supervision of lag distance via the position control, and the realization of software end-switches which execute monitoring of the operating field of NC axes. Such procedures are specifically made for every function program. They are usually implemented into the NC control systems that are available on the market today.

Another method is to keep the monitoring and diagnosis functions separate from the real functions. A monitoring and diagnosis unit (Fig. 9.10) made for a special function or for different functions (Fig. 9.11) can receive the data of one or several functions via adequate interfaces [8, 9].

With the development of control functions, the diagnosis interface has to be considered as a separate entity. The diagnosis function is separately coded and is therefore portable on to additional hardware.

All in all, one can say that the method mentioned first does not cover control-internal hardware fault sources. If the second procedure is used, faults of the control hardware can also be recognized and a certain standardization of the diagnosis

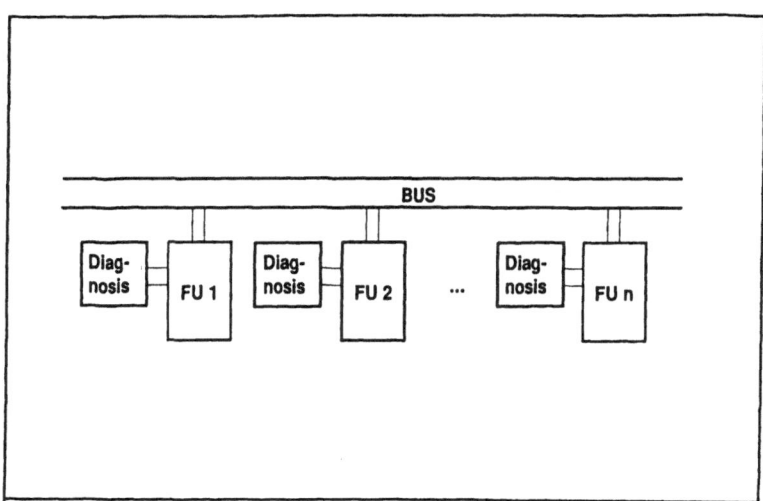

Fig. 9.10. Separate monitoring and diagnosis modules used for single functional units.

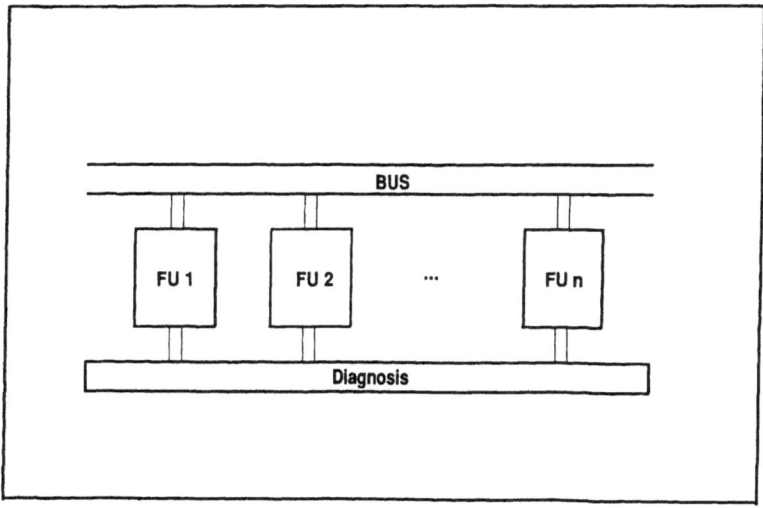

Fig. 9.11. Separate monitoring and diagnosis systems used for overall control.

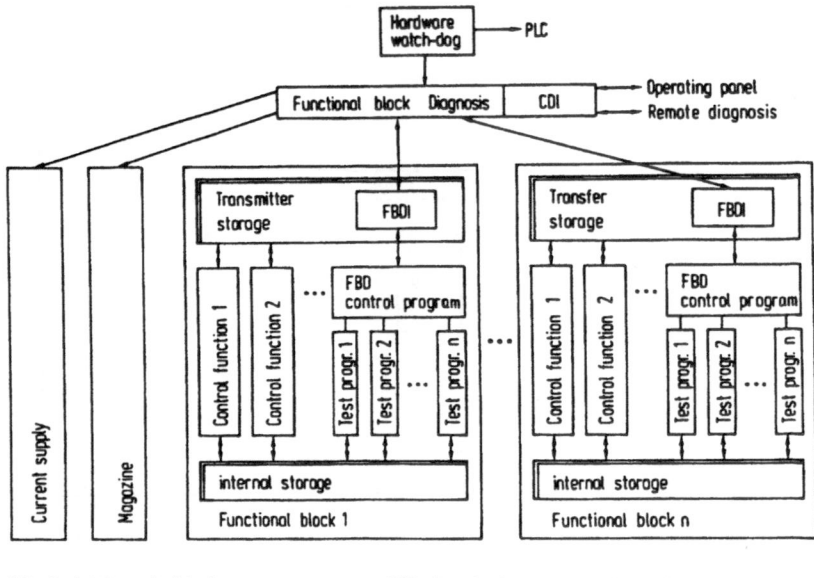

CDI: Central diagnosis interface   FBDI: Functional block diagnosis interface
    Interfaces for diagnosis systems   FBD: Functional block diagnosis

Fig. 9.12. Structure of an integration diagnosis system for control according to the MPST principle.

components is possible. Furthermore, diagnoses become possible which can only be derived from an analysis of variables of the different function components [10].

9.2.3   An Example of the Implementation of Test Functions into a Control

For the implementation of test functions into an NC control which is structured according to the MPST conception [5] the structure shown in Fig. 9.12 has been laid down. Each single functional block of the control is enlarged by addition of a function which can be called up for the diagnosis [4, 5, 7].

In the control function blocks, a diagnosis function block is placed over functions which can be called up for the diagnosis. This diagnosis function block controls the diagnosis processes and executes the inputs and outputs for the operator. It also contains the callable functions.

The control program of the callable function "diagnosis" is realized as a task on its own.

9.2.4   Special Methods for Software Control

Normally faults occur only with components that are liable to fail, so that software which has been tested once and which does not age can be looked on as fault free. Experience shows, however, that software faults can be uncovered even after it has been in operation for a very long time, because the enormous number of combinations cannot be tested during the test phase by simply using the input

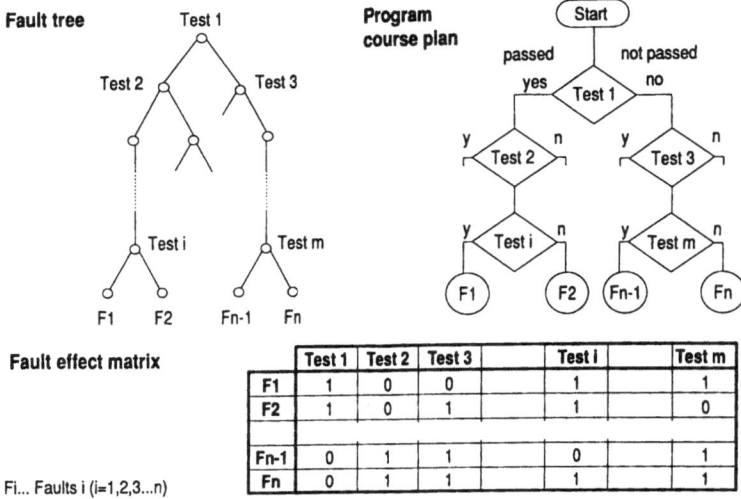

Fault effect matrix

	Test 1	Test 2	Test 3		Test i		Test m
F1	1	0	0		1		1
F2	1	0	1		1		0
Fn-1	0	1	1		0		1
Fn	0	1	1		1		1

Fi... Faults i (i=1,2,3...n)

Fig. 9.13. Connection between fault tree and fault effect matrix.

parameter. Furthermore, control software is now configurable, and therefore there are always new untested combinations to be made with other system components, so complete elimination of faults before the actual operating period is often infeasible. Therefore, it is reasonable to supervise on-line both the system software and the operator software. If spare hardware is unavailable, hardware and software faults cannot be definitely distinguished. Software diagnosis is done either by tests using specific pattern behaviour and given input parameters, or by plausibility supervision of the function result (e.g. monitoring the limit value), or simply by specifying a time of operation. A further method compares results on software diversity, i.e. a multiple realization of one and the same facts, during which various pieces of hardware should be used (diversified redundancy) if possible.

With function-oriented diagnosis technology, one can use the fact that functions need different hardware and software modules according to the tasks they are called upon to perform, so that faulty modules can be located by various tests. These tests lead to a typical fault pattern for every fault, which is shown in a fault effect matrix. A fault search programme with a tree structure can be used to analyse the location of the fault. The generating pattern of the fault effect matrix shows the corresponding fault. See Fig. 9.13.

The matrix does not need to be symmetrical. The tests can also be made with different tree structures, which means they are independent of one another. The method is in general use to locate faults – both internal and external [8].

9.2.5   Diagnosis of Control-External Faults

The diagnosis of control-external faults (see Fig. 9.14) covers all modules and components, which, according to their inputs and outputs, can be assigned as actuators or signal transmitters. These include, for example, systems to execute numerically controlled movements, systems to execute protection functions, and other mechanical components.

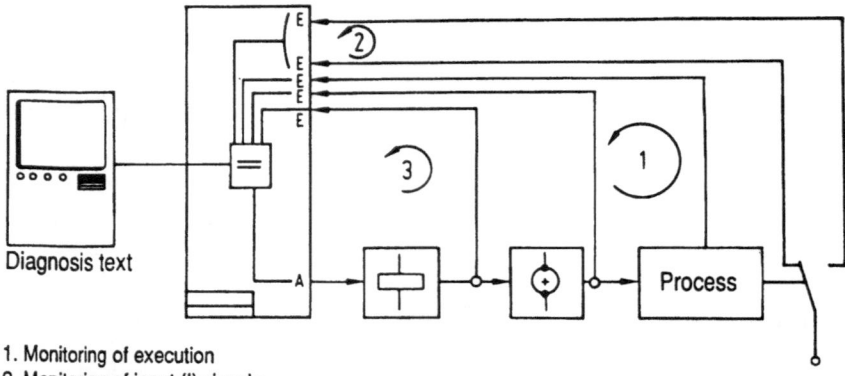

1. Monitoring of execution
2. Monitoring of input (I) signals
3. Monitoring of output (O) signals

Fig. 9.14. Control via diagnosis of external faults.

As far as monitoring is concerned, two different tasks are involved:

1. Monitoring of modules with logical functions by means of Boolean variables;
2. Monitoring of modules with process behaviour by means of continuous variables.

9.2.6 Supervision of Modules with Logical Functions

A first important task is the monitoring of switching functions (in accordance with Fig. 9.16). There the following procedures can be especially applied [6]: (1) monitoring of outputs by feedback; (2) monitoring of execution, for example by time observation; (3) monitoring of end-switch signals by antivalent circuits; (4) monitoring of the function of end-switches by software end-switches via measuring systems.

Another important method of diagnosing external faults is the detection of unallowed value combinations. If one takes the overall control system of a PLC as a switching system with m inputs and n outputs, this system can theoretically have

$$C = 2^m \, 2^n \qquad (9.1)$$

states. To get a realizable diagnosis, the number of states to be tested has to be as low as possible. This number is reached by dividing the system into k process chains and treating every output separately (see Fig. 9.15).

These considerations lead to a control concept which allows the automatic search of conditions:

1. The control program is divided into program chains. These program chains run practically simultaneously.
2. Every program chain is divided into separate steps. These steps are worked off sequentially. Only if a step is worked off completely can one continue to the next.

This program structure ensures that at any point in time only one step is active in

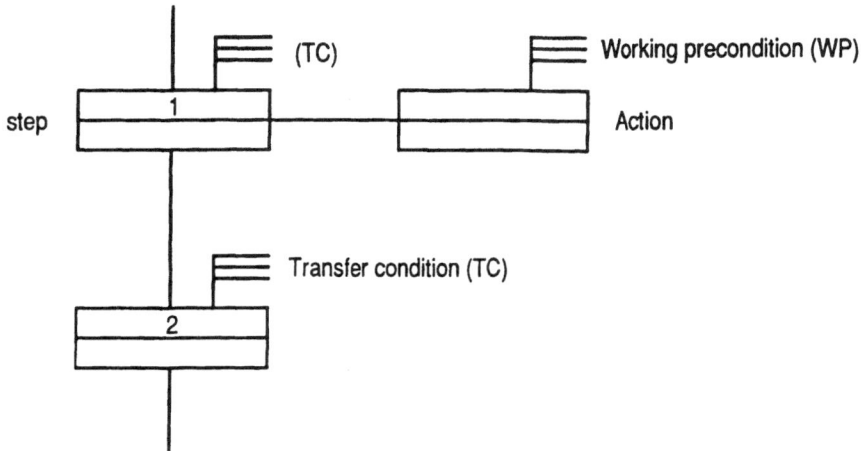

Fig. 9.15. Process chain with input conditions.

every chain. Thus, the operating system can be put in a position to determine automatically the active step of a chain, and to indicate it. To adapt the inputs in a diagnosis-oriented way to this structure, each program step is executed only if two conditions are fulfilled: the working pre-condition (WP) and the transfer condition (TC) [14, 16, 17].

Thus, there are 4 possible states for every step (S):

S	WP	TC
0	0	0
1	1	0
2	0	1
3	1	1

The purpose aimed at in the control program can be defined as follows: an attempt has to be made to reach condition $S = 3$, i.e. always to switch to the next step.

Every other condition is a hindrance and is looked on as a diagnosis cause. Should a disturbance occur, the intended step is shown on the screen with its logical conditions, so that the maintenance personnel are in a position to identify the source of the fault.

Another very convenient method of diagnosis evaluation is the programming of state graphs. During this programming, the definite transfer condition for a following condition is registered into a transfer matrix for every condition. Accordingly, a fault effect matrix for unallowed state conditions can be made, which gives an indication of the location of faults by a comparison with the actual state. See Fig. 9.16.

9.2.7   Supervision of Functions for Process Behaviour by Means of Continuous Variables

Continuous variables can be subject to both simple measures for the monitoring of constraints, and complicated signal analysis procedures – so that faults in, for

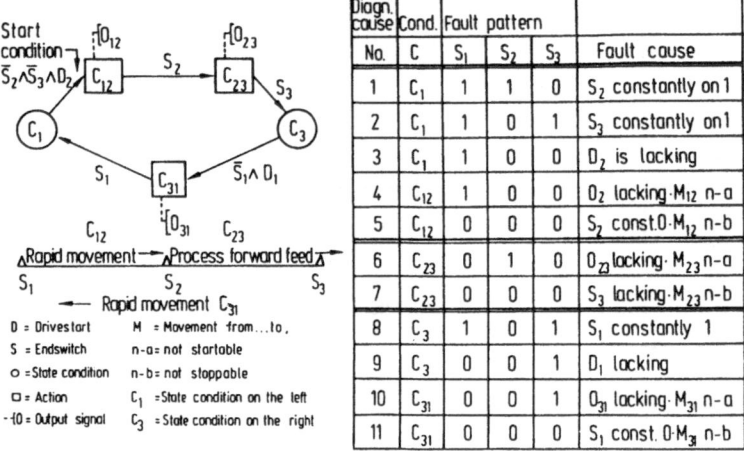

Diagn. cause No.	Cond. C	Fault pattern S_1	S_2	S_3	Fault cause
1	C_1	1	1	0	S_2 constantly on 1
2	C_1	1	0	1	S_3 constantly on 1
3	C_1	1	0	0	D_2 is locking
4	C_{12}	1	0	0	D_2 locking·M_{12} n-a
5	C_{12}	0	0	0	S_2 const.0·M_{12} n-b
6	C_{23}	0	1	0	D_{23} locking· M_{23} n-a
7	C_{23}	0	0	0	S_3 locking·M_{23} n-b
8	C_3	1	0	1	S_1 constantly 1
9	C_3	0	0	1	D_1 locking
10	C_{31}	0	0	1	D_{31} locking·M_{31} n-a
11	C_{31}	0	0	0	S_1 const. 0·M_{31} n-b

Fig. 9.16. Fault effect matrix for the drilling unit of a transfer line.

example, the electrical drive system, and wear and mechanical defects (unacceptable levels of friction, loss) can be recognized in a reasonable time. A pre-condition for this method is the generation of a model of the functions to be examined, whereby different structures can be looked at. In the field of control technology, function and parameter models are important for diagnosis purposes.

9.2.8   Function Model

This system can be described as a block structure of sub-functions in which every sub-function has several input variables and one output variable. Here, the input and output variables are supervised on limit or plausibility values. Figure 9.17 shows an example of the function model for position control of a numerical axis, with lag distance being used to monitor and detect faults via numerically controlled movements. With this system, the lag distance is controlled by the actual processing speed and the parameters of the control loop. If this monitoring is done along several axes at the same time, it is also called "path monitoring" or "contour monitoring".

9.2.9   Parallel Model

The description of the dynamic behaviour of a machine and its elements and components is made by differential and difference equations, condition matrixes and various other mathematical techniques. The detection of the parameters (coefficients of the differential equations or the transfer function) of this model can be made either theoretically or experimentally, both in the time and the frequency domain, and serves as a parallel model to observe the process. The transfer function is continuously estimated from the input and output variables of the path to be observed, in order to compare the physical parameters (e.g. damping, friction, specified values) with these of the parallel model, and possibly to derive a fault detection system or trends of changes, from deviations. See Fig. 9.18.

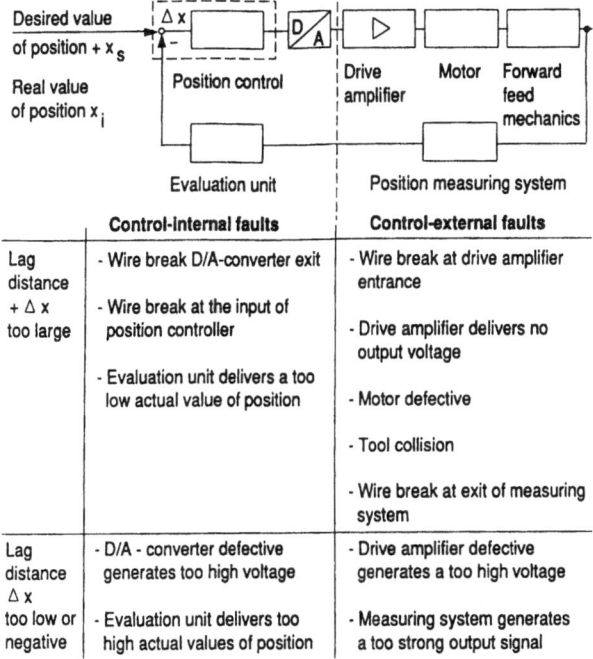

Fig. 9.17. Examples of faults detectable with lag distance.

	Control-internal faults	Control-external faults
Lag distance $+ \Delta x$ too large	- Wire break D/A-converter exit - Wire break at the input of position controller - Evaluation unit delivers a too low actual value of position	- Wire break at drive amplifier entrance - Drive amplifier delivers no output voltage - Motor defective - Tool collision - Wire break at exit of measuring system
Lag distance Δx too low or negative	- D/A - converter defective generates too high voltage - Evaluation unit delivers too high actual values of position	- Drive amplifier defective generates a too high voltage - Measuring system generates a too strong output signal

Since, for efficient monitoring, a calculation of parameters which can keep step with the process is required, algorithms with short calculation times are necessary. Here the parameter estimation methods are especially important because of their rapid convergence. Methods of this kind are not used for machine tools at present, but they are the subject of many research projects all over the world and may soon be introduced in industry.

9.3 Teleservice System for Machine Tools with ISDN

9.3.1 Problems

Complex machine tools impose strict requirements on their operators, in initial operation and re-tooling or in case of disturbances. In many cases, neither adequately educated personnel nor sufficient test and diagnosis aids for the machine are available in these situations. Until manufacturing can restart, high costs are the result for customers owing to the downtime of the machine, and for machine manufacturers owing to after-sales service. Analyses of after-sales service have shown that these downtimes can be reduced in many cases and a part of the mobile after-sales service can be saved if, besides verbal communication between customer and manufacturers, there is also the possibility of direct access to control data from

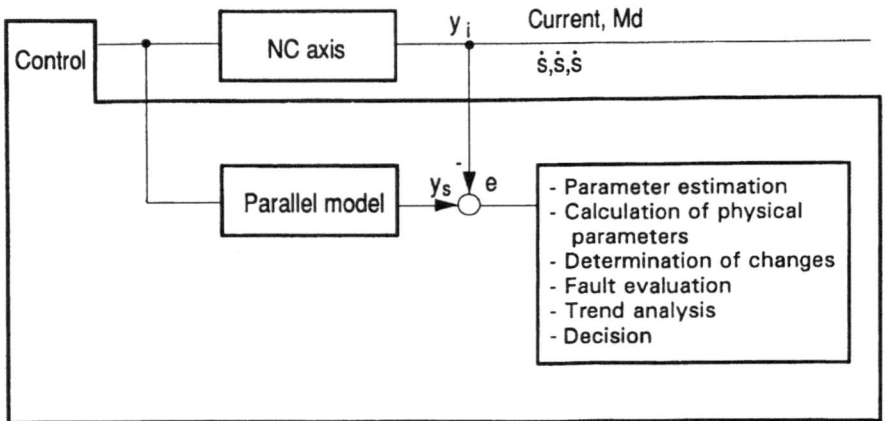

Fig. 9.18. Observation with the help of a parallel model.

the service central station or of a dialogue between the manufacturer's diagnosis system and the customer [15].

Thus, new functions for the machine tool have to be initiated frequently, and changes of control-internal data have to be analysed at the same time to locate faults. For this, verbal instructions to the machine operator with simultaneous data availability are necessary.

To achieve this with existing resources, the installation of a second separate telephone line is necessary. A provisional solution is sometimes offered using just one line, but then synchronization problems arise because of the manual changing over between language and data at the modem.

In the past these deficiencies led to the reluctant acceptance of such teleservice systems in practice, both on the part of service technicians and customers.

With the introduction of the Integrated Services Digital Network (ISDN) at the end of 1988, an efficient new network became available which, for functional as well as for financial reasons, is especially suitable for setting up teleservice systems.

9.3.2 General Characteristic Data of ISDN

The ISDN operates in time multiplex, and supplies the user at the basic telephone connection with two so-called B channels with a signal capacity of 64 kbit s^{-1} each and a D channel with a signal capacity of 16 kbit s^{-1}. The two B channels are line-connected, and form useful channels which are independent of each other and can be used simultaneously: they transmit language, text, pictures and data. The D channel transmits information for controlling and supervising purposes. See Fig. 9.19.

The introduction in series of ISDN into the German Federal Republic began in eight big cities at the end of 1988. In the intervening period, many other cities have followed this example. According to the German Federal Postal Administration, the whole country should be supplied with this system by 1993 [5]. In other Western

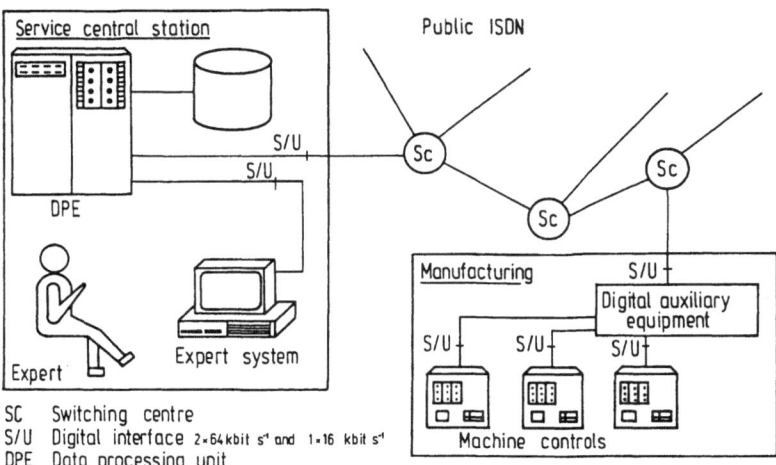

SC   Switching centre
S/U Digital interface 2×64 kbit s⁻¹ and 1×16 kbit s⁻¹
DPE Data processing unit

Fig. 9.19. Remote diagnosis.

European states, similar networks will be established during the same period. The network of the Netherlands is already connected to that of Germany. In almost all Western industrial nations, such as the USA and Japan, national networks are being built up, which in the near future will be connected with one another, so that a worldwide teleservice will be available soon.

9.4 Use of Expert Systems

Simple logical systems can generally be divided into modules in such a way that a completely automatic diagnosis can be carried out. A machine tool, however, is very complex, so that it becomes impossible to supervise all existing elementary modules for a justifiable expenditure (Fig. 9.20).

Furthermore, there are elements which cannot be diagnosed automatically, such as the communication interface for operation in the form of entry keyboards or display output. In this case a specialist who is a diagnosis expert is required. Because of their capabilities of "thinking", and their capabilities also to learn, compare, recognize patterns and associate, humans are extremely efficient diagnosis systems which for the time being can not be matched by computers; thus the term "artificial intelligence" is completely misleading. As far as data processing with its searching and algorithmical processes is concerned, computers are unbeatably fast. If such properties are helpful in diagnosis, so-called "Expert Systems" can be used as a computer-aided tool which, for a very limited field of general tasks, can be described as a structured knowledge memory with conclusion rules.

Today several methods, such as rules, frames, semantical networks and object/attribute/value relations, are used to represent the knowledge. They are quite suitable for recording empirical knowledge. For the description of a machine tool, a function-oriented knowledge representation, which is structured and frame-oriented, is useful – as mentioned earlier. This kind of knowledge representation corresponds to the well-known block structure of control technology. For generation

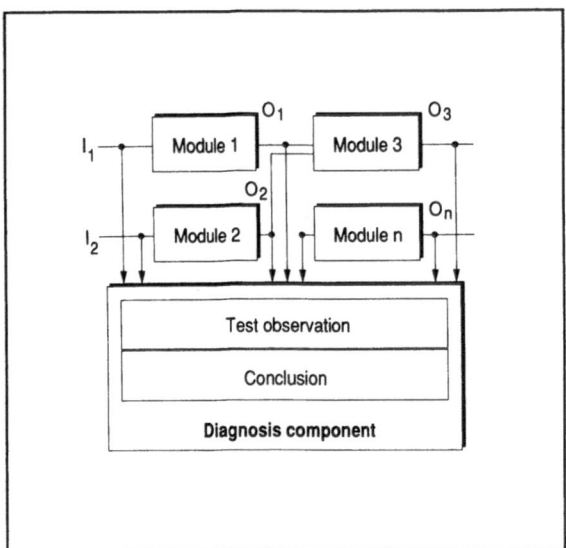

Fig. 9.20. Completely automatic diagnosis by division into elementary modules.

of a knowledge base, facts and rules which relate to both functional and practical knowledge have to be recorded for every module of the machine.

By means of the conclusion component (inference), the dialogue between the service technician and the Expert System is controlled, whereby these components can be structured differently. With the functional inference, an approach to faults is made by separating those modules in which the fault cannot be found. This is shown by the correct input and output values. A rule-oriented inference initiates the empirical method of an expert by if–then relations. Strategies for this are forward and backward chainings.

An essential requirement for an Expert System is a short response time at the consultation to avoid acceptance problems by the operator. Experience shows that these requirements are not yet fulfilled for larger knowledge bases and medium databases. Newer approaches with procedural languages anticipate that this problem will also be solved in due course, and Expert Systems should be able to play an important role in fault diagnosis in the future [15].

References

1. Leonards F. Diagnosesysteme bei Transferstraßen und Industrierobotern in der Automobilindustrie. VDI/IPA-Seminary, Stuttgart, 1984
2. Weck M. Verbesserung der Verfügbarkeit komplexer Fertigungssysteme. In Proceedings "Automatische Produktionssysteme", München, 1985, pp 232–250
3. Balzert H. Die Entwicklung von Software-Systemen. B.I.-Wissenschaftsverlag, Mannheim, 1982
4. DIN 66264. Mehrprozessor-Steuersystem für Arbeitsmaschinen (MPST). Part 1: Parallelbus. Part 2: Regeln für den Informationsaustausch. Beuth-Verlag, Berlin, 1986
5. Stute G *et al*. Test-und Diagnosesystem für modulare Mehrprozessorsteuersysteme (MPST). KfK-PFT 68, Karlsruhe, 1983

6. Röcker F. Automatisierte Diagnose durch strukturierte Programmierung. VDI Bericht, Düsseldorf, 1983
7. Holtz K. Multitaskingbetriebssystem für eine Mehrprozessorsteuerung (MPST). Girardet-Verlag, HGF-Kurzbericht, Essen, Blatt 84/16
8. Möller H. Integrierte Überwachungs- und Diagnosesysteme für numerische Steuerungen. Springer-Verlag, ISW 60, Berlin, 1986
9. Grimm W. Diagnosesystem für steuerungsperiphere Fehler an Fertigungseinrichtungen. Springer-Verlag, ISW 65, Berlin, 1986
10. Pritschow G, Gaukler J. Eigendiagnose von Funktionsprogrammen einer numerischen Steuerung. Girardet-Verlag, HGF-Kurzbericht, Essen, Blatt 88/19, 1988
11. Knörschild E. Speicherprogrammierbare Steuerungen für den sicherheitstechnischen Einsatz. VDI-Bericht 1983; 481: 43–45
12. N.N. VDI-Richtlinie 2880 Teil 5 (Draft). Speicherprogrammierbare Steuerungsgeräte; Sicherheitstechnische Grundsätze. VDI-Verlag, 1985
13. N.N. VDI/VDE Richtlinie 3541 (Draft). Steuerungseinrichtungen mit vereinbarten gesicherten Funktionen, Blatt 1–3. VDI-Verlag, 1983
14. Isermann R, Freyermuth B. Process fault diagnosis based on process model knowledge. Intl Computers in Engineering Conf, ASME, Anaheim, California, 1989
15. Grimm W. Expertensysteme zur Fehlerdiagnose an Drehmaschinen. Vortrag zum 5. Darmstädter Fertigungstechnischen Symposium "Frühdiagnose", Darmstadt, 1990
16. Isermann R. Process fault detection based on modeling and estimation methods – a survey. Automatica 1984; 20(4): 387–404
17. Isermann R. Identifikation dynamischer Systeme, Bd 1, 2. Springer-Verlag, Berlin, 1988

10 Signal Processing for Automatic Supervision

D. Dornfeld

10.1 Background on the Developing Needs of Sensors and Signal Processing in Manufacturing Automation

Consumers are now demanding products that are reasonably priced, reliable and tailored to their needs. As a result, manufacturers have had to develop manufacturing systems that are flexible and can accommodate a variety of products promising high performance. According to Ayres [1] high performance demands precision and complexity to differing degrees and, since complexity is often accompanied by a higher likelihood for defects and errors, increased attention to monitoring and screening devices during production.

The expense of automating manufacturing operations is high enough for operation often to be needed around the clock. These processes are necessarily untended or minimally tended owing to the expense of labour or lack of trained personnel to monitor the process. Thus there is also a great demand for monitoring systems to ensure the safe and efficient performance of these systems during untended operation. If we add to this the great diversity of materials, operating conditions and tooling likely to be experienced over the range of workpieces produced it is highly likely that malfunctions will occur. In the absence of good models of these processes to predict performance, sensors have been utilized in these systems to reduce the likelihood of unexpected malfunctions. Tönshoff et al. [2] have made an excellent review of the monitoring trends in machining processes and point out the improvements that have already been achieved by the introduction of monitoring to manufacturing processes. Studies by Ito [3] have pointed out the pivotal role that sensing technology will play in the development of future factory systems. As pointed out above, both processing and system conditions must be monitored to ensure optimum performance. Tönshoff identifies three major strategies at present used for implementing process monitoring and control using time-critical and non-time-critical situation sensing techniques. These are:

1. Open-loop monitoring systems that measure some condition of the machine tool or process and then display or activate an alarm to prompt human intervention;
2. Open-loop diagnostic systems that attempt to determine a functional or causal relationship between a machine failure and its cause; and
3. Closed-loop adaptive control systems that automatically adapt machining

conditions to changes in the process environment according to pre-determined strategies.

Improvements have been suggested by Tönshoff to make these systems more effective. One of the suggestions is related to improved sensors and sensor data-handling techniques. This is the topic of this chapter.

Ito [3] predicts that compact multiple-purpose sensors and sensors for "ambiguity factors" will be developed. It is also clear that more than one sensor will be utilized to improve the reliability of the monitoring system. There have been strategies developed using multiple sensors in the past but these are essentially sets of sensor systems operating independently of each other, each sensor alerting to a different phenomenon. Of interest here is the integration of sensors to provide an environment that uses the combined information from a number of sensors to render a decision on the state of a manufacturing process, tool or machine. This is referred to as sensor fusion.

More recently Dornfeld [4] extended the review of Tönshoff by focusing on intelligent sensor systems and emphasizing the recent research in multi-sensor systems for process monitoring. In this case a variety of sensors are used to provide a range of coverage of process characteristics with the goal of ensuring a higher reliability. This multi-sensor approach then requires more attention to feature extraction, information integration and decision-making in real time to be effective. Sensing systems for manufacturing processes must balance a number of options if they are to be effective. For example:

1. Is the application process monitoring and control or machine diagnostics (that is, to control the output of the process or ensure correct functioning of the machine components' lubrication, cycle etc.)? This will determine the type of signal processing/models, type of sensor and its location.

2. Does the nature of the application require a sensor response to a detected phenomenon that is slow or fast (i.e. will the workpiece be damaged in the next few milliseconds or within the next few cycles if no action is taken as a result of this detected information)? This will determine the type of sensor to be used and, most significantly, the amount of digital signal processing/hardware needed to meet time demands.

3. Does the sensing technique require that the sensor be in contact or not in contact with the component under surveillance (the tool, machine or workpiece)? This will determine the type of sensor that can be employed as well as the degree of modification of the machine, tooling and process needed to implement the sensor.

4. Does the application require a direct or indirect sensor measurement? That is, is the transducer measuring the phenomenon itself (for example, the diameter of a workpiece during a turning operation) or something affecting the diameter from which the change in diameter can be inferred (such as the length of the cutting tool creating the diameter)? This determines the type, location, required signal treatment and performance (in terms of meeting the monitoring objectives) for the sensor employed.

5. Does the measurement made by the sensor need to be done in real time, i.e. during the process, or can it be done before and after the process, or perhaps between steps in the process? This, along with the time response of the sensor, will dictate the type of sensor that can be used.

There have been many sensor methodologies suggested for process, tool or machine monitoring in the manufacturing environment. Hoshi [5] reviewed many differing sensing applications in machining processes. Included by Hoshi are reviews of the following measurements or sensors (the most common "competing technologies") for edge chipping, fracture and wear (related to cutting tools), and poor hole quality (drilling): touch sensors, load amperage, vibration, torque/force limiter and acoustic emission (AE). Most of the processes monitored involve drilling, and the more straightforward approaches of monitoring, tool touch and load amperage are most frequently used. Vibration sensing and acoustic emission sensing are the next most often applied but, according to Hoshi, only touch sensing was reported to be 100% successful. Load amperage, vibration and AE were reported to be 80%, 75% and 33% successful, respectively. This indicated that much additional research is needed to make these techniques useful. Often the addition of enhanced signal processing methodologies can make these sensing techniques more reliable.

Achieving untended manufacturing has been described as perhaps "the biggest obstacle" confronting the development of computer integrated manufacturing or, on the smaller scale, flexible manufacturing systems [6]. Sensors function to collect information for the evaluation of the performance for the system and its consistency with analytical predictions. Further, sensors also play an important role in the diagnostic capabilities of systems and methodologies for predictive maintenance. The Machine Tool Task Force Study [7] identified the lack of reliable, rugged sensors as a main stumbling block to the real-time control and monitoring of manufacturing systems. Sensors must often operate in hostile environments, and existing sensors are either limited in accuracy, reliability, range or response, or are inappropriate for some of the phenomena under observation. An excellent review of sensors for untended machining operations can be found in [7, 8]. In addition, Iwata [9] has reviewed the need for on-going research on sensing technologies to improve the machine tool function.

It is important to ensure that sensing methodologies applied to process monitoring yield data that can be quantitatively related to the process being investigated or monitored. This is often not the case and the result is sensor data of limited reliability. Here the word reliable is used in the sense of the *dependency* of the data on the basic process parameters as well as the *repeatability* of performance of the sensor in real environments. A sensing technique that has application in data collection for process modelling and analysis as well as real-time monitoring is most desirable.

Interest in developing a capability for untended manufacturing systems is growing along with the implementation of advanced flexible manufacturing systems. To remove the need for observation of the manufacturing processes and minimize the time lost due to repair or correction of unexpected failures in the system, new sensing methodologies and sensor-based control schemes are being proposed and evaluated. Clearly, one of the major blocks to implementation of true untended manufacturing is the lack of suitable sensors for process monitoring. In spite of the many sensing technologies in existence today, few, if any, totally untended operations exist. The sensing tasks for untended manufacturing operations are substantial. In addition to being economical, sensors must be non-intrusive, rugged and resistive to the hazardous environment in which they operate, accurate, exhibit a high sensitivity to the features or phenomena being monitored, have an output that is linear (or linearizable) with respect to the sensed feature if possible, and be highly repeatable and reliable. It is also desirable to have sensors that have diagnostic capabilities as well as being self-calibrating. Few sensors can meet all of these specifications.

It seems reasonable that, to be effective in untended manufacturing, combinations of sensors will be needed to provide corroborative information on the state of the manufacturing operation. This often includes integration with other sensors and the co-processing of data from several basic different sources, since one sensor alone may be unable to determine the process state adequately. This will require efficient methodologies for the integration of sensor information. Easily available information on the operation of the process (such as speeds, feeds etc. in machining) will need to be considered as well. Some early efforts by Matsushima and Sata [10] have demonstrated the potential of combining information from several sources in process monitoring.

10.2   Signal Processing, Feature Extraction and Sensor Fusion

10.2.1   Introduction

Human monitoring of manufacturing processes can attribute its success to the ability of humans to distinguish, by nature of the physical senses and experience, the "significant" information in what is observed from the meaningless. In general, humans are very capable as process monitors because of the high degree of development of sensory abilities, essentially noise-free data (unique memory triggers), parallel processing of information and the knowledge acquired through training and experience (Rangwala [11]). Limitations are seen when one of the basic human sensor specifications is violated: something happening too fast to see or out of range of hearing or visual sensitivity as a result of the frequency content. These limitations have always served as some of the justification for the use of sensors. Sensors, of course, are also limited in their ability to yield an output that is sensitive to an important input. Thus we need to consider the use of signal processing and along with that feature extraction. In most cases the utilization of any signal processing methodology has as its goal one or more of the following: the determination of a suitable "process" model from which the influence of certain process variables can be discerned; the generation of features from sensor data that can be used to determine process state; and the generation of data features so that changes in the performance of the process can be "tracked".

An overview of signal processing and feature extraction is conveniently summarized in Fig. 10.1, taken from Rangwala [11]. Here the measurement vector extracted from the signal representation from the sensor (basic signal conditioning) is the "feedstock" for the feature selection process (local conditioning), resulting in a feature vector. The characteristics of the feature vector include signal elements that are sensitive to the parameters of interest in the process. The "decision-making" process is characterized in the lower portion of Fig. 10.1. Based on a suitable "learning" scheme which maps a teaching pattern (i.e. the process characteristics we desire to recognize) onto the feature vector, a pattern association is generated. The "pattern association" contains a matrix of associations between the desired characteristics and the features of the sensor information. In application, the pattern association matrix operates on the feature vector and extracts correlations between features and characteristics – these are taken to be "decisions" on the state of the

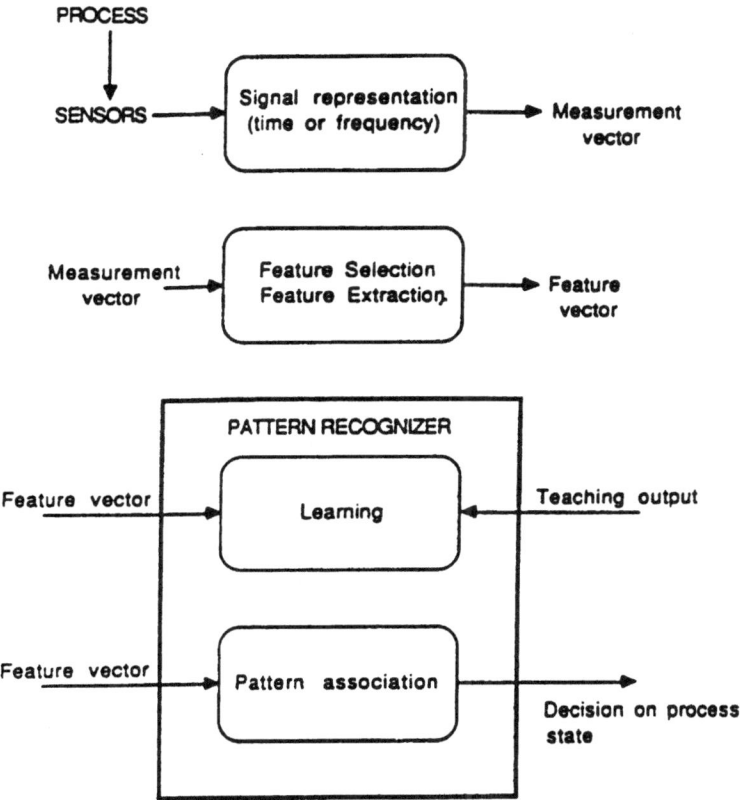

Fig. 10.1. Structure of a typical pattern recognition system.

process if the process characteristics are suitably structured (e.g. tool worn, weld penetration incomplete, material flawed etc.).

There is a close relationship between sensor fusion methodologies and signal processing/feature extraction so some overlap in methodologies will exist. Advanced information processing and fusion methodologies are most often embodied in what could be referred to as an "intelligent sensor". Hence, we shall first discuss, generally, the concept of an intelligent sensor and then described the use of neural networks as a sensor fusion strategy with multi-sensor inputs including acoustic emission. An example of the application of recursive autoregressive models for feature extraction from acoustic emission signals for tool-wear monitoring will also be presented.

10.2.2 Intelligent Sensor Defined

The philosophy of implementation of any sensing methodology for diagnostics or process monitoring can be divided into two simple approaches. In the first, one uses a sensing technique for which the output shows some relationship to the characteristics of the process. After determining the sensor output and behaviour for the "normal" machine operation or process, one observes the behaviour of the signal until it deviates from the normal thus indicating a problem. In the second approach, one attempts to determine a model linking the sensor output to the process mechanics and then, with sensor information, use this model to predict the behaviour of the process. Both methods are useful according to prevailing circumstances. The first is, perhaps, the more straightforward but liable to misinterpretation if some change in the process occurs that was not foreseen (that is, "normal" is no longer normal.) To ensure against this type of misinterpretation, intelligence has been added to the sensors to give sophistication to the feature extraction and decision-making processes. Intelligent sensing systems most commonly have been associated with robot systems operating in unstructured environments. This has been motivated by the need to integrate multiple sensors for flexibility in control of the robot – see Ruokangas *et al.* [12]. In these applications, information from only one sensor is generally insufficient to allow complete specification of the environment for task planning and execution. Multiple sensors are often employed for object location and recognition, for example, and use cameras, infra-red, ultrasonic and tactile sensing devices. The integration of the data from all of these sensors operating simultaneously is the major challenge for sensor fusion methodologies in robot applications. Yamasaki [13] also identifies intelligent sensors as part of the lowest layer in a multi-layer control system for complex processes. In the case described in that work, feature extraction for speech pattern recognition, an intermediate layer provides some control and optimization at the lower level as well.

The development of an intelligent sensor for monitoring a manufacturing operation generally requires the following three hierarchical stages [11]:

1. Determining the sensitivity of a sensor signal to the process parameter or parameters to be monitored;
2. Developing an appropriate in-process real-time signal processing method for extracting signal parameters that are rich in information regarding the process parameters being monitored but relatively insensitive to other parameters; and
3. Developing a decision-making scheme that can make a decision on the process state based on the data obtained from all previous experiences as well as current sensor information.

This strategy yields both hardware and software elements. Figure 10.2, from [14], is a schematic of an intelligent sensor system showing the usual sensor element and associated signal amplification and transmission electronics found in all basic sensors as well as additional elements. These include signal conditioning (conversion of sensor output to information relative to a process parameter), local conditioning and decision-making (feature extraction and state estimation) as well as self-calibration and diagnostic functions (to ensure that the sensor system is functioning properly and that any faults can be found easily). The sensor system provides a high-level signal to the process controller or for further fusion with the output of other sensors.

10.2.3   Sensor Fusion Defined

With a specific focus for the monitoring in mind, researchers have developed over the years a wide variety of sensors and sensing strategies, each attempting to predict

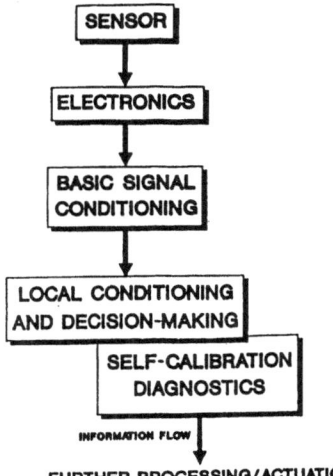

Fig. 10.2. Schematic of an intelligent sensor.

or detect a specific phenomenon during the operation of the process and in the presence of noise and other environmental contaminants. Although able to accomplish the task for a narrow set of conditions, these specific techniques have almost uniformly failed to be reliable enough to work over the range of operating conditions and environments commonly available in manufacturing facilities. Thus, researchers have begun to look at ways of collecting the maximum amount of information about the state of a process from a number of different sensors (each of which is able to provide an output related to the phenomenon of interest, although at varying reliability). The strategy of integrating the information from a variety of sensors with the expectation that this will "increase the accuracy and ... resolve ambiguities in the knowledge about the environment," in the words of Chiu *et al.* [15], is called sensor fusion. Sensor fusion is able to provide data for the decision-making process: that has a low uncertainty due to the inherent randomness or noise in the sensor signals; includes significant features covering a broader range of operating conditions; and because of redundancy accommodates changes (due to calibration, drift etc.) in the operating characteristics of the individual sensors. In fact, perhaps the most advantageous aspect of sensor fusion is the richness of information available to the signal processing/feature extraction and decision-making methodology employed as part of the sensor system.

Sensor fusion is best defined in terms of the "intelligent sensor" (Dornfeld [14, 16]) as introduced above, since that sensor system is structured to utilize many of the same elements needed for sensor fusion. These elements include the basic sensor hardware (transducer element and electronics such as the piezo element and charge amplifier in a load-cell or bimetallic and reference junctions with a Wheatstone bridge in a thermocouple), as well as basic signal conditioning, decision-making and self-calibration and diagnostic capabilities. Middlehock and Hoogerwerf [17] define an integrated sensor along similar lines, as a device with one or more transducer elements, signal conditioning and signal processing electronics, microcontroller and communication circuitry integrated into the same package.

10.2.4   Fusion Methodologies

10.2.4.1   Introduction

This section reviews, briefly, several techniques for integrating the information from several sources. These are discussed in light of the requirements for monitoring manufacturing processes. The objective of sensor fusion is to increase the reliability of the information so that a decision on the state of the process is reached. This tends to couple fusion techniques closely with feature extraction methodologies and pattern recognition techniques. The problem here is to establish the relationship between the measured parameter and the process parameter. There are two principal ways to encode this relationship – Rangwala [11]:

1. Theoretical – the relationship between a phenomenon and the measured parameters of the process (say tool wear and the process); and
2. Empirical – experimental data is used to tune the parameters of a proposed model.

As mentioned earlier, reliable theoretical models relating sensor output and process characteristics are often difficult to develop because of the complexity and variability of the process and the problems associated with incorporating large numbers of variables in the model. As a result, empirical methods which can use sensor data to tune unknown parameters of a proposed relation are very attractive. These types of approaches can be implemented either by: (a) proposing a relationship between a particular process characteristic and the sensor outputs, and then using experimental data to tune the unknown parameters of a model; or (b) associating patterns of sensor data with an appropriate decision on the process state without consideration of any model relating sensor data to the state. The second approach is generally referred to as pattern recognition and involves three critical stages – Ahmed and Rao [18]:

Sampling of input signal to acquire the measurement vector;

Feature selection and extraction;

Classification in the feature space to permit a decision on the process state to be made.

These have previously been illustrated in Fig. 10.1. The pattern recognition approach provides a framework for machine learning and knowledge synthesis in a manufacturing environment by observation of sensor data and with minimal human intervention. More importantly, such an approach allows for integration of information from multiple sources (such as different sensors), which is our principal interest here.

Sata et al. [10, 19] were among the first researchers to propose the application of pattern recognition techniques to machine process monitoring. They attempted to recognize chip breakage, formation of built-up edge and the presence of chatter in a turning operation using the features of the spectrum of the cutting force in the 0–150 Hz range. Dornfeld and Pan [20] used the event rate of the rms energy of an acoustic emission signal along with feedrate and cutting velocity in order to provide a decision on the chip formation produced during a turning operation. Emel and Kannatey-Asibu [21] used the spectral features of the acoustic emission signal to classify fresh and worn cutting tools. Balakrishnan et al. [22] used a linear discriminant function technique to combine cutting force and acoustic emission information for cutting tool monitoring.

The manufacturing process may be monitored by a variety of sensors and, typically, the sensor output is a digitized time-domain waveform. The signal can then be either processed in the time domain (for example, extract the time-series parameters of the signal) or in the frequency domain (power spectrum representation). The effect of this is to convert the original time-domain record into a measurement vector. In most cases, this mapping does not preserve information in the original signal. Usually, the dimension of the measurement vector is very high and it becomes necessary to reduce this dimension owing to computational considerations. There are two prevalent approaches at this stage: select only those components of the measurement vector which maximize the signal-to-noise ratio, or map the measurement vector into a lower-dimensional space by means of a suitable transformation (feature extraction). The outcome of the feature selection/extraction stage is a lower-dimensional feature vector. These features are used in pattern recognition techniques and as inputs to sensor fusion methodologies.

10.2.4.2   Linear Discriminant Functions

The discussion here focuses on linear discriminant pattern classifiers trained using the perceptron learning algorithm. These are often referred to as perceptrons. A detailed discussion of the perceptron learning algorithm and its application to machine learning can be found in Nilsson [23].

A state can be characterized by a set of measured numbers, x_1, x_2, ..., x_d, which form the pattern X for the state. A pattern classifier assigns the patterns into one of R different categories. A pattern can be viewed as a point in d-dimensional pattern space with its components as the co-ordinates. The vector extending from the origin to the point can also be used to represent the pattern, called a pattern vector. Each pattern classifier can also be viewed as a set of surfaces, called decision surfaces, which partitions the pattern space into R different regions. The decision surfaces are implicitly defined by a set of functions containing R members, $g_1(X)$, $g_2(X)$, ..., $g_R(X)$, called discriminant functions, which are scalar and single-valued functions of the pattern X. The discriminant functions are chosen such that for all X in category i, $g_i(X) > g_j(X)$ for all $i \neq j$, and $i,j = 1,2, ..., R$. The classification works as follows. A pattern, X, is presented to R discriminators, each of which computes the values of a discriminant function. The outputs of the discriminators, called discriminants, are compared by a maximum selector, which indicates the largest discriminant, i. Then the pattern X is classified into category i.

Discriminant functions can be selected in a variety of ways. In the simplest form, the discriminant functions are linear and the decision surfaces are hyperplanes. This is the easiest for calculation and is especially good for real-time application but, thus, does not guarantee that it is the most effective way for pattern classification. A linear discriminant function has the following form:

$$g_i(X) = W_i(X) = \sum_{k=1}^{k=d+1} w_k x_k \qquad i = 1, 2, ..., R \qquad (10.1)$$

where W_i = weight vector $(w_1, w_2, ..., w_{d+1})$
and X = pattern vector $(x_1, x_2, ..., x_{d+1})$

The threshold margin, which is the minimum difference required between the largest discriminant and all others for making a classification, is embedded in w_{d+1}.

The weight vectors of all the discriminant functions are obtained by an adjustment process known as training. A set of typical patterns similar to those that the discriminant functions must ultimately classify are chosen to form a training set. All the classifications of the training pattern are known. We have adopted the error-correction training procedure, a non-parametric training method, in this study. The training procedure starts with arbitrary initial values for W_i. Every pattern in the training set is presented one at a time in any sequence to the discriminant functions, and each pattern can recur infinitely often. Adjustments are made whenever the discriminant function responds incorrectly to any pattern. Suppose that a pattern, X_k, belonging to category i is the ith pattern presented in the training sequence. If $g_i(X_k)$ = W_iX_k exceeds all other discriminants, i.e. $g_i(X_k)>g_j(X_k)$, for all $j \neq i$, the response is correct, and no correction of the weight vector is made. Suppose, however, that $g_i(X_k) \leq g_j(X_k)$ for some $j \neq i$ and $j \in J$, which is an integer set containing all the indices of weight vectors that need to be corrected for X_k. Then the following rule is applied to obtain a new set of weight vectors from the present one (W_p):

$$W_p = \begin{cases} W_i + a_kX_k & p = i \\ W_j - \alpha_{jk}a_kX_k & p = j \in J \\ W_p & \text{otherwise} \end{cases}$$

The correction increments, a_k and α_{jk}, are allowed to vary with the number of corrections, k. There must exist constants a_{min} and a_{max}, such that

$$0 < a_{min} \leq a_k \leq a_{max} \quad k = 1, 2, \ldots$$

and α_{jk} must satisfy

$$\alpha_{jk} \geq 0; \text{ and } \sum_{j \in J} \alpha_{jk} = 1$$

The effect of the adjustment is to increase the value of ith discriminant by $a_k |X_k|^2$, while the value of jth discriminant is reduced by $\alpha_{jk}a_k|X_k|^2$. If the training set is "linearly separable", i.e. there exists a set of linear discriminant functions that can classify all patterns correctly, then the training method guarantees that the weight vectors are yielded after a finite number of corrections.

If only two categories exist, i.e. $R = 2$, the classification is made by deciding which one, $g_1(X)$ or $g_2(X)$, is larger. It turns out that this decision can be implemented by evaluating the sign of a single discriminant function: $g(X) = g_1(X) - g_2(X)$ The single linear discriminant function has the same form as defined in Equation (10.1). If $g(X)>0$, X is placed in category 1; if $g(X)<0$, X is placed in category 2. The two regions in the pattern space are separated by a decision surface $g(X) = 0$. The pattern classifier can be implemented according to the block diagram in Fig. 10.3, from [23]. Such a structure, consisting of weights, a summing device and a threshold element, is called a threshold logic unit (TLU). Desirable properties of perceptrons are that they offer a means for machine learning based on observations of training data, and are suited for parallel operation since the weighted values of the different components of the feature vector can be computed in parallel. Thus, information from different features is determined.

Once a pattern classifier has been trained it should be able to associate incoming sensory patterns with a decision on process status. The degree to which correct associations can be made depends on the following factors [11]: (1) quality of information available in the sensory features; and (2) type of processing that the raw

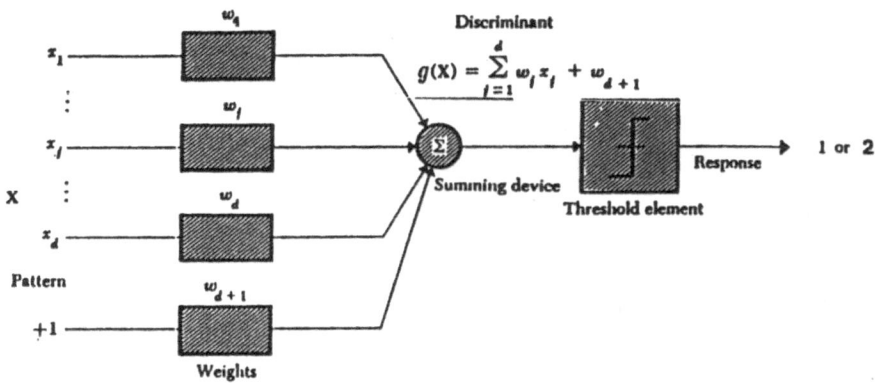

Fig. 10.3. The threshold logic unit (TLU).

sensory features are subjected to during the training and pattern recognition phases. The first factor reflects the amount of discriminatory information contained in the raw sensor signal features presented to the pattern classifier. Ideally, good information implies that the features are very sensitive to the process characteristics of interest and insensitive to the effects of noise and process variables. In many situations, even though the raw signal features contain the necessary information for discriminating between two distinct states, it may be encoded in a complex manner or coloured as a result of the effects of noise and process variables. In such cases it becomes necessary to process the raw sensor information in order to extract features suitable for classification purposes and to reject noise. Some of the techniques in Section 10.3 do exactly that.

Although valuable information may be contained in the entire measurement vector, from practical considerations only a few of these components can be used for training and pattern association. This is because in training a pattern classifier such as a perceptron the minimum number of training samples to be used is

$$N = 2(d+1) \tag{10.2}$$

where N is the number of training samples and d is the number of features used. This constrains the training procedure so that generalization behaviour of the classifier is acceptable.

The approach for reducing the dimension of the measurement vector is to retain only those components of the spectra that show a high sensitivity to the process characteristics of interest and low sensitivity to noise or process parameters. Considering that the measurement vector is D-dimensional, the objective is to select d features which maximize a criterion representing the signal-to-noise ratio of the features. The selected d features are the components of a d-dimensional feature vector. The criterion function in this case uses the concept of inter-class Euclidean distance measures and is discussed in greater detail in Devijver and Kittler [24]. A typical criterion is

$$J = \text{trace}(S_w^{-1} S_b) \tag{10.3}$$

where S_w is the within-class scatter matrix and S_b is the between-class scatter matrix of the d-dimensional feature vector. S_w measures the scatter of data points within a class representing a process state and S_b measures the distance between clusters representing data points in the d-dimensional feature space of different states. Intuitively, the value of J represents the signal-to-noise ratio of the feature vectors. Adding new features increases the J value since additional features cause the distance between mean values of the clusters representing states to increase. A high value of J indicates that the clusters corresponding to two different process states are far apart and that the scatter within the cluster is small.

10.2.4.3   Neural Networks

It is preferable if the feature extraction and learning activities occur at the same time, since this would allow the extraction of optimal information from the features along with noise rejection. Such an approach is possible using neural network pattern recognizers. A neural network is a collection of simple, interconnected processors which operate in parallel and store knowledge about the strength of the connections between the individual processors. Such parallel networks of computing elements crudely resemble processing activity in the brain and have been successfully applied to intelligent tasks such as learning and pattern recognition.

Artificial neural networks are an attempt to mimic the computational architecture of the human brain in electronic hardware, the objective being to incorporate intelligent functions such as learning and pattern recognition in computers. The human brain consists of a large number of interconnected neurons, each possessing very simple computational abilities. However, the interactions between the neurons allow for parallel processing of information, which greatly enhances the speed of computation and causes a large amount of knowledge to be brought to bear in processing this information – Hinton and Fahlman [25]. A typical neuron consists of three components: the cell body, the input lines into the neurons (dendrites) and output lines emerging from the cell body (axons). The axons of a neuron are connected to the dendrites of other neurons at points called synapses. This forms a highly interconnected system of neurons which communicate with each other via synapses. The synapses determine the strength of the connection between two neurons. A typical neuron receives inputs from various other neurons via the dendrites. The time-averaged sum of these inputs causes biochemical reactions inside the cell body, which result in pulses of electrochemical activity being transmitted over the axon lines of the neuron. The pulse rate depends on the magnitude of the input excitation to the neuron, and is usually assumed to be a sigmoid function of the input. This is because the output saturates at extreme values of the input (the output pulse rate lies between 0 and 500 Hz for a typical neuron).

The synapse between two neurons may be excitory (in which case, a high activity in one neuron causes a high activity in the other neuron) or inhibitory (in which case, a high activity in one neuron causes activity in the other neuron to be suppressed). The synapses develop through learning processes in the brain, although the exact mechanisms are not known at present. The information processing tasks that the human brain excels at, such as pattern recognition and the ability to retrieve from memory based on partial or incorrect cues (content addressable memory) are due to the collective processing activity of a large number of interconnected neurons operating in parallel and with the knowledge stored in the strength of the synapses between these neurons – Hopfield [26].

Artificial neural networks can be implemented by using amplifiers with a sigmoid input–output relationship as the "neuron" element and resistive connections

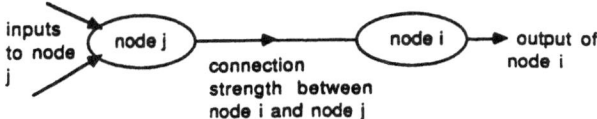

Fig. 10.4. Behaviour of nodes in a neural network.

between the amplifiers representing the synapses between the neurons. The conductance values of the resistive elements represent the strength of the connection between the individual processors and the sign of the conductance determines whether it is excitory (positive conductance) or inhibitory (negative conductance). The voltage output of the amplifier represents the activity level of a given processor whereas the input to the processor is simply a current-summing operation on the outputs of the other processors. A schematic of two such interconnected processors is shown in Fig. 10.4. The connection strength between the jth and the ith neuron is represented as w_{ij}, and is simply the conductance value of the connection between these processors. The input to the ith processor is given by

$$\text{INPUT}_i = \sum_{j=1}^{N} w_{ij} \text{ OUT}_j \tag{10.4}$$

where N is the total number of processors connected to the ith processor and OUT_j is the output of the jth processor. The output of a processor is usually assumed to be a sigmoid function of the input or, in some cases, a threshold function may also be used. The conductance values or the "weights" are the learning parameters of the system and encode the knowledge in the system. These learning parameters are acquired through learning. Physically, this implies that the resistors should be able to change their resistance values in response to the learning process, i.e. adaptive or programmable resistors are required.

Computation in neural networks occurs by propagation of signals through the connections between the processors. Each processor is capable of very simple functions such as current summing and elementary arithmetic operations. In a general network, all the processors may be connected to each other, with the knowledge of the system encoded in the "strength" of connections between the individual processors. Some of the processors may receive inputs from the external world (input nodes) whereas others may transmit their outputs to the external world (output nodes). The remaining nodes which are not directly connected to the external world are called "hidden" nodes. These nodes are important since they are responsible for feature extraction and internal representation of the knowledge acquired through the learning process. One important feature of neural networks is that the knowledge in the system directly determines how the processors interact in contrast to being stored in a separate knowledge base, waiting to be accessed sequentially by a CPU. Also, the knowledge is distributed over a large number of connections which results in a fail soft operation: a failure of some of the processors will cause graceful degradation in performance rather than a complete loss of the knowledge base.

Pattern recognition tasks typically involve associating sets of patterns. In a neural network, the knowledge required to associate the correct sets of patterns is encoded in the values of the weights or connection strengths between the processors (Fig. 10.4). Neural architectures proposed by Hopfield [26, 27] have been demonstrated to implement content addressable memories successfully, whereby a system of interconnected processors can retrieve a complete memory given only a partial knowledge of the memory. Hopfield networks consist of a system of N processors, each processor being connected to all other processors. The operation of such a network can be described as a dynamic system to which the network state converges. The state of the network is represented by the activation levels of the individual processors at a given instant of time, and is an N-dimensional vector. The important point here is that the location of the attraction basins (or equilibrium points) can be controlled by suitable selection of the connection strengths between the processor states (which represents an incomplete or incorrect memory), and so the network will converge to one of the attraction basins of the system. If the initial state is close enough to the desired state, the correct memory (represented by the converged state) will be retrieved. Hopfield networks have also been used to solve combinatorial optimization problems (Hopfield and Tank [28] and Tank and Hopfield [29]).

One class of neural networks which have been shown to be successful at pattern recognition tasks are multi-layered, feedforward networks of the type shown in Fig. 10.5. There are three kinds of processing units in such networks: input layer nodes which accept patterns from the external world, output layer nodes which generate outputs to the external world, and hidden nodes which do not directly interact with the external world. The role of the hidden nodes is to form internal representations of the patterns presented at the input layer. These networks perform pattern association tasks in which a pattern presented at the input layer of the network is associated with a pattern at the output layer. In these networks, information propagates from the bottom to the top layer, with connections existing only between processors in adjacent layers. Let:

$w_{i,j,k}$ = weight between the jth processor in $(k-1)$th layer and the ith processor in the kth layer

$\text{net}_{i,k}$ = input to the ith node in the kth layer

$\text{out}_{i,k}$ = output of the ith node in the kth layer

$t_{i,k}$ = threshold value associated with the ith node in the kth layer

The input to a processor is then given as:

$$\text{net}_{i,k} = [\sum_j W_{i,j,k}\text{out}_{j,k-1}] + t_{i,k} \tag{10.5}$$

The output of a given processor is a sigmoid function of the input and can be expressed as:

$$\text{out}_{i,k} = F(\text{net}_{i,k}) = \frac{1}{e^{-\text{net}_{i,k}}} \tag{10.6}$$

Although recognizing that multi-layered neural networks possess many attractive properties, one of the obstacles to their development is the absence of an efficient learning algorithm for training such networks. The generalized delta rule, developed independently by Rumelhart and McClelland [30] and by Le Cun [31], fills this gap and has been shown to work efficiently on pattern association tasks. This is a supervised learning procedure in which examples of input and output patterns (representing the patterns to be associated) are used to train the network. The rule consists of presenting an input pattern to the network, propagating activity among the various processors according to Equations (10.5) and (10.6), and computing the

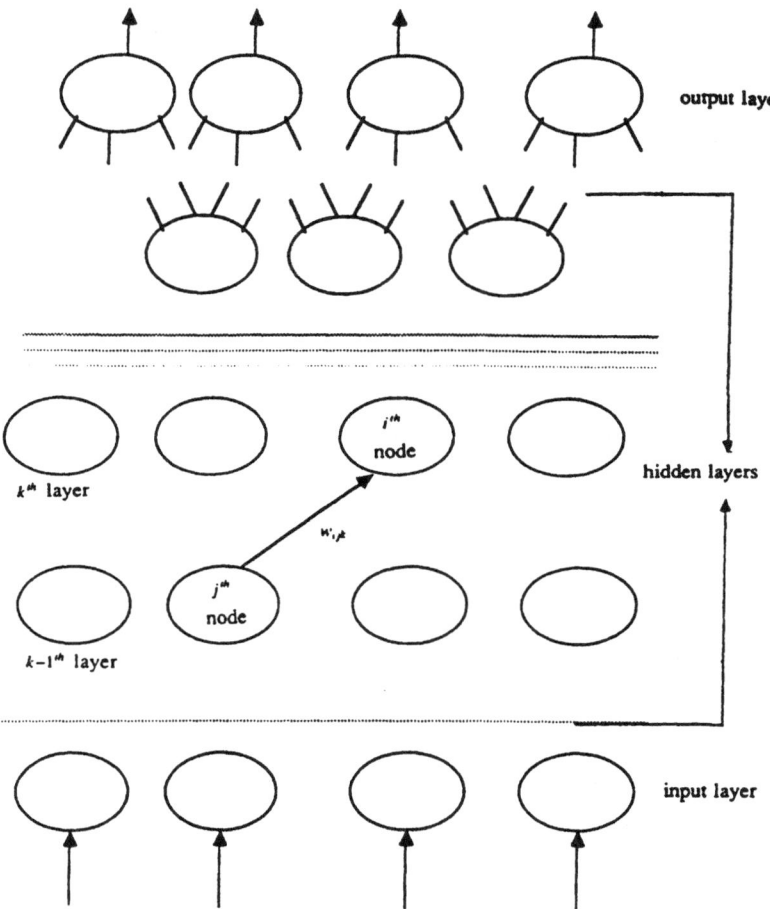

Fig. 10.5. Structure of a feedforward neural network.

pattern at the output nodes with the current set of learning parameters (thresholds and weights of the network). The actual output pattern is then compared with the desired output pattern and the error is calculated as:

$$E = \frac{1}{2} \sum_{i=1}^{q} (d_i - a_i)^2 \tag{10.7}$$

where d_i is the desired output at the ith output layer node, a_i is the actual output and q is the total number of nodes i in the output layer. The procedure of calculating the error is repeated for all sets of training input–output patterns and the individual errors are added to compute the total error. This constitutes the forward pass through the network.

Next, the error is propagated from the top to bottom layer through the network to modify the weights and thresholds in such a way that the error term is minimized.

This involves minimization of a non-linear error function with respect to the threshold and weight values, using gradient descent. Computation of the error term with respect to each weight and threshold is accomplished using local information at each node, so that gradient calculations at each layer can be accomplished in parallel. Rumelhart and McClelland [30] use the gradient information to adjust the weights and thresholds as follows:

$$\Delta w_{i,j,k} = -\eta \, \frac{\partial E}{\partial w_{i,j,k}} \qquad (10.8)$$

$$\Delta t_{i,k} = -\beta \, \frac{\partial E}{\partial t_{i,k}} \qquad (10.9)$$

where η and β are the step sizes in the minimization process.

A linear network is one in which the output of a processor is a linear function of its input, with the input to a processor defined as in Equation (10.5). For a linear system, the error surface is bowl shaped and has only one minimum point, so that convergence is guaranteed. In the present case, however, the error function is a non-linear function of the learning parameters, so that any gradient descent scheme for error minimization is prone to termination in a local minimum. There is no guarantee that a gradient descent procedure will find a set of thresholds and weights so that the error term is zero. However, as pointed out by Rumelhart and McClelland [30], this does not seem to present difficulties in practical implementations, since the number of hidden layers and number of nodes in each hidden layer can be so chosen that a set of weights and thresholds which drive the error to zero can usually be found, for a given implementation. The final values of the learning parameters are randomized and usually lie between −1 and 1.

Once such a network has been trained using a set of training patterns, it can be used to associate patterns presented at the input layer with appropriate patterns at the output layer. The advantage of using these networks for pattern recognition tasks is that learning can be accomplished purely by observation of sensor data, learning and pattern recognition is accomplished using parallel computation, and such networks can form internal representations of the raw sensory information independently. Internal representations are necessary because the raw information may be noisy, redundant and not suited for making pattern classification decisions. Another advantage of using such networks is that knowledge is stored in the connection strengths between the processors and directly determines how the network operates, rather than being stored in a database, waiting to be accessed by the CPU.

There are similarities between the perception type network discussed in Section 10.2.4.2 and the multi-layered networks discussed here. In fact, multi-layer networks have also been referred to as multi-layered perceptrons – Nilsson [23]. The perceptron performs pattern classification by computing a weighted sum of sensor inputs and comparing it with a threshold value. In the case of a perceptron, there are only two layers of units with the outer layer unit implementing a threshold type function. The perceptron implements a hyperplane in feature space, with the hyperplane surface representing the decision surface. No processing of raw information is carried out. With multi-layered neural nets, however, the raw input patterns undergo further processing to filter out noise in these patterns. The training procedure forces the hidden nodes to perform feature extraction on the raw features so that as information propagates through the network, the noise is filtered out. The feature extraction capability of the hidden nodes is developed during the training procedure so that the extracted features are better suited for the classification task. The last two layers of the neural network essentially implement a perceptron, however the features used in this case are the internal features which are relatively noise free. This leads to better generalization abilities of the classifier. Further, another difference between the perceptron and the multi-layered neural network is that the perceptron can perform only a linear separation of the sensor features whereas the feature extraction abilities of the hidden nodes in a multi-layered neural network allow the network to perform arbitrary mappings between input and output patterns.

10.3   Applications of Signal Processing and Sensor Fusion

10.3.1   Introduction

Three examples of applications of intelligent sensor signal processing and sensor fusion are presented here to illustrate the practical implementation of the methodologies discussed in earlier sections. The examples are drawn from research work related to process monitoring for manufacturing – specifically metal cutting. However, the principles behind the applications are appropriate for a wide range of processes and problems. The methodologies presented include the linear dis-criminant function applied to metal-cutting chip-form determination, adaptive autoregressive models for feature extraction for tool-wear monitoring, and neural network sensor fusion for tool-condition monitoring. All of these examples rely on acoustic emission generated by the machining process as one of the sensor outputs observed during process monitoring. Acoustic emission from metal cutting is described briefly below.

10.3.2   Background on Acoustic Emission from Metal Cutting

Acoustic emission (AE), i.e. stress waves spontaneously released in materials undergoing deformation, fracture (or both), friction and rubbing, and impact are usually of two distinct types – continuous and burst. Continuous-type signals arrive at the transducer in large numbers (i.e. distinct events cannot be distinguished). Burst-type signals appear as series of bursts (related to distinct events such as growth of a crack or a fracture) that are distinctly separable. In the machining process, continuous AE signals are generated in the shear zone, at the chip–tool interface and at the tool–work interface. Burst AE signals are generated as a result of chip breakage during or after chip formation, or as a result of tool cracking or fracture.

   Research over the past several years has established the effectiveness of AE-based sensing methodologies for machine-tool condition monitoring and process analysis. The problems of detecting tool wear and fracture of single point turning tools motivated much of this early work. In addition, the sensitivity of the AE signal to the various contact areas and deformation regions in the cutting and chip formation process has led to the analysis of AE signals as a basic tool for analysis of the cutting process. Dornfeld [32] presents an overview of AE in manufacturing with additional background on the subject.

   The major advantage of using acoustic emission to detect the condition of tool wear is that the frequency range of the acoustic emission signal is much higher than that of the machine vibrations and environmental noises. Therefore, a relatively uncontaminated signal can be easily obtained by the use of a high-pass filter. In addition, acoustic emission can be measured by simply mounting a piezo-electric transducer on the tool holder. It does not interfere with the cutting operation and thus allows for continuous monitoring of the tool condition. However, owing to its high-frequency nature and the sensitivity to the microstructural behaviour of the material, acoustic emission signals often have to be treated with additional signal processing schemes so that the most useful information can be extracted. The detection of tool wear using acoustic emission has been attempted by several researchers. The signal analysis methodologies reported include rms (root-mean-square) measurement, event count analysis and frequency analysis.

10.3.3   Determination of Chip-Forming States using a Linear Discriminant Function Technique with Acoustic Emission

10.3.3.1   Introduction

In automated factories untended machine tools are required to operate for long periods of time with little or no operator supervision. Under these circumstances it is crucial that chip control, i.e. ensuring that discontinuous chips are produced, be maintained. Continuous chips can cause catastrophic failure of tooling, entanglements and damage to the workpiece, hinder the efficient operation of mechanized chip disposal systems, and interfere with automated tool handling equipment. The problem of chip form detection has been characterized by Nakayama and Ogawa [33] as one of the most serious in machine tool automation.

As a result of this challenging problem, many sensing devices have been proposed for the detection of chip form. One promising technique uses the analysis of acoustic emission signals generated by the chips as an indication of chip form. The generation of acoustic emission from the fracture of discontinuous chips was suggested by Dornfeld [34] and has been experimentally confirmed in a number of studies (Iwata and Moriwaki [35] and Dornfeld and Lan [36]). These studies have shown that there is an excellent correlation between the event rate of the AE signal and the frequency of chip formation for discontinuous chip machining of steel and aluminium work materials.

The "burst" AE activity due to discontinuous chip formation is superimposed over the continuous AE due to the plastic deformation and rubbing in the machining process. A methodology for analysis of this AE has been developed by Yee et al. [37], based on the event rate of the root-mean-square (rms) voltage of the AE above a pre-determined threshold, to distinguish between continuous and discontinuous chip forms. It is difficult in some cases to identify the exact chip-forming condition near the transition from a continuous to a discontinuous chip based on any single factor. Further, under differing machining conditions, the definition of a desirable discontinuous chip form may vary. Thus, additional signal analysis capabilities are required.

The linear discriminant function as described in Section 10.2.4.2 is adopted here to help distinguish the cutting states and identify the continuous/discontinuous chip-forming boundary. Discriminant functions, including the event rate of the rms of AE, are obtained through a training process using a set of pattern vectors, the classifications of which are known. The discriminant functions are then applied to the pattern vectors of test cuts. A more efficient discriminator, the threshold logic unit (TLU), is also successfully applied to the classification of chip form. The significance of the event rate of the rms of AE as a component of the pattern vector for characterizing cutting states is evaluated. The criterion for selecting cutting patterns for the training set is also discussed.

10.3.3.2 Classification of Cutting States

Tests were done to distinguish the cutting states into either continuous or discontinuous chip formation – Dornfeld and Pan [20]. The following parameters were selected to characterize a cutting pattern:

1. Cutting speed (x 1000 fpm (m s^{-1}));
2. Feedrate (x 0.001 ipr (mm rev^{-1}));
3. Depth of cut (x 0.01 inch (mm));
4. Event rate of the rms of AE using 1.1 times average rms level as threshold (Hz).

The training patterns were selected so that they represented machining at a variety of cutting speeds, feedrates and depths of cut. Cutting conditions near the boundary of continuous or discontinuous chip formation for this tool geometery and work material were especially included. Seven training pattern vectors were used to determine the weight vectors. The correction increment was selected to be equal to the inverse of the number of corrections, i.e. $1/k$, with a minimum of 0.002. The following weight vectors (i.e. cutting speed weight, feedrate weight, depth of cut weight, ...) were obtained after 48 iterations:

continuous chip (7.50, 10.30, 11.17, –4.70, 4.76)
discontinuous chip (-5.50, –8.30, –9.17, 6.70 –2.76)

Cutting experiments were carried out under various conditions to check the validity of the recognition system for the cutting-state classification. Cutting speeds ranged from 1250 fpm to 1800 fpm (6.35 to 9.24 m s^{-1}); feedrates from 0.002 ipr to 0.020 ipr (0.05 to 0.51 mm rev^{-1}); and depths of cut from 0.040 in. to 0.060 in. (1.02 to 1.52 mm).

An analysis was carried out to identify the significance of each component of the pattern vector in the classification. New pattern vectors were formed by eliminating one component at a time, e.g. using cutting speed, feedrate and depth of cut (without AE event rate), or using cutting speed, depth of cut and event rate of the rms of AE (without feedrate) as parameters each time. The seven new training patterns were used to search for the discriminant functions. Again, these classifiers were applied to the 16 cutting tests. Table 10.1 shows the number of iterations needed to obtain solutions, the weight vectors and the successful rate of the classifications done on different types of pattern vectors. It shows that the classification is not reliable without the inclusion of the event rate of the rms of AE as a pattern component, indicating the effectiveness of the AE technique in distinguishing chip form. Based on the number of iterations needed to obtain weight vectors, more effort is needed to get a set of discriminant functions without feedrate as a component. This verifies that feedrate is the second most important factor in chip-form classification. It was also the most effective parameter for changing the chip form during test cuts.

Table 10.1. Results of LDF classification for cutting tests excluding one component at a time

Item Case	Number of iterations to get weight vector	Discriminant function for	Weight of cutting speed	Weight of feed rate	Weight of depth of cut	Weight of event rate of the rms of AE	Weight of constant	Successful rate of test
No cutting speed in pattern	524	Cont. chip	—	8.62	12.61	-3.85	5.93	100%
		Disc. chip	—	-6.62	-10.61	5.85	-3.97	
No feed-rate in pattern	2200	Cont. chip	8.26	—	2.10	0.23	5.29	100%
		Disc. chip	-6.26	—	-0.10	1.77	-3.29	
No depth of cut in pattern	49	Cont. chip	10.00	7.24	—	-2.24	6.11	100%
		Disc. chip	-8.00	-5.24	—	4.24	-4.11	
No event rate of the rms of AE in pattern	53	Cont. chip	4.65	0.57	-4.43	—	3.02	87.5%
		Disc. chip	-2.65	1.43	2.04	—	-1.02	

However, the absolute value of the weight for each component does not really indicate the relative importance among the parameters. This is due to a variety of reasons including differences in selection of correction increments and in the relative scale factors (instrumentation gains, for example) among the components. The analysis above is a more suitable way to judge the relative importance of each component in the classification.

Since the basic classification needed is between continuous and discontinuous chip form in this study, the performance of the TLU was evaluated to simplify the process. With the five-component pattern vectors, we assumed that the discontinuous chip state generated a positive discriminant, and the continuous chip state generated a negative one. Using the same training set as above with all parameters included, the weight vector was obtained after 69 iterations as:

$$(-6.98, -11.87, -12.28, 7.04, -3.34)$$

The discriminant function was applied again to the cutting tests and the classifications determined were correct every time, as shown in Table 10.1. In the two-category classification, the calculation is straightforward using a TLU because of the single discriminant function involved. This also implies that the classification scheme can be implemented in hardware for real-time applications.

In most machining processes the exact depth of cut is not always known without human observation or other sensor input, whereas cutting speed and feedrate are more easily obtained. Although the depth of cut has some influence on the chip formation, the effect is much less prominent in the pattern than feedrate or event rate of rms of AE, as seen above. An analysis was done to see whether or not depth of cut could be eliminated from the cutting pattern without jeopardizing the performance of the discriminator. First, the two-member discriminant functions and TLU were obtained through the training patterns without depth of cut as a component. Then, classifiers were applied to the new patterns for the 16 cutting tests above, as shown in Table 10.1. The test data indicates that the classification is still highly reliable without the depth of cut as a pattern component. When the difference between two discriminants in the TLU is larger, the classification is more definite and less ambiguous. Examining Table 10.1, it can be seen that the different and absolute values of discriminants decrease for the classification without depth of cut included. It is apparent that depth of cut has some positive contribution for the classification and the new discriminant functions are less effective than the original ones. However, using only those easily obtained parameters (i.e. no depth of cut) to distinguish chip form can work well in on-line monitoring according to the analysis.

10.3.4 Tool Wear Detection using Time Series Analysis of Acoustic Emission

10.3.4.1 Introduction

Acoustic emission is primarily generated by the deformation in the primary shear zone and the sliding friction in the secondary shear zone and tool flank/work surface contact area. As the cutting tool wears, additional frictional action between the tool flank and the workpiece creates additional acoustic emission. The portion of AE that is attributed to friction on the wear-land becomes more important when the flank/

workpiece contact area increases as a result of tool wear. Acoustic emissions generated from shearing and friction exhibit different signal characteristics since the mechanisms by which AE is produced in these processes are fundamentally different. As a result, the signal characteristics of acoustic emission are expected to change when the tool wear-land progresses.

The detection of tool wear using acoustic emission has been attempted by several researchers, as discussed in the previous section. The signal analysis methodologies reported include rms (root-mean-square) measurement, event count analysis and frequency analysis. Using the first method, the rms voltage level of AE is shown to increase in proportion to the amount of flank wear. In the second method, correlation is found between the amount of flank wear and the number of accumulated counts of the AE signal whose amplitude exceeds some pre-selected threshold value. The rms voltage level and event count based on a properly selected threshold are both representative of the signal power content, which increases with wear-land in general. However, the crater wear, because of its influence on the tool effective rake angle, tends to reduce the sensitivity of AE energy content to progressive flank wear. Tool coatings may also influence AE energy. Moreover, cutting parameters will change the AE energy content independently since the AE power released is proportional to the strain rate in the cutting process. As a result of using a single parameter to characterize the signal, power level analysis often fails to distinguish between the change of AE source mechanisms (such as the flank wear and the growth of crater wear on tool rake) and the change of cutting parameters (such as feedrate, cutting speed and depth of cut). Therefore, calibrations of AE power at all combinations of cutting parameters will be necessary actually to implement the monitoring technique. In the third method, spectral characteristics are found to be a function of the tool-wear condition – Emel and Kannatey-Asibu [38]. However, difficulties usually encountered in the frequency analysis are the "signal colouring" effects caused by the propagation media, sensor frequency response and instrumentation system function. Furthermore, the discrete Fourier transform introduces a large number of orthogonal parameters to characterize the AE signal, therefore a parameter selection process has to be implemented before spectral analysis can be practically carried out.

In this example, a time-series analysis technique is used to characterize the acoustic emission rms signal with an autoregressive model. The model parameters are constantly updated according to the rms signal dynamics which are strongly dependent on the AE source mechanism. As a result, these time-varying parameters are expected to contain information about the condition of cutting tool wear. A detailed description of the background and implementation of time-series modelling is given below. The detection scheme illustrated monitors the time series model parameters of the acoustic emission rms signal generated during metal cutting. When the parameter pattern exceeds the pre-set allowable range, a severely worn tool condition is deduced, and tool change should take place. The merits of this technique are that the number of parameters to be monitored is much less than in spectral analysis, and it provides a larger degree of freedom to describe the acoustic emission signal than by just studying its DC power level.

10.3.4.2   Time Series Analysis

It is sometimes possible to derive the mathematical model for a dynamic system based on physical laws which then allows us to calculate the value of some time-varying quantity at any particular time. This type of model would be entirely deterministic. However, very few dynamic systems are totally deterministic because changes arising from unknown or unquantified effects may take place during the process. Thus, it is convenient to construct stochastic models that can describe the dynamics from a probabilistic point of view. In this way the underlying system physics or system characteristics can be studied or ascertained from experimental data.

A time series is generally defined as a sequence of observed data ordered in time (or other variables such as space). In manufacturing process monitoring we often encounter such ordered sets of data corresponding to the output of a force transducer during a machining process, the temperature of a sheet during forming or the vibration amplitude of a grinding wheel. The statistical methodology associated with the analysis of these sequences of data is referred to as time series analysis (Wu and Pandit [39]) and this approach has been applied to the analysis of a wide range of manufacturing processes. Any statistical dependence between the data is seen in the correlation or autocorrelation between successive observations. It is impossible to discuss time series analysis in detail here. We will address one simple form, the autoregressive model, using a stochastic gradient algorithm. This method does not require any prior knowledge of the signal statistical properties and the modelling algorithm is adaptive in the sense that the parameters are updated at every time step, as opposed to batch algorithms which estimate the parameters only after a whole data set has been collected. In the application of the time-series modelling approach to in-process characterization of systems or processes that have fast time-varying dynamics or features, the modelling technique has to be adaptive such that the information contained in the measured data can promptly be used to reflect instantaneous system dynamics or features. Not only is the adaptability needed for real-time analysis but it also reduces the memory size required since the oldest data point is discarded at each time step.

In an Nth-order autoregressive (AR) model for a time series $y(k)$, where k is the discrete time index, $y(k)$ can be predicted (called $\hat{y}(k)$) based on the current N previous values as

$$\hat{y}(k) = n(k) + \sum_{i=1}^{N} a_i(k)y(k-i) \qquad (10.10)$$

where $n(k)$ designates a white noise. In the present case, $y(k)$ is the measured signal at the transducer site and the a_i values are the model parameters.

If the measured data value of $y(k)$ is different from the value predicted by using Equation (10.10), some error will occur. The error signal, $e(k)$, is defined to be the difference between the model-predicted value $\hat{y}(k)$ and the actual sampled value $y(k)$:

$$e(k) = y(k) - \hat{y}(k) \qquad (10.11)$$

The behaviour of output signal $y(k)$ is closely related to the model parameters a_i as well. During a cutting operation, when the acoustic emission signal changes as a result of tool wear progression, the model parameters become time varying and are utilized to track the tool wear. The model parameters are calculated from the measurement of the acoustic emission sensor signal, $y(k)$, using the stochastic gradient algorithm. With the stochastic gradient algorithm, the model parameters are adjusted every time a new data point is sampled. Each adjustment is an effort to minimize the square of the error signal at that instant, $e^2(k)$. The selection of the adaptation gain is critical since it governs both the stability of adaptation and the speed of convergence.

If some of the model parameters do not vary significantly with process state or if the model order is so high that the real-time implementation of signal analysis is difficult, it becomes necessary to ignore some less important model parameters. The importance of a model parameter is evaluated according to its ability to discriminate different process states. A discrimination index associated with each parameter is formulated here to quantitatively describe the relative importance of that model parameter. For any process state specified by A, an "ith parameter mean" can be defined for the ith model parameter as its mean value with respect to time:

$$a_{i,A}^o = \frac{1}{M} \sum_{k=1}^{M} a_{i,A}(k) \qquad (10.12)$$

where M is the number of total adaptation time steps. One natural way to evaluate the capability of a parameter a_i in separating two different states, A and B, is through the observation of its between-class variation $Q_i[A,B]$ defined as:

$$Q_i[A,B] = |a_{i,A}^o - a_{i,B}^o| \qquad (10.13)$$

When the variation of $a_i(k)$ between process states A and B is of the same order as its parameter mean, $Q_i[A,B]$ will decrease with increasing i. That is

$$Q_i[A,B] \geq Q_j[A,B] \qquad i,j = 1-N \text{ and } i \leq j \qquad (10.14)$$

Therefore, $Q_i[A,B]$ alone cannot represent the relative importance of a_i in distinguishing process states. To formulate a better index for the relative importance of model parameters, a within-class variation in the ith parameter for state A is defined as:

$$S_{i,A} = \left[\frac{1}{M} \sum_{k=1}^{M} (a_{i,A}(k) - a_{i,A}^o)^2 \right]^{1/2} \qquad (10.15)$$

Similarly, $S_{i,A}$ decreases with respect to i as the variation of $a_i(k)$ between conditions A and B is on the same order as its parameter mean:

$$S_{i,A} \geq S_{j,A} \qquad i,j = 1-N \text{ and } i \leq j \qquad (10.16)$$

A discrimination index $J_i[A,B]$ between the two conditions A and B based on the ith parameter can then be obtained by normalization of $Q_i[A,B]$ with $S_{i,A}$ and $S_{i,B}$. That is

$$J_i[A,B] = \frac{Q_i[A,B]}{[S_{i,A} S_{i,B}]^{1/2}} \qquad (10.17)$$

The discrimination index defined in this way does not necessarily decrease with i. A greater discrimination index implies that the difference in this specific parameter for two conditions is more pronounced, and that the parameter varies less within either of the conditions. Therefore, the discrimination index, $J_i[A,B]$, is a suitable indication of how successfully the two conditions, A and B, can be separated through the observation of the ith model parameter. The most important parameter is the one that maximizes the discrimination index. Parameter reduction can then be achieved by ignoring the parameters with small discrimination indices.

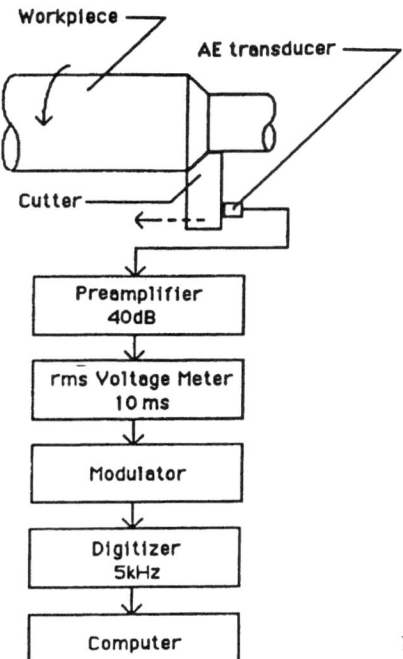

Workpiece

AE transducer

Cutter

Preamplifier 40dB

rms Voltage Meter 10 ms

Modulator

Digitizer 5kHz

Computer

Fig. 10.6. Experimental set-up and signal processing flow diagram.

The rms signal is a function of the AE low-frequency components and the total power contained in the individual bursts of AE generated during the machining. To make the technique sensitive only to changes in the AE source rather than to the change of cutting parameters (rms is proportional to cutting speed), the DC component of the rms voltage is filtered out. Thus the autoregressive model tracks only the dynamic properties but not the mean energy level of the AE signal. In this way, an off-line parameter calibration procedure is needed only once under any cutting condition to map out the allowable range for the parameters, since the parameters will be affected by tool wear but not by the change of cutting velocity, depth of cut, or feedrate.

10.3.4.3 Experimental Evaluation

A series of experiments were conducted to test the performance of the proposed technique, (Liang and Dornfeld [40]). Figure 10.6 shows the experimental set-up and the signal processing schematic. A Kennametal K68 tungsten carbide insert tool was mounted on a tool holder with a 5° rake. The workpieces used were low carbon steel bars of 2 inch (5.08 cm) diameter. Experiments were performed under five different cutting conditions as listed in Table 10.2. Conditions 2 and 3 were the same as condition 1 except for different amounts of flank wear. Conditions 1, 4 and 5 were all conducted using fresh cutting tools but with different cutting parameters. No chatter, built-up edge, or rake face crater was observed during cutting.

Table 10.2. Experimental conditions

No.	Tool condition	Cutting parameters
1	Fresh	1000 rpm, 0.008 ipr (0.203 mmpr) 0.02 in. (0.508 mm) depth of cut
2	0.0078 in. (0.19 mm) flank wear	Same as above
3	0.0197 in. (0.50 mm) flank wear	Same as above
4	Fresh	1000 rpm, 0.008 ipr (0.203 mmpr) 0.03 in. (0.762 mm) depth of cut
5	Fresh	800 rpm, 0.005 ipr (0.127 mmpr) 0.03 in. (0.762 mm) depth of cut

A typical time series of acoustic emission rms signals recorded over 0.4 seconds is shown in Fig. 10.7a. The DC level of the signal, normally about 300 mV, has been removed. Also plotted on this diagram is a broken line which shows the predicted values of the AE rms from a 6th-order autoregressive model with an adaptation gain of 1.0e–5. Some signal dynamics at the beginning of cutting were lost for just a short period while the adaptation began. However, the predicted values agree with the original signal extremely well, as seen from the figure. Figures 10.7b and 10.7c show the time variation of the error signal and the first model parameter respectively. The decreasing trend in error signal with time is clearly seen. After 400 ms of adaptation, the error signal is confined within ±5% of the measured signal. The first model parameter takes about 300 ms to achieve its optimal value. The time then required to track the time-varying optimal value will be much shorter than 300 ms since the initial condition will always be the current value rather than zero.

If an autoregressive signal is modelled by an autoregressive model of sufficiently high order, the error signal should be white since all the parameters are optimized in the sense of minimizing the mean-square error. The error signal $e(k)$ shows no major autocorrelation. The existence of some minor correlated components indicates that: (1) the acoustic emission rms signal generated during the metal cutting operation is not a purely autoregressive process; (2) the model order, 6 in this case, is not sufficiently high, so the truncated higher-order terms have to be incorporated into the error signal; and (3) based on the formulation of the stochastic gradient algorithm, the model parameters are not exact solutions to the problem of minimizing the mean-square error but rather only approach the solution ever closer with time. However, these limitations are not significant here.

The modelling algorithm works on a pre-selected model order and adaptation gain, which are kept constant throughout the whole process. The guideline to the selection of the model order and gain are based on the minimization of the sum of square errors (SSE), defined as

$$\text{SSE} = \sum_{k=1}^{M} e^2(k) \tag{10.18}$$

where M is a large number, 2048 in this study. The correlation between model order, adaptation gain and SSE suggests that a larger adaptation gain in the region of low model order or a higher model order in the region of small adaptation gain will result in a smaller sum of square errors. However, a large adaptation gain associated with a high model order will cause unstable adaptation. An optimal order of 6 and an adaptation gain of 0.6e–5 were selected since they minimized the SSE.

The parameter vectors $[a_1, a_2, a_3, a_4, a_5, a_6]$ exhibit the distinction between parameters for various tool wear conditions and the agreement of parameters under

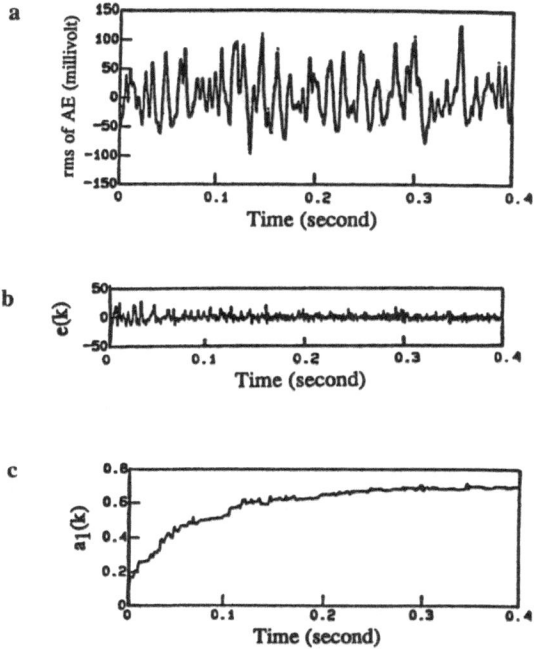

Fig. 10.7. a Root mean square of acoustic emission signal (solid line – experimental data; dots – model predicted data). **b** Error signal. **c** First model parameter of a 6th-order (AR(6)) model.

various cutting speeds, feedrates and depths of cut. The discrimination indices between classes 1 and 2, 1 and 3, 1 and 4, and 1 and 5 were calculated for each model parameter. The results of these tests show that the separation attributed to changes in cutting parameters is very low, whereas the separation resulting from various states of tool wear is comparatively high. All data collected under conditions 1, 2 and 3 are then combined into a single class, 1+2+3, and the discrimination index calculated. The resulting discrimination indices are lower than they are in cases when fresh tool conditions are not combined. This is expected from the fact that the combined class has a higher within-class variation. Based on the discrimination analysis, parameters 1 and 4 can be shown to be the two most important features in terms of detecting and tracking the amount of tool wear while maintaining insensitivity to machining tool condition.

A two-dimensional parameter plane is constructed by plotting the fourth parameter against the first one, as shown in Fig. 10.8. Owing to the non-stationarity of the acoustic emission signal, the model parameters are time-varying and form clusters in the plane. However, these clusters, each representing a different tool wear condition, are clearly separable. As a result, the condition of tool wear can be effectively detected by monitoring the time trajectories of the model parameters. The parameter plane in Fig. 10.8 suggests that a tool change should be called for when $[a_1, a_4]$ goes beyond $[0.46, 0.07]$ since a 0.0197 inch (0.5 mm) wear-land tool is considered damaged for these studies.

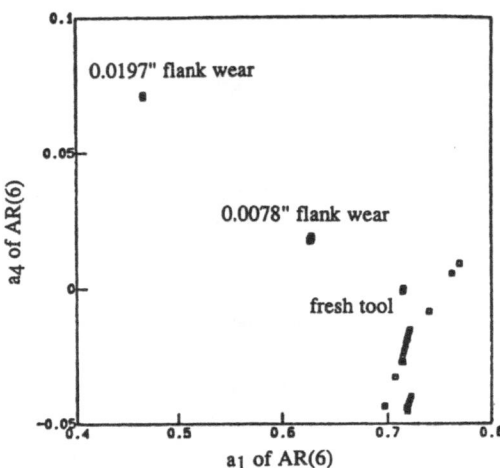

Fig. 10.8. Parameter plane of the first and fourth parameters.

10.3.5 Sensor Integration using Neural Networks for Intelligent Tool Condition Monitoring

10.3.5.1 Introduction

In this example, a technique for intelligent tool condition monitoring which employs information from multiple sensors is presented. This information is integrated via a neural network, a parallel computing architecture which can learn to recognize patterns of sensor information and associate them with decisions on the tool wear state. Initial efforts by Rangwala [11] and Rangwala and Dornfeld [41] demonstrated the feasibility of using neural networks for sensor integration in tool wear monitoring tasks. The networks were used as learning and pattern recognition devices, and were able to associate sensor signal patterns successfully with the appropriate decision on tool wear. Chryssolouris and Domroese [42] performed simulations in order to study the learning capabilities of these networks. Based on the simulation results, they proposed the use of neural networks as the decision-making component in an intelligent tool condition monitoring system. As shown in this example, neural networks are able to filter out noise in the sensor data and this enhances their ability for successful pattern association tasks. These aspects are experimentally evaluated for tool wear monitoring in a turning operation, under a range of machining conditions, to determine both discrete as well as continuous tool wear states.

10.3.5.2 Use of Multiple Sensors

In the application discussed here, it was decided to use AE and cutting force information in order to develop an intelligent tool condition monitoring system. The primary and secondary shear zones are important sources of AE when cutting with a fresh tool. Kannatey-Asibu and Dornfeld [43] have presented a comprehensive analysis for AE generation during orthogonal cutting with a fresh tool. In the

Table 10.3. Effect of velocity and wear on acoustic emission

Condition	Total power	Low freq. power	High freq. power	Mean frequency	Standard dev. of frequency
Increased velocity	+	+	+	−	+
Increased wear	+	+	−	−	−

presence of flank wear, the tool–work interface becomes an additional zone of significant AE generation owing to intense friction between the tool and workpiece surfaces which move past one another at high relative velocities. The effects of tool wear on AE generation in the primary and secondary zones must also be considered. Kobayashi and Thomsen [44] conducted experiments with artificially ground worn tools and concluded that the presence of flank-land did not have an observable effect on the shear angle. This implies that flank wear does not affect the AE characteristics in the primary and secondary shear zones. However, the presence of crater wear affects the effective rake angle of the tool, and this could affect the generation of AE from the primary and secondary shear zones.

The root-mean-square (rms) level of the AE signal (V_{rms}) measures the total power level of the signal and has been found to be sensitive to the degree of flank wear in a turning operation. Experiments conducted by Lan [45] for machining of SAE 4340 steel with carbide tools indicate that V_{rms} increases with machining time owing to increased flank wear. However, in cases where the crater wear is significant, V_{rms} tends to decrease or remains constant. Since the presence of flank wear is expected to increase V_{rms}, Lan concluded that the effect of crater wear is to cause a drop in V_{rms}. The fact that V_{rms} remains constant with increased tool wear as a result of the opposing effects of flank and crater wear makes it difficult to design an AE-based tool wear monitoring system which uses only information on the rms level of the signal.

Emel and Kannatey-Asibu [38] present experimental data which shows that the power spectrum is sensitive to tool wear and process conditions. Results for machining of AISI 1060 with carbide inserts [11] also indicated that the AE power spectrum was sensitive to the level of flank wear and process parameters such as the cutting velocity. Table 10.3, reproduced from [11], summarizes the qualitative effects of tool wear and cutting velocity on the AE power spectrum. The mean frequency divides the total power of the spectrum into two equal parts, whereas standard deviation of frequency indicates the spread in power content around the main frequency. Table 10.3 shows that increases in flank wear and cutting velocity cause an increase in the low-frequency (100–300 kHz) power of the AE signal. Other effects such as feedrate and depth of cut change as well, since chip tangling and chip breakage processes are also expected to affect the AE spectral characteristics. An important consideration for tool wear monitoring is that appropriate schemes should be used in order to identify those spectral regions which show maximum sensitivity to tool wear under a range of process conditions.

The performance of an AE-based tool wear monitoring system can be enhanced by complementing the AE information with information from other sensors mounted on the machine tool (for example, force or power sensors). The magnitude of the cutting force is sensitive to the occurrence of tool wear in a turning operation (Andrews and Tlusty [8]). According to Wright [46] however, cutting force information by itself is inadequate for tool wear detection because its magnitude is also dependent on the cutting velocity. Another problem is that although flank wear tends to increase the cutting force, the accompanying crater wear tends to reduce it, so that the magnitude of the cutting force may not show any sensitivity to tool wear. Cook [47] and Martin *et al.* [48] have shown, along with others, that the cutting force spectrum (which reflects the dynamic characteristics of the cutting force) is sensitive to tool flank wear. Vibrations in the direction of the cutting force are induced because of flank wear in high-frequency regions (> 5 kHz) and lower-frequency regions (< 300 Hz). The former is due to vibrations of the tool holder whereas the latter is attributed to workpiece vibrations. The force spectrum is also dependent on process variables such as cutting velocity, feedrate and oscillations in the shear angle during chip formation.

The AE and cutting force information relate to different effects of tool wear. Acoustic emission is sensitive to the microscopic activities (and the resulting stress waves) related to plastic deformation and friction in the cutting zone. The cutting force spectrum is sensitive to the vibrations induced in the tool and workpiece as a consequence of the effects of flank wear. The advantage of using AE and cutting force sensors is that they provide information relating to microscopic (stress waves) and macroscopic (vibrations) effects of tool wear. This helps provide better signal features to the pattern classifier, allowing a greater reliability in making decisions on the state of tool wear.

10.3.5.3   Experimental Evaluation

To apply the neural network-based machine learning approach discussed in Section 10.2.4.3, a series of machining tests were conducted on a Tree lathe. The experimental set-up is similar to that shown in Fig. 10.6. The work material was case hardened AISI 1060 bars (hardened workpieces were used in order to induce faster tool wear). A Kennametal TPGF-322 insert of grade K68 was used. The bars were of nominal diameter 2 inches (50.8 mm) and 12 inches (305 mm) in length. An acoustic emission transducer (type D9201) was mounted on the tool shank. The tool shank was mounted in a fixture instrumented with a Kistler force dynamometer (type 9251A). The fixture was mounted in the tool turret. The AE sensor output was passed through a preamplifier (with a fixed gain of 40 dB) which high-pass-filters the incoming signal above 50 kHz. The preamplified signal was passed through an amplifier (5 dB gain) and recorded on the video channel of a modified Sony recorder. The cutting force signal was passed through a charge amplifier, the audio channel of the Sony recorder. The process variables were varied in the following range:

Feedrate: 0.002–0.008 ipr (0.05–0.20 mm rev^{-1})
Depth: 0.01–0.03 inch (0.25–0.75 mm)
Velocity: 278–556 sfpm (85–170 m min^{-1})

No signals were collected while machining the hardened layer (approximately 1.5–2 mm thick) of the workpiece. Signals were collected only when the workpiece diameter was 45 mm (1.75 inches) diameter or less. The tool flank-land was measured using an optical comparator. The procedure used was to measure the flank-land width after every two passes through the soft section of the bar and after every pass through the hardened layer. The tool wear was recorded at flank wear levels of 0.1 mm (0.004 inch), 0.125 mm (0.005 inch), 0.25 mm (0.01 inch), 0.5 mm (0.02 inch) and 0.75 mm (0.03 inch). Signals collected between these wear levels were ascribed to the wear value at the lower end of the interval. For example, signals collected between 0 and 0.1 mm flank wear were assumed to be generated during cutting with a fresh tool. Between wear levels of 0.25 mm and 0.5 mm, no signals were recorded, although cutting proceeded. Signals collected during cutting below 0.25 mm flank wear were assumed to belong to fresh tool cutting, whereas signals associated with a flank wear level of 0.5 mm were assumed to belong to worn tool category.

During post-processing, the signals recorded on video tape were played back, filtered and digitized on a HP waveform recorder. The digitized AE signals had a record length of 1024 points, sampled at 5 MHz, and the digitized force signals were of record length 512 points, sampled at 1 KHz. The sampled AE and force records were synchronized as closely as possible using the tape counter number as a reference. A total of 65 samples of fresh tool cutting and 58 samples of worn tool were collected for the purposes of training and testing. The force time-domain record is of length 512 (sampled at 1 kHz) and the AE time-domain record length is 1024 (sampled at 5 MHz). Using a Fast Fourier Transform (FFT) program yields the power spectrum representations of the time-domain records. Consider the power spectrum as a vector whose components are the signal power at various discrete frequencies. The cutting force spectrum is of dimension 256 (256 discrete frequencies with a resolution of 2 Hz) and the AE spectrum is of dimension 512 (512 discrete frequencies with a 5 kHz resolution). Combining the AE and cutting force spectra yields a vector of dimension 768, each component of the vector representing the signal power at a discrete frequency in either the cutting force or the AE signal. This vector is referred to as the measurement vector.

Although valuable information may be contained in the entire measurement vector, from practical considerations, only a few of these components can be used for training and pattern association purposes. In this example, 30 measurement vectors (equally divided between fresh and worn tool states) corresponding to various machining conditions were used to estimate S_w and S_b. The final d features were selected using the Sequential Forward Search (SFS) algorithm, developed by Whitney [49]. The algorithm works as follows: out of the D features in the measurement vector, select the one feature which maximizes J. Call this feature ζ_1. Next, pair each of the remaining $D-1$ features with ζ_1 and compute J according to Equation (10.3) for each of these pairs. The pair which maximizes J is selected as the new feature set. This procedure is repeated until all d features have been selected. It should be pointed out that the SFS algorithm is sub-optimal in the sense that it does not guarantee that the best feature set is selected. However, it is computationally viable and yields feature sets whose signal-to-noise ratio is reasonably close to the optimal case [24]. It was decided that 30 samples (equally divided between fresh and worn tool cutting) would be used for purposes of training. According to the criterion in Equation (10.2), the dimension of the feature vector was chosen to be 6. Three feature sets were selected using the procedure discussed above. Set 1 features were selected using a combined measurement vector of the AE and force spectra. Application of the SFS algorithm yielded 4 AE and 2 force features in this case. Set 2 features were selected by considering the AE and force spectra as separate measurement vectors and selecting three features from each. In this case, the feature vector consists of three AE and three force features. Set 3 features were selected considering only the AE spectrum as the measurement vector. The selected features for each set and the corresponding J values as each additional feature is added are shown in Tables 10.4 to 10.6. It is seen that adding new features

Table 10.4. Set 1 features

No.	Feature	J^*
1	AE (88 kHz)	0.89
2	Force (43 Hz)	3.22
3	AE (161 kHz)	5.33
4	Force (10 Hz)	8.76
5	AE (122 kHz)	12.76
6	AE (151 kHz)	20.10

* Based on 30 training samples

Table 10.5. Set 2 features

No.	AE features		Force features	
	Feature	J^*	Feature	J^*
1	88 kHz	0.89	10 Hz	0.56
2	504 kHz	1.79	33 Hz	1.17
3	493 kHz	2.54	39 Hz	2.00

* Based on 30 training samples

Table 10.6. Set 3 features

No.	Feature	J^*
1	AE (88 kHz)	0.89
2	AE (504 kHz)	1.79
3	AE (493 kHz)	2.54
4	AE (68 kHz)	3.31
5	AE (293 kHz)	4.7
6	AE (449 kHz)	5.86

* Based on 30 training samples

increases J, since additional features cause the distance between the mean values of fresh and worn tool clusters to increase. Note that these values of J are based on estimates of S_w and S_b, computed using 30 samples of the measurement vector.

Since measurement vectors corresponding to different process conditions are used, the selected features should show a low sensitivity to changes in process variables. However, some sensitivity may still be present, so that it makes sense to use the process conditions as additional features. Information such as the feedrate and cutting velocity is easily available from the machine controller and can be used as additional features. Depth of cut information is difficult to obtain on-line, and is not used as a feature. Thus a change in sensor feature values arising from a change in the depth of cut has the effect of noise corrupting the sensor feature value. Therefore a total of 8 features in the feature vector (input to the network) were used: 6 features from force and AE (2 force and 4 AE), 1 feature corresponding to cutting velocity and 1 feature corresponding to feedrate.

Various design parameters affect the performance of the tool wear monitoring system. These include factors such as the number of training samples, the number

Table 10.7. Description of cutting conditions

Machining condition	Tool status	Feed (ipr)	Vel. (fpm)	Depth of cut (in.)	Flank-land (in.)	Flank-land (mm)
F1	Fresh	0.008	450	0.03	0.000	0.00
W1	Worn	0.008	450	0.02	0.020	0.50
W2	Worn	0.008	450	0.02	0.030	0.75
W3	Worn	0.005	370	0.02	0.020	0.50
W4	Worn	0.008	278	0.02	0.020	0.50
W5	Worn	0.004	370	0.02	0.030	0.75
W6	Worn	0.002	278	0.02	0.030	0.75
W7	Worn	0.005	556	0.01	0.030	0.75

of sensors and sensor features used and the structure of the neural network. The effect of these factors on the performance of the tool wear monitoring system is evaluated next and a design which yields the best performance is shown. Although the design detail will vary according to the exact situation, the methodology presented here will yield practical design strategies for implementing on-line process monitoring systems.

10.3.5.4 Discussion of Results

The perceptron training algorithm was used to train a linear classifier, using set 1 features. In order to see the effects of sensor fusion, perceptrons were also trained using set 2 and set 3 features. Unless otherwise specified, all training sets contained 30 samples, equally divided between fresh and worn tool cutting. The trained classifiers were then tested on the remaining 93 samples (of which 50 corresponded to fresh tool cutting and the remainder to worn tool cutting). Sets 1 and 2 yield comparable performance (88% and 87% classification success rates, respectively) whereas the performance of set 3 is lower (80% success rate). This indicates that feature sets composed of multiple sensor information provide better classification performance. To see the effect of process variables on the sensor features, the relative influence of changes in depth of cut and tool wear on the first AE feature (88 kHz) was studied. We consider one group of samples (group F1) corresponding to fresh tool cutting and seven groups of samples (groups W1–W7) corresponding to worn tool cutting. The exact machining conditions for each sample group is shown in Table 10.7. The value of J for each fresh-worn sample group was calculated. It was seen that in all 7 cases, the AE feature has zero sensitivity to tool wear (a J value of 0.001 or less is considered as zero sensitivity). Group F1 samples correspond to a higher depth of cut than do group W1–W7. Thus, simply looking at the AE feature would cause increases in depth of cut to be mistaken for a "tool worn" condition, and hence lead to classification errors. The sensitivity of the force feature (10 Hz) to tool wear was seen to be reasonably high, regardless of changes in depth of cut. Including the force feature would, in this case, reduce classification errors. Of course, additional AE features may also provide sensitivity to tool wear under these operational conditions (in fact, this is the motivation for using a large number of features). However, as a larger number of features from one sensor are used, the information provided by them becomes highly correlated, so that a loss of sensitivity to tool wear in one feature is accompanied by a loss of sensitivity in other features of the same sensor signal.

Feature sets 1, 2 and 3 were used to train and test multi-layered neural networks. Sensor feature values were normalized in order to prevent saturation of the sigmoid function. This was done by dividing the feature value by its maximum value in the training set. Neural networks with a single hidden layer and three nodes in the hidden layer were used. The number of nodes in the input layer is equal to the number of input features, which in the current case is eight (six sensor features and two process features). The output layer contains a single node, whose output level associates the current input pattern with a decision on tool wear. This yields a network with an 8–3–1 structure. During the training phase, the target state of the output node was fixed at 0.01 for fresh tool patterns and 0.99 for worn tool patterns. The minimization of the error was achieved by using conjugate gradient optimization, which adjusts the weights and thresholds in a direction which minimizes the error. The weights and thresholds were initialized to uniformly distributed random values lying between −1 and 1. During the testing stage, a pattern presented at the input layer was associated with a "fresh tool" decision if the output node activity was between 0 and 0.5, otherwise the pattern was associated with a worn tool state.

A threshold value is associated with all nodes in the input, hidden and output layers. The role of the threshold is to compare the weighted sum of inputs to the node and generate an output which depends on the difference between this sum and the node threshold. The threshold value thus acts as a filter for incoming signals. Theoretically, the learning procedure maps worn tool samples to an output node activity of 1 whereas fresh tool samples are associated with zero activity of the output node, so that the signal-to-noise ratio (measured by the value of the discriminant index, J) of the output node feature approaches infinity. In practice, this does not occur because the output node error does not converge exactly to zero, however, since it is sufficiently close to zero, each filtering step in the network is expected to suppress noise and increase the signal-to-noise ratio as the signal propagates through the network. In order to observe the noise suppression behaviour discussed above, the trained 8–3–1 network (set 1 features) was presented with all 123 samples and the variations of the J value of the features at each layer were calculated. The value of J increases at every layer, implying higher separability of the fresh and worn tool patterns at the decision layer.

The classification success rate of the 8–3–1 network used above, based on 93 test samples, was found to be 94%. For comparison purposes, a perceptron network using the same normalized input features as the neural network was also trained and tested. The classification success rate in this case was found to be 88% (similar to that obtained with non-normalized features). The superior performance of neural networks is attributed to their noise suppression abilities. It is of interest to see how the classification performance is affected when the number of features presented to the input layer is varied. To observe this, the least significant feature in the input feature vector was dropped sequentially. Process features were always included as part of the feature vector. The modified vectors were used to train and test the performance of a perceptron and a neural network. An increase in the number of features used at the input layer generally improves classification performance. For a given number of features, the performance of the neural network is seen to be superior to that of the perceptron. The effect of not using the process features is that the performance is adversely affected when the process features are not included. In this case, it is possible that changes in process conditions are wrongly considered to be due to changes in tool wear state, so that the classification error rate increases. In this case also, the neural network performs better than the perceptron network. One aspect that is not fully explained is that although increase in the number of sensor features generally improves performance, in some cases, the use of an additional

Table 10.8. Performance of neural network

Features	J value (sensor features)	J value (input node features)	J value (hidden node features)	Success rate (%)
Set 1	1.02	2.3	5.5	94
Set 2	1.70	3.6	11.5	97
Set 3	0.72	1.4	2.1	84

feature causes a deterioration in the performance level. A possible reason may be that the training and test data statistics for that particular feature may be very different, so that the trained network may not be able to respond correctly to test data. Training anomalies may also contribute to this sort of behaviour.

The performance of the 8–3–1 networks trained and tested using sets 1, 2 and 3 features are compared in Table 10.8. Set 2 features show the best performance (97%), and this correlates with the fact that the signal-to-noise ratio of the sensor features as well as the internal features is highest for set 2. On the other hand, set 3 features (which are very noisy) show poorer performance (84%). The data presented in Table 10.8 suggests that the noise suppression ability of a neural network is a function of the amount of noise present in the sensor features. Noisy sensor features cause the learning ability of the network to degrade, consequently the internal features are noisy. On the other hand, noise suppression is enhanced when the incoming sensor features have a high signal-to-noise ratio. In order to see if the initial values of the learning parameters or network structure have significant effects, set 3 features were used to train different networks with various initial values. In all cases, the classification performance ranged from 80% to 85%. In some cases, the signal-to-noise ratio of the hidden node features was found to be lower than that of the input node features. This probably occurs because noise in the sensor features degrades the learning ability of the network. Consequently, even though feature extraction on training set samples increases the signal-to-noise ratio, performance deteriorates when new samples are propagated through the network.

10.3.5.5   Continuous Tool Condition Monitoring

The traditional application of neural networks in process monitoring has concentrated on rather discrete estimations of the process state. That is, determining whether or not a condition (or state) exists. This would indicate whether a cutting tool was new or worn, and the user must determine the threshold for the worn state. A more interesting application is the real-time estimation of the state over a period of operation. This would avoid the problem of only classifying the process in terms of two extreme states and indicate the gradual change in process conditions. Some progress was made towards this goal by Moriwaki and Hino [50] who demonstrated the use of statistical features of the AE signal to characterize correctly three states of tool wear – unworn, medium wear and worn.

The work described here is based on Wang and Dornfeld [51] and demonstrates how a multi-layered feed forward neural network can be used to integrate information from multiple parameters of the AE signal to monitor the gradual increase of tool wear during a turning operation. The root-mean-square AE signal was measured during the machining operation and this signal was passed through a signal processing block in which the mean rms AE, the integrated rms AE, the skew and kurtosis of an assumed ß distribution for the rms AE and the autoregressive

Tool Wear Level

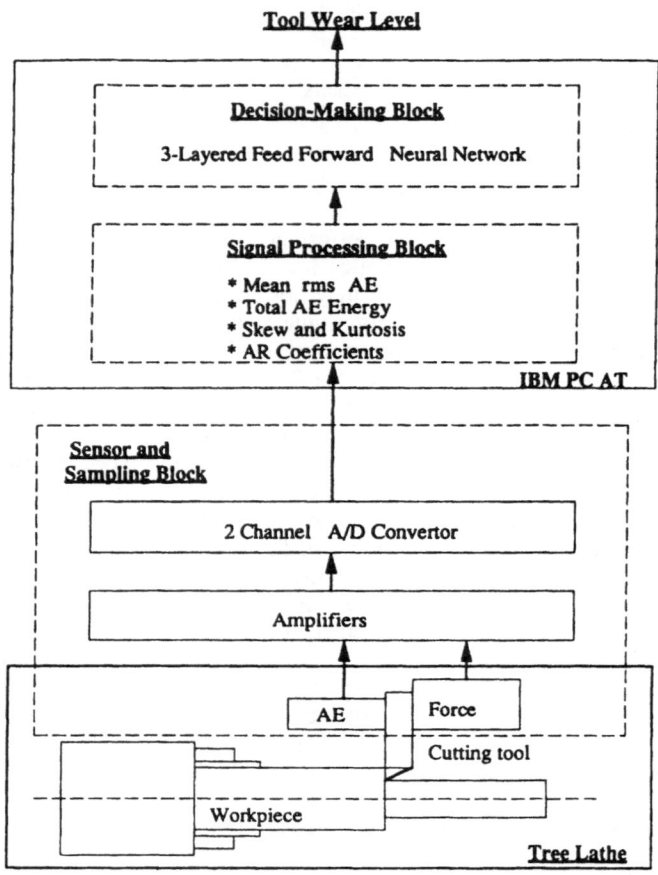

Fig. 10.9. Tool wear monitoring system.

(AR) coefficients of the rms were calculated. These parameters were then presented to a previously trained neural network (three layer with five nodes in the hidden layer and six inputs as described) to identify the increase of tool wear. To learn the necessary input/output mapping for tool wear monitoring, a group of training samples is presented to the network and the weights and thresholds of the network are adjusted accordingly using back propagation (BP).

The effectiveness of the monitoring strategy was evaluated in machining tests on a lathe with a single point carbide insert tool; the tests showed a continuous growth in flank wear-land. This wear-land was measured at intervals (approximately every 90 seconds of machining). The monitoring system is shown schematically in Fig. 10.9. The system consists of a Tree lathe, a sensor and signal sampling block, a signal processing block and a decision-making block. The AE sensor output was passed through a preamplifier and then an rms meter. The raw rms AE signal was sampled by an IBM/PC AT computer through a 12-bit A/D converter and saved as an ASCII file for further processing. The sampling frequency for the raw AE rms

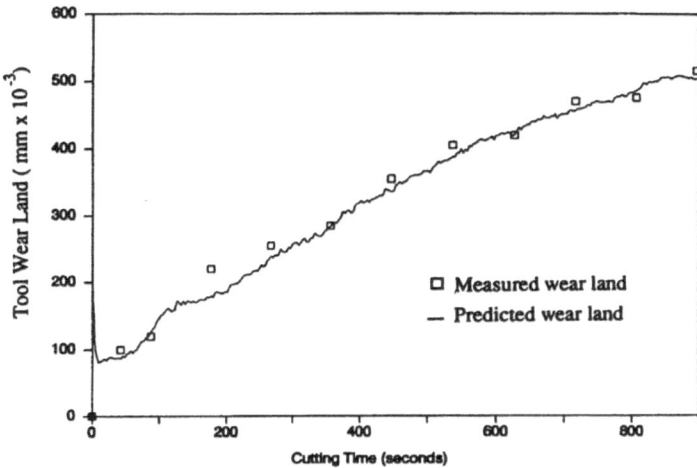

Fig. 10.10. Measured and predicted wear-land using neural network output.

was 1 kHz. A total of 300 samples of raw rms AE signal were collected at regular intervals (3 seconds) for training and testing. Each rms AE sample consisted of 1024 points. The machining condition was chosen so that there was no crater wear and the flank wear was clearly seen. The raw AE rms signal was then passed through the signal processing block which calculated all the AE parameters needed by the decision-making block. All AE parameters (mean rms AE, total AE energy, skew, kurtosis and AR coefficients) were then fed to the pre-trained neural network to make the final decision on tool wear level. Figure 10.10 shows typical results of measured and predicted wear-land growth for a machining test in which the neural network was trained on wear and sensor data from various other machining tests. In these cases the system was able to predict the progressive flank wear accurately for the unknown tool with an error of approximately 10%.

10.4 New Signal Processing Techniques for Automatic Supervision

Research on sensor fusion using neural networks for tool condition monitoring is evolving to include the results of earlier feature extraction studies and to take advantage of unsupervised learning algorithms for "self-learning" networks. Choi *et al.* [52] and Dornfeld [53] detail the results of research utilizing the parameters of an autoregressive model of acoustic emission and tool force as inputs to a neural network of similar structure to that described here. Figure 10.11 from [53] shows a schematic of the system. The unsupervised learning capabilities of adaptive resonance networks have potential for tool condition monitoring as well because of their "self-learning" ability. Drawbacks of the BP algorithm described here are that, in some cases, the training process used may restrict the range of conditions over which the system performs successfully. With adaptive resonance networks, features of sensor measurement vectors that are significantly different from those

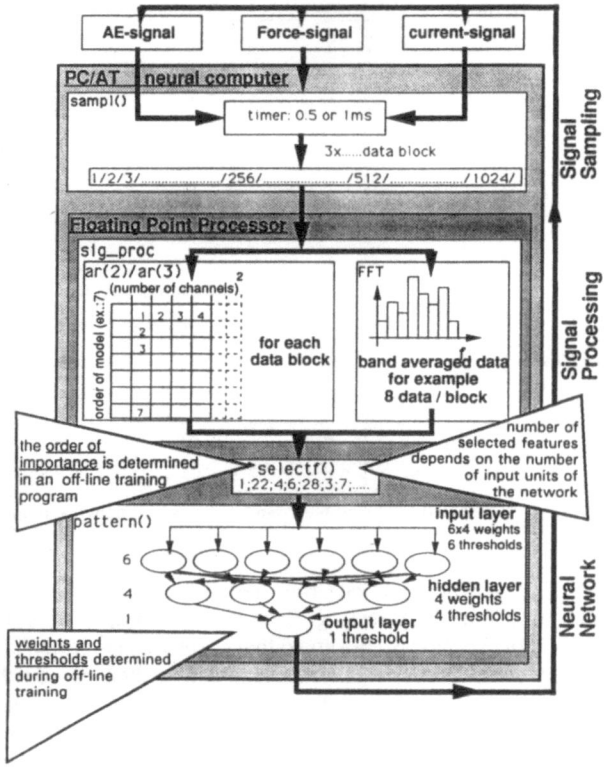

Fig. 10.11. Schematic of operation of monitoring system.

previously "learned" during operation will trigger the establishment of a new category or process state. Thus, the dependence on "pre-trained" networks is removed and the opportunity to update the network in-process exists.

A method of recognizing tool wear states in a turning operation from the integrated information of cutting force and acoustic emission signals is presented in [54]. The approach, which employs gradient adaptive lattice analysis and pattern recognition techniques, is fast and yields accurate recognition of tool wear states under a wide range of cutting conditions. The gradient adaptive lattice algorithm is applied to compute the autoregressive as well as the partial correlation coefficients recursively for both cutting force and AE signals with high computational efficiency. Those coefficients are chosen to characterize the sensing signals and used as the feature inputs for pattern recognition. A stepwise search procedure was applied to select the most useful features for the recognition process. Both unsupervised learning technique (fuzzy C-means algorithm) and supervised learning technique (linear discriminant analysis) are used in the pattern recognition analysis. Characteristics of each method are discussed and performance in recognizing the states of the cutting tools (fresh or worn) is evaluated. The results show that this approach was successfully applied to distinguish a fresh from a worn tool state with a high percentage (over 90%) of correct classification.

Finally, Langari *et al.* [55] have proposed, as a basis for intelligent control of metal cutting, a methodology for integration of sensing and control. In particular, they present a strategy for the modelling and design of a robust control algorithm in the H infinity (H_∞) framework that addresses some of the intricate issues associated with control of machining processes, such as structural resonance modes and modelling inaccuracies. This strategy, when coupled with sensor feedback from the process, here acoustic emission from a finishing operation, promises more robust control and insensitivity to external disturbances.

10.5   Conclusion

In a recent report, Dornfeld [4], research on multi-sensor or intelligent sensor systems was reviewed. A wide variety of sensors are proposed for inclusion in these systems, including acoustic emission, force and torque, low-frequency structural vibration, motor current and power, surface profilometry, computer vision, displacement probes, acceleration, temperature and audio microphones. Researchers have developed and evaluated these systems for tool condition monitoring, process control, process monitoring and process optimization – often attempting combinations of tasks using signal analysis in both the time and frequency domain. Some of these systems have provision for feedrate or speed control in response to sensor feedback, and employ a variety of advanced signal processing methodologies for feature extraction and decision-making (e.g. zero crossing, skew, kurtosis, fast Fourier transformation, linear and quadratic discriminant functions, finite element analysis, group method of data handling, band-pass filtering and autoregressive moving average time series modelling.) The work in this area will only increase as researchers realize the potential of sensor systems with advanced signal processing for built-in feature extraction and decision-making.

Advances in both software and hardware for signal processing for feature extraction and sensor fusion, combined with improvements in existing sensor technology and new micro-sensor technology, will make efficient and reliable tool condition monitoring based on multi-sensor systems more realizeable. Unfortunately, it is not possible to review here all of the critical aspects of this area – especially the hardware features. The reader is referred to the *Handbook of Intelligent Sensors for Industrial Automation* [56], especially Chapter 16 on sensor fusion, for a discussion of hardware issues in sensor fusion. These software and hardware advances will be applicable to a wide range of manufacturing supervision problems, making true automatic supervision possible.

It should be evident that the potential for process monitoring and process investigation in manufacturing is immense. The use of intelligent sensing systems coupled with advanced process controllers offers the possibilities of increased reliability and productivity of complex manufacturing processes. This is equally true of "traditional" processes and materials as well as the new processes and materials recently developed. Although one prefers to be able to describe the behaviour and performance of processes quantitatively, the availability of sensors can aid in the development of those quantitative process models as well as enhance their application in process control. It is expected that these intelligent sensors will see increasing application in developing and realizing *smart* tools and machines, *flexible* machine tools with increased ranges of performance, and *flexible* tooling and fixtures which support the recent thrust in the development of so-called agile manufacturing systems.

Acknowledgements

Research on intelligent sensors in the Laboratory for Manufacturing Automation at the University of California at Berkeley has been supported by the Industrial Affiliates of the Laboratory as well as the National Science Foundation, Ford Motor Company, Fanuc Ltd, Citroen Automobiles, Renault Automobiles, and Alcoa.

References

1. Ayres RU. Complexity, reliability, and design: manufacturing implications. Manufacturing Review 1988; 1(1): 27
2. Tönshoff HK, Wulfsberg JP, Kals HJJ, Koenig W. Developments and trends in monitoring and control of machining processes. Annals of the CIRP 1988; 32(2): 611–622
3. Ito Y. Conceptualizing the future factory system. Manufacturing Review 1988; 1(4): 252–258
4. Dornfeld DA. Monitoring of machining process – literature review. Presented at the CIRP STC "C" meeting, Paris, January 1991
5. Hoshi T. Automatic tool failure monitoring in drilling and thread tapping. In: Proc 3rd Intl Conf Automatic Supervision, Monitoring and Adaptive Control in Manufacturing, AC'90, CIRP, Rydzyna, Poland, 1990 pp 41–58
6. Schaffer G. Sensors: the ears and eyes of CIM. American Machininst, Special Report No 765, July 1983
7. Birla S. Sensors for adaptive control and machine diagnostics. In: Technology of Machine Tools – Machine Tool Task Force Report, Machine Tool Controls, vol 4, Miskell RV ed, Lawrence Livermore National Laboratory,UCRL-52960, 1980, pp 7.12-1–7.12-70
8. Andrews GC, Tlusty J. A critical review of sensors for unmanned machining. Annals of the CIRP 1983; 32(2): 563–572
9. Iwata K. Sensing technologies for improving the machine tool function. In: Proc 3rd Intl Machine Tool Engineers Conf, JMTBA, Tokyo, Japan, 1988, pp 87–109
10. Matsushima K, Sata T. Development of the intelligent machine tool. J Faculty of Engineering, University of Tokyo (B) 1980; 35(3): 395–405
11. Rangwala S. Machining process characterization and intelligent tool condition monitoring using acoustic emission signal analysis. PhD thesis, Department of Mechanical Engineering, University of California, Berkeley, California, 1988
12. Ruokangas CC, Blank MS, Martin JF, Schoenwald JS. Integration of multiple sensors to provide flexible control strategies. In: Proc IEEE International Conference on Robotics and Automation, IEEE, San Francisco, California, 1986, pp 1947–1953
13. Yamasaki H. Intelligent sensing technology. J Japan Society of Precision Engineering 1989; 55(9): 14–19 (in Japanese)
14. Dornfeld DA. Acoustic emission process monitoring for untended manufacturing. In: Proc Japan–USA Symp Flexible Automation, JAACE, Osaka, Japan, July 1986
15. Chiu SL, Morley DJ, Martin JF. Sensor data fusion on a parallel processor. In: Proc 1987 IEEE International Conf Robotics and Automation, IEEE, Raleigh, North Carolina, 1987, pp 1629–1633
16. Dornfeld DA. Intelligent sensors for monitoring untended manufacturing processes. In: Proc 2nd International Machine Tool Research Forum, NMTBA, Chicago, Illinois, September 1987
17. Middlehock S, Hoogerwerf, AC. IEEE Transducer '85 Digest 1985; pp 2–7
18. Ahmed N, Rao KK. Orthogonal transforms for digital signal processing. Springer, New York, 1975
19. Sata T, Matsushima K, Nagakura T, Kono E. Learning and recognition of the cutting states by spectrum analysis. Annals of the CIRP 1973; 22(1): 41–42
20. Dornfeld DA, Pan CS. Determination of chip forming states using a linear discriminant function technique with acoustic emission. In Proc 13th North American Manufacturing Research Conference, SME, University of California, Berkeley, May 1985, pp 285–303
21. Emel E, Kannatey-Asibu Jr E. Characterization of tool wear and breakage by pattern recognition analysis of acoustic emission signals. In: Proc 14th North American Manufacturing Research Conference, SME, University of Minnesota, Minneapolis, 1986, pp 266–272
22. Balakrishnan P, Trabelsi H, Kannatey-Asibu Jr E, Emel E. A sensor fusion approach to cutting tool monitoring. In: Advances in Manufacturing Systems Integration and Processes, Proc 15th NSF

Conf Production Research and Technology, SME, University of California, Berkeley, California 1989, pp 101–108

23. Nilsson NJ. Learning machines – foundations of trainable pattern-classifying systems. McGraw-Hill, New York, 1965
24. Devijver PA, Kittler J. Pattern recognition – a statistical approach. Prentice-Hall, Englewood Cliffs, New Jersey, 1988
25. Hinton G, Fahlman S. Connectionist architectures for artificial intelligence. IEEE Computer 1987, pp 100–109
26. Hopfield J. Neural networks and physical systems with emergent collective computational abilities. In: Proc National Academy of Sciences 1982; 79: 2544–2588
27. Hopfield JJ. Neurons with graded response have collective computation properties like those of two-state neurons. In: Proc National Academy of Sciences 1984; 81: 3088–3092
28. Hopfield JJ, Tank D. Computing with neural circuits: a model. Science 1986; 233: 625–633
29. Tank D, Hopfield JJ. Simple neural optimization networks: an A/D converter, signal decision circuit and a linear programming circuit. IEEE Trans Circuits and Systems 1986; CAS-33(5): 533–541
30. Rumelhart D, McClelland J. Parallel distributed processing, vol 1. MIT Press, Cambridge, Massachusetts, 1986
31. Le Cun Y. A learning procedure for assymetric threshold networks. In: Proc Cognitiva, Paris, 1985
32. Dornfeld DA. Application of acoustic emission techniques in manufacturing. In: Proc 11th International Acoustic Emission Symposium, JSNDI, Fukuoka, Japan, 1992, pp 1–15
33. Nakayama K, Ogawa M. Basic rules on the form of chip in metal cutting. Annals of the CIRP 1978; 27(1): 17–21
34. Dornfeld DA. Acoustic emission and metalworking – survey of potential and examples of applications. In: Proc 8th North American Metalworking Research Conference, SME, University of Missouri, Rolla, Missouri, 1981, pp 207–213
35. Iwata K, Moriwaki T. Cutting state identification and in-process tool wear sensing by acoustic emission analysis. Bull Japan Society of Precision Engineering 1978; 12: 213–215
36. Dornfeld DA, Lan MS. Chip form detection using acoustic emission. In: Proc 11th North American Manufacturing Research Conference, SME, University of Wisconsin, Madison, Wisconsin, 1983, pp 386–389
37. Yee KW, Blomquist D, Dornfeld DA, Pan CS. An acoustic emission chip-form monitor for single point turning. In: Proc 26th Intl Machine Tool Design and Research Conf, University of Manchester, 1986
38. Emel E, Kannatey-Asibu E. Tool failure monitoring in turning by pattern recognition analysis and AE signals. Trans ASME J Engineering Industry 1988; 110(2): 137–145
39. Wu SM, Pandit SM. Time series and systems analysis with applications, Wiley, New York, 1983
40. Liang S, Dornfeld DA. Tool wear analysis using time series analysis of acoustic emission. Trans ASME J Engineering in Industry 1989; 111(3): 199–205
41. Rangwala S, Dornfeld D. Integration of sensors via neural networks for detection of tool wear states. In: Proc Winter Annual Meeting of the ASME, PED 25, 1987, pp 109–120
42. Chryssolouris G, Domroese M. Sensor integration for tool wear estimation in machining. In: Proc Winter Annual Meeting of the ASME, Symp Sensors and Controls for Manufacturing, 1988, pp 115–123
43. Kannatey-Asibu E, Dornfeld DA. Quantitative relationships for acoustic emission from orthogonal metal cutting. Trans ASME J Engineering for Industry 1981; 103: 330–340
44. Kobayashi S, Thomsen E. The role of friction in metal cutting. Trans ASME Engineering for Industry 1960; 82: 324–332
45. Lan M. Investigation of tool wear, fracture and chip formation in metal cutting using acoustic emission. PhD dissertation, University of California at Berkeley, Department of Mechanical Engineering, 1983
46. Wright PK. Physical models of tool wear for adaptive control in flexible machining cells. In: Computer integrated manufacturing, PED-Vol 8, ASME (Martinez MR, Leu MC eds), New York, 1983, pp 19–31
47. Cook N. Self excited vibrations in metal cutting. Trans ASME J Engineering for Industry 1959; 183–186
48. Martin P, Mutels B, Drapier J. Influence of lathe tool wear on the vibrations sustained in cutting. In: Proc 15th MTDR Conference, Birmingham, Alabama, 1974, pp 251–257
49. Whitney A. A direct method of non-parametric measurement selection. IEEE Trans Computers 1971; 20: 110–1103

50. Moriwaki T, Hino R. Application of neural networks to AE signal for automatic detection of cutting tool life. In: Proc 1st Intl Conf Automation Technology, CSIAAI, Taipei, Taiwan, 1990, pp 811–818
51. Wang ZX, Dornfeld DA. In-process tool wear monitoring using neural networks. In: Proc Japan–USA Symp Flexible Automation, ASME, San Francisco, California, 1992, pp 263–270
52. Choi GS. Wang ZX, Dornfeld DA, Tsujino K. Development of an intelligent on-line tool wear monitoring system for turning operations. In: Proc Japan–USA Symp Flexible Automation, ISCIE, Kyoto, Japan, 1990
53. Dornfeld DA. Neural network sensor fusion for tool condition monitoring. Annals of the CIRP 1990; 39(1)
54. Jiaa CL, Dornfeld DA. Detection of tool wear using gradient adaptive lattice and pattern recognition analysis. Mechanical Systems and Signal Processing 1992; 6(2): 97–120
55. Langari R, Dornfeld DA, Wang ZX. Intelligent sensing and control in metal cutting. In: Proc 4th Intl Symp Robotics and Manufacturing (ISRAM '92), ASME, Sante Fe, New Mexico, 1992, pp 749–755
56. Zuech N. ed Handbook of intelligent sensors for industrial automation. Addison-Wesley, Reading, Massachusetts, 1992, Chapter 16

11 Geometrical Adaptive Control of Manufacturing Systems

J. Peklenik

11.1 Introduction

In the manufacture of discrete products in the mechanical, electrical and other industries there are three major objectives (1) good performance and reliability of the product; (2) competitive price related to comparable products on the market; and (3) high-quality products.

We shall deal in this chapter with the problems of how to achieve and maintain the quality of mechanical parts during production. Particular attention will be paid to automatic control of the geometrical parameters of parts produced in flexible manufacturing systems (FMS) or cells (FMC).

The quality of the geometrical parameters of parts is basically related to the following features: dimensional deviations, shape deviations and surface roughness. Information on these geometrical parameters is essential, and it is usually specified by the designer. The selection of allowable tolerances is closely related to the functional requirements of the product.

The quality specification must be carried out under conditions where fabrication of the parts may take place on various machine tools, as a function of the production time. There are many uncontrollable disturbances affecting the stability of the geometrical quality of parts [1]. These may have their origin in the dynamics of the machine tools and/or the manufacturing process itself or in the surface interface dynamics.

Our discussion will focus, therefore, on the conceptual development of the geometrical quality model of the part, which will result in a definition of dimensional accuracy and its transfer function (DATF). The stability of the geometrical quality is a time-dependent function, describing the state of geometrical accuracy of the parts manufactured as a function of time t.

In order to maintain the stability of the geometrical quality, it is necessary to develop a geometrical adaptive control (GAC), enabling the machining system to accomplish the objectives $\{Z^*\}$ related to the geometrical quality of the parts, as specified by the designer.

There are three areas that the designer of the GAC systems has to consider. Firstly, the means and methods of measuring geometrical deviations from the theoretical parameters. Secondly, the compensation systems enabling, for example, the change of relative position between the tool and the surface generated in order to maintain the tolerances, or change of the feedrate, cutting speed or some other machining parameter contributing to the stability of the quality parameters of the part.

11.2   Quality of Machined Parts

The quality of a mechanical part is a complex characteristic related to its geometrical and material properties.

The designer specifies the quality of the mechanical components in relation to their functional performance required during the exploitation of a product, of which a particular component is a part. In addition, the useful life of the product depends strongly on the quality of its parts.

These specifications represent the reference values, expressing the objectives $\{Z^*\}$, set by the designer, which have to be accomplished during fabrication of the parts.

Because of the various influences that affect both the material and its geometrical properties during the manufacturing process, the actual geometrical and physical values characterizing the quality of the parts may not be realized within the acceptable limits. Therefore there is a need for on-line quality identification and consequently adaptive control.

There are two sets of potential problems in developing on-line quality control during automated production:

1. The control of those material properties affected by the conditions under which the manufacturing process is implemented; and
2. The control of the geometrical parameters of the parts, in relation to the length and shape tolerances, the surface roughness, deviations from perpendicularity, parallelism etc.

We will now briefly discuss both of these areas of quality control.

11.2.1   Quality of Material Properties

The material of a blank represents an input into a manufacturing process in which a part is generated to within the designer specifications. Physical and chemical properties affect to a great extent not only the functional behaviour of the mechanical part under load, but also its fabrication by various manufacturing processes. There are two considerations to be taken into account:

1. The material properties affect the tool wear, which is one of the major parameters in optimizing the machining conditions. On the other hand to a great extent the tool wear influences the geometrical quality of the parts.
2. The forces and the temperatures at the machining interface cause structural damage to the surface of the part, as Fig 11.1 indicates.

Material properties can be changed during fabrication only in certain cases (e.g. heat treatment, plastic deformation of the surface, preliminary treatment before painting etc.). For this reason these properties cannot be the object of control in machining. However, it is necessary to establish effective quality control of the incoming materials from suppliers, in order to establish whether their properties are within the required specifications, and to minimize the uncontrollable effects on the

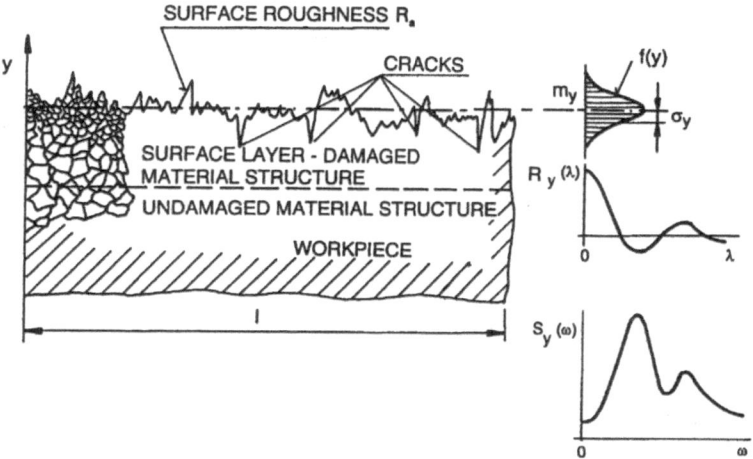

Fig. 11.1. Characteristics of surface integrity.

machining processes related to dimensional and shape accuracy, as well on surface integrity.

In this chapter we will not discuss the problem of quality control of materials, despite the fact that the management and control of the material inputs required for automatic production are significant if a successful CIM operation is to be achieved.

11.2.2 Quality of the Geometry and Surface of the Parts

The term "quality of a mechanical part" describes a complex characteristic defined by the following geometrical parameters:

Actual dimensional tolerances T,

Actual shape tolerances T_s (e.g. roundness, straightness etc.),

Actual surface roughness R_a

describing the quality to be accomplished in the final fabrication process of a part. We shall deal therefore with finished parts, intended solely for the assembly of a product. The intermediate stages in the fabrication processes are also important. However, the methods of control are basically the same. The significance of the final quality is evident and is directly correlated to the functional behaviour of the parts under load.

In addition to these parameters of basic quality, there are a number of others assessing such factors as parallelism, perpendicularity etc.

We shall consider problems related to the automatic control of the three basic parameters, T, T_s and R_a, limiting the discussion to these questions only.

The designer specifies the quality of mechanical parts in relation to the functional performance of a device. These specifications of quality represent the reference values which must be achieved and maintained during the fabrication of the parts.

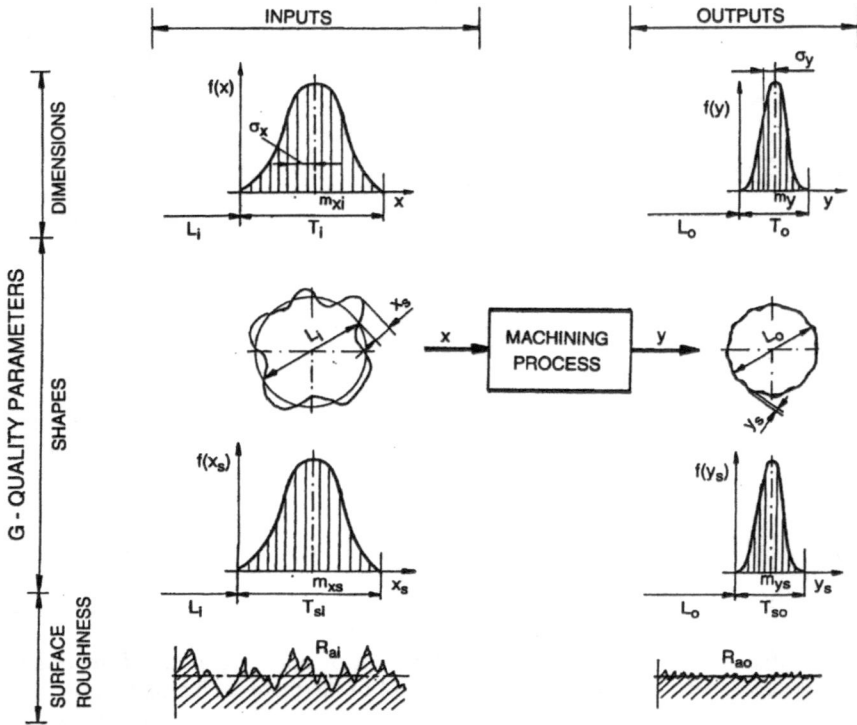

Fig. 11.2. Transformation of the parameters of geometrical quality in machining.

11.3 Transformation of the Parameters of Geometric Quality during Machining

The machining process, responsible for the generation of parts with desired shapes, dimensions and surface quality, can be considered as a material removal process in which a part (or parts) of low quality is transformed into a part (or parts) of higher quality. Figure 11.2 illustrates this schematically. For instance, a set of turned parts with a diameter of L_i yields an input tolerance of T_i. The parts will be ground in order to achieve an output diameter of L_o with a tolerance of T_o, which is tighter than the input value T_i and must be within the reference value T_R specified by the designer. Similar analysis applies also to errors in shape. The roundness deviations x_s of cylindrical parts at input are distributed within the tolerance limits T_{si} and transformed into tighter shape tolerance T_{so} at output. Therefore the accuracy of shape has been improved by machining, as required by the specification. Also the surface roughness at input, specified as a value of R_{ai}, is a parameter of quality that is transformed during a machining operation into a smoother surface R_{ao} at output.

WITH REFERENCE TOLERANCES T_L and T_D

THEORETICAL GEOMETRY

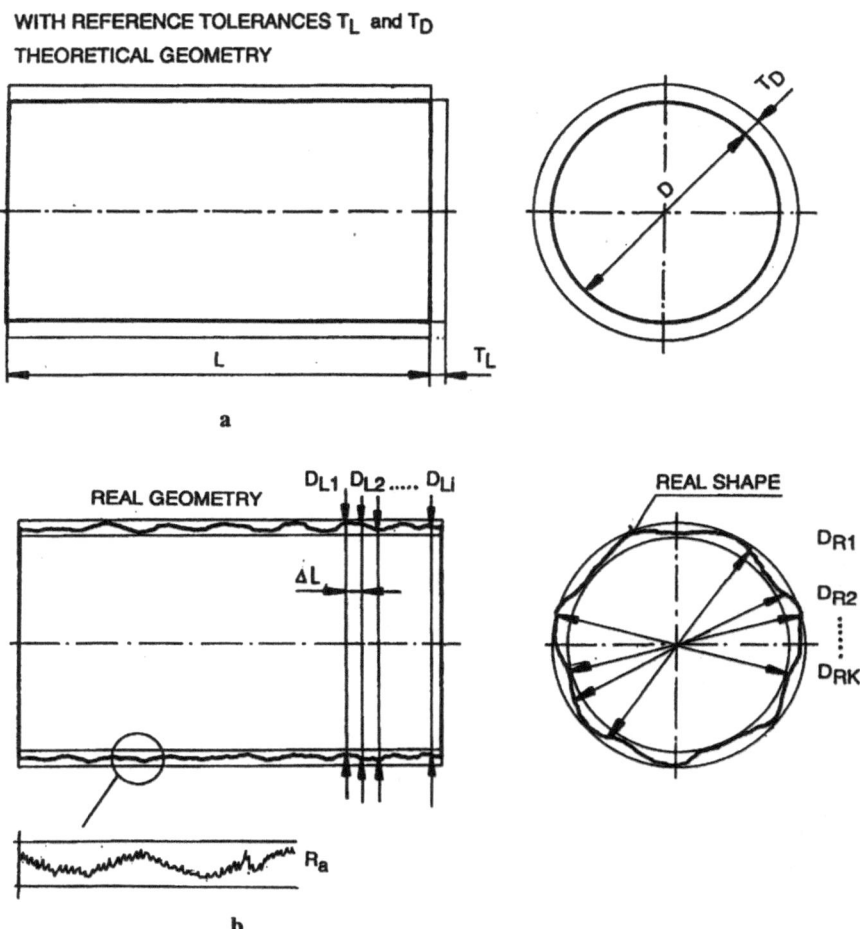

Fig. 11.3. Theoretical and actual geometry of a part with reference tolerances.

11.4 Quantitative Description of the Geometrical Quality of a Part

11.4.1 Dimensional Characteristics

A part (e.g. a rotational component) is defined by the designer as a cylinder of diameter D and length L, with tolerances T_D and T_L and the surface roughness R_a – see Fig. 11.3a. If these specifications are met, the part is accepted as "good" and may be assembled in the product.

Analysing the actual geometry of the part, we may find that the straightness as well as the roundness of the cylinder fluctuate (Fig. 11.3b). In addition, the surface roughness, specified by the designer as a value R_a, represents variations of micro-

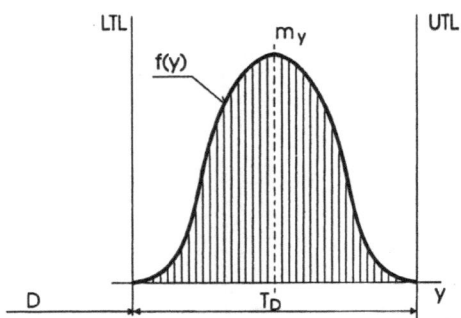

Fig. 11.4. Gaussian probability density distribution $f(y)$ containing the dimensional deviations y_{Li} and y_{Ri}.

geometry in the plane normal to the rotational axis, ignoring the frequency content of the surface profile along the cylinder.

Now let us consider the geometrical aspects. Figure 11.3b shows measurements of the diameters D_{Li} made by positioning measuring instruments parallel to the rotational axis. The positioning distance ΔL is small. The deviations y_{Li} are calculated as the difference between the measured and the theoretical diameter y_{Li} = $D_{Li} - D$. Repeating the process by measuring the diameters D_{Ri} (Fig. 11.3b), we obtain the dimensional errors y_{Ri} as the difference between the measured diameters D_{Ri} and the theoretical value D, hence $y_{Ri} = D_{Ri} - D$.

The probability density $f(y)$, as shown in Fig. 11.4, consists of the deviations y_{Li} = y_i and $y_{Ri} = y_i$, and can be assumed as Gaussian.

In the case of the distribution $f(y)$ falling within the tolerance limits (UTL and LTL), the part is considered as "acceptable" if the surface roughness is also within the allowed R_a specifications. It is, however, quite possible that a part within the limits of $f(y)$ lies outside the tolerance field.

Existing geometrical adaptive control systems usually measure a dimension of a part only once. The selection of the measuring position is arbitrary. Therefore, it is highly probable that the deviation y will fall within the tolerance limits. On the other hand, it is quite possible, because of various errors in shape, that the measurements will not show the effect of these deviations on the dimensional errors of a part. Hence a part, deemed "acceptable" may in reality be just the opposite.

Therefore assessment of geometrical quality, as shown in Fig. 11.3, must include both aspects – dimensional and shape deviations – as an integral dimensional characteristic of a part, as described by the probability density $f(y)$.

To extend the discussion, we will assume that measurements of the length deviations y yield only one probability density function $f_p(y)$ for the population, corresponding to a normal distribution for the estimated parameters of the population n. Then the mean value

$$m_{yp} = \frac{1}{n} \sum_{i=1}^{n} y_i \qquad (11.1)$$

and the variance

Table 11.1 ISO system of tolerances

Quality IT*		from 1 up to 3	over 3 up to 6	over 6 up to 10	over 10 up to 18	over 18 up to 30	over 30 up to 50	over 50 up to 80	over 80 up to 120	over 120 up to 180	over 180 up to 250	over 250 up to 315	over 315 up to 400	over 400 up to 500	Tolerances (in i units)
1	IT 1	0.8	1	1	1.2	1.5	1.5	2	2.5	3.5	4.5	6	7	8	—
2	IT 2	1.2	1.5	1.5	2	2.5	2.5	3	4	5	7	8	9	10	—
3	IT 3	2	2.5	2.5	3	4	4	5	6	8	10	12	13	15	—
4	IT 4	3	4	4	5	6	7	8	10	12	14	16	18	20	—
5	IT 5	4	5	6	8	9	11	13	15	18	20	23	25	27	≈7
6	IT 6	6	8	9	11	13	16	19	22	25	29	32	36	40	10
7	IT 7	10	12	15	18	21	25	30	35	40	46	52	57	63	16
8	IT 8	14	18	22	27	33	39	46	54	63	72	81	89	97	25
9	IT 9	25	30	36	43	52	62	74	87	100	115	130	140	155	40
10	IT 10	40	48	58	70	84	100	120	140	160	185	210	230	250	64
11	IT 11	60	75	90	110	130	160	190	220	250	290	320	360	400	100
12	IT 12	100	120	150	180	210	250	300	350	400	460	520	570	630	160
13	IT 13	140	180	220	270	330	390	460	540	630	720	810	890	970	250
14	IT 14	250	300	360	430	520	620	740	870	1000	1150	1300	1400	1550	400
15	IT 15	400	480	580	700	840	1000	1200	1400	1600	1850	2100	2300	2500	640
16	IT 16	600	750	900	1100	1300	1600	1900	2200	2500	2900	3200	3600	4000	1000

*Basic tolerances.
Values in the table are in μm = 0.001 mm.
Tolerance unit $i = 0.45 \sqrt[3]{L} + 0.001 L$ [μm], where L = length in mm.

$$D_{yp} = \frac{1}{n-1} \sum_{i=1}^{n} (y_i - m_{yp})^2 \qquad (11.2)$$

and the standard deviation

$$\sigma_{yp} = \sqrt{D_{yp}}$$

The estimator m_y determines the position of $f(y)$ in the tolerance field, and the standard deviation σ_y is a measure of the variations of y around m_y.

The designer selects the dimensional tolerances from reference tables, representing the quality classes as a function of length L.

Table 11.1 presents the ISO system of basic tolerances, to be used for the development of accuracy criteria in order to estimate the geometrical quality of the parts, objectively.

11.4.2 Shape Characteristics

The roundness, straightness and/or flatness of parts are usual specifications related to the required shape characteristics. Figure 11.5 shows the general form of shape deviations appearing on rotational parts (see Fig. 11.3) and/or on flat surfaces.

The assessment of the actual shape can be done in two ways [2]:

1. Estimate the deviations $y(L)$ in the direction perpendicular to the theoretical shape, which results in the determination of m_y and σ_y (see Fig. 11.5a).

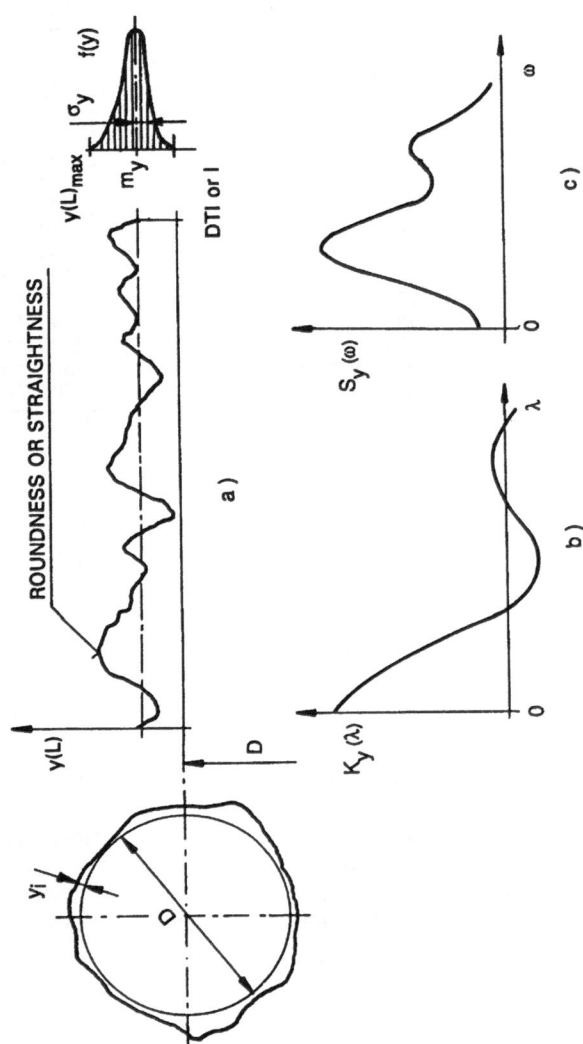

Fig. 11.5. Shape characteristics of a part: **a** actual shape $y(L)$ against $f(y)$; **b** correlation function $K_y(\lambda)$; **c** power spectrum $S_y(\omega)$ of $y(L)$.

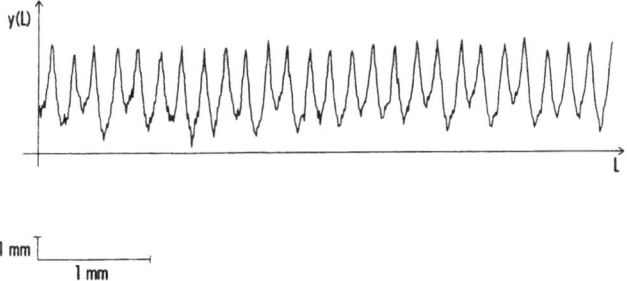

y(L)

0.1 mm

1 mm

Fig. 11.6. Surface roughness of a part.

2. Estimate the autocorrelation function $K_y(\lambda)$ of $y(L)$ (Fig 11.5b) and/or its Fourier transform, the power spectrum $S_y(\omega)$ (see Fig 11.5c). These two estimates make it possible to assess the frequency content of the actual shape $y(L)$.

The estimate of autocorrelation function is expressed as follows:

$$K_y(\lambda) = \frac{1}{n-1} \sum_{i=1}^{n} (y_k - m_y)(y_r - m_y) \tag{11.3}$$

and the power spectrum by:

$$S_y(\omega) = \frac{2}{\pi} \int_0^{\infty} k_y(\lambda) \cos \omega \lambda \, d\lambda \tag{11.4}$$

The estimate of the normalized autocorrelation function is

$$k_y(\lambda) = \frac{K_y(\lambda)}{D_y} \tag{11.5}$$

The designer usually specifies only the maximum allowable shape deviations $y_{max}(L)$, based more or less on experience (Fig. 11.5a).

The estimates of m_y and/or σ_y and the estimates of $K_y(\lambda)$ and/or $S_y(\omega)$ are at present not used under fabrication conditions. There are no standardized reference data available for errors of shape. A detailed account of these matters is given in [3].

11.4.3 Surface Characteristics

The surface quality also represents a complex characteristic that defines both the micro-geometrical configuration of the surface and the state of the surface layer. The structure of this layer (Fig. 11.1) is in close correlation with the changes in material properties that result from the mechanical and thermal loads imposed during machining.

Estimates of the surface roughness (Fig. 11.6) of parts after fabrication are made using R_a (or CLA) and R_z. Computations of the correlation function $K_y(\lambda)$ and/or the

Table 11.2 Roughness classes [4]

Roughness class	Grade	Roughness parameters (μm)		Basic length L (mm)
		R_a	R_z	
1	—	—	320–160	
2	—	—	160– 80	80
3	—	—	80– 40	
4	—	—	40– 20	2.5
5	—	—	20– 10	
6	a	2.5 –2.0	—	
	b	2.0 –1.6	—	
	c	1.6 –1.25	—	
7	a	1.25–1.0	—	
	b	1.00–0.80	—	0.8
	c	0.80–0.63	—	
8	a	0.63–0.50	—	
	b	0.50–0.40	—	
	c	0.40–0.32	—	
9	a	0.32 –0.25	—	
	b	0.25 –0.20	—	
	c	0.20 –0.16	—	
10	a	0.160–0.125	—	
	b	0.125–0.100	—	
	c	0.100–0.080	—	
11	a	0.080–0.063	—	0.25
	b	0.063–0.050	—	
	c	0.050–0.040	—	
12	a	0.040–0.032	—	
	b	0.032–0.025	—	
	c	0.025–0.020	—	
13	a	—	0.100–0.080	
	b	—	0.080–0.063	
	c	—	0.063–0.050	0.03
14	a	—	0.050–0.040	
	b	—	0.040–0.032	
	c	—	0.032–0.025	

power spectrum $S_y(\omega)$ according to Equations (11.3) and (11.4) have not yet been used in a production environment.

The designer specifies the R_a (CLA) and/or R_z values according to the functional requirements of the parts, using reference values for surface roughness. Table 11.2 shows one of the standardized sets of R values [4] which designers can use for surface roughness reference specification.

Assessment of structural changes in the surface layer is a very complex physical and metallographical problem and can only be accomplished by using highly sophisticated instrumentation.

We shall now describe how geometrical adaptive quality control can be used to maintain the quality of the parts in automatic production. The dimensional and shape characteristics are assessed by the distribution function $f(y)$ and described by its estimators m_y and σ_y.

11.5 Modelling of Dimensional Accuracy

The development of the control of dimensional parameters (length, shape, surface roughness) requires clarification of the term "dimensional accuracy". This is necessary owing to the nature of the problem. Figure 11.2 indicates that the system approach has been used in dealing with this problem. That means that transformation of the parameters of geometrical quality of a part entering a machining process produces at the output a higher-dimensional accuracy than the part had at the input. We will limit accuracy modelling, however, to dimensional aspects only.

The first comprehensive model of dimensional accuracy was proposed by J. Peklenik in 1961. It is based on the ISO tolerance system (see Table 11.1) as the reference structure, and the estimation of the statistical parameters m_y and σ_y of the probability density function $f(y)$ which is affected by systematic and random influences (tool wear, thermal deformations etc.).

The Index of Dimensional Accuracy (IDA) is defined as the ratio between the tolerance T prescribed by the designer and the generated position and width of the probability density $f(y)$ for dimensional deviations y in this tolerance field. The length equivalent φ of the ISO tolerance system and the quality class IT in which the parts must be fabricated [5, 6] has also to be taken into account.

This basic relation is evident from Fig. 11.7. It may be expressed as

$$\text{IDA} = \frac{T}{k\,\sigma_y + c} \cdot \frac{1}{\varphi^n} \qquad (11.6)$$

$$c = (\sigma_{yp} - \sigma_y)\,\frac{k}{2} + m_y \qquad (11.7)$$

Substituting Equation (11.7) into Equation (11.6) we obtain

$$\text{IDA} = \frac{2T}{2m_y + k\,(\sigma_{yp} + \sigma_y)} \cdot \frac{1}{\varphi^n} \qquad (11.8)$$

where:
m_y = mean value of the dimensional errors y;
σ_y = standard deviation of the dimensional errors y;
σ_{yp} = standard deviation of the population;
k = coefficient characterizing the type of distribution;
  = 6 for normal distribution, which comprises 99.998% of all y values (in practical cases a value of $k = 4$ can be considered, where approximately 99% of all y values are included);
  = 4.4 for Maxwell distribution;
φ = 1.3 (length equivalent of the ISO tolerance system)
n = IT − 1 [exponent (IT1, IT2, ..., IT16)].
IT = quality class number of the ISO tolerance system.

Investigations into GAC systems have indicated that Equation (11.6) must include the possibility of variations in dimensional accuracy, which occur owing to systematic errors caused by tool wear, thermal deformations of the machining tool, etc. This means that the mean value m_y of the real distribution $f(y)$ is oscillating

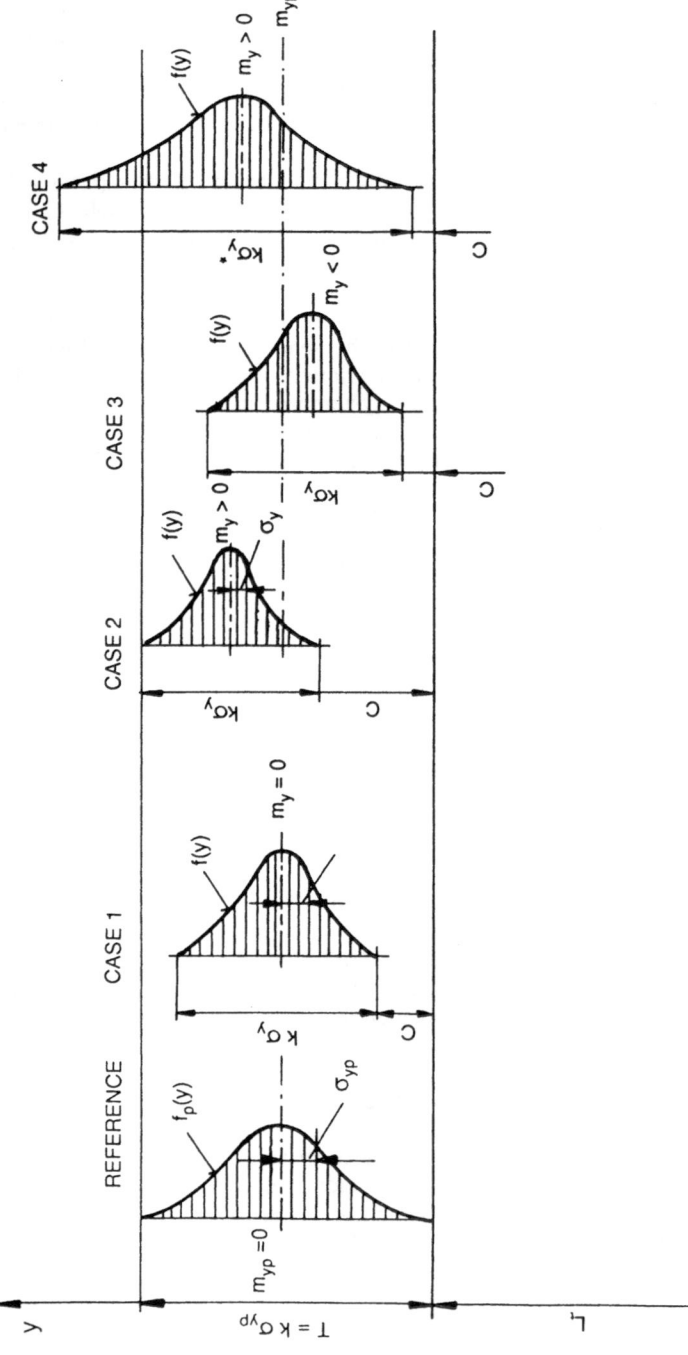

Fig. 11.7. Statistical relations in modelling the IDA.

about the centre of the tolerance field, thus influencing the dimensional quality of the part. It is also true that dimensional control in a GAC system can only be realized by compensating for systematic deviations, described by m_y, as a function of the machining time.

Figure 11.7 illustrates the relationships between the statistical parameters for improved modelling of the index of dimensional accuracy (IDA). The designer selects a tolerance T for a theoretical dimension L of a part. Between L and T there is a relationship defined in the ISO basic tolerancing system: parts with increasing length L and same quality class (IT1, IT2, ..., IT16) exhibit an increase of tolerance T. This property is assessed by the length equivalent φ^n, where the exponent is calculated as $n = IT-1$, and the value $\varphi = 1.3$ represents the steps between the length classes [5].

On the other hand, the part or parts to be machined must satisfy the tolerance specification given by the designer. However, the distribution density function $f_p(y)$ of the entire population of y exerts a real influence on the systematic and random errors of the machining system and the machining interface. The position of the $f_p(y)$ function in the tolerance field determined by the mean value m_y, as well as the range $k\sigma_{yp}$, should satisfy the condition

$$T = k\sigma_{yp} \tag{11.9}$$

where the centre line of the tolerance field T corresponds to the value $m_{yp} = 0$ and the width of $f_p(y)$, expressed as a product $k\sigma_{yp}$, satisfies equation (11.9).

The Index of Dimensional Accuracy (IDA) is in this case a function of the quality class only, depending on the value φ^n, and can be expressed as

$$IDA = \frac{2T}{2m_y + k\sigma_{yp}} \cdot \frac{1}{\varphi^n} = \frac{1}{\varphi^n} \tag{11.10}$$

The machining of parts produces various situations that affect the IDA values. There are four characteristic cases (Fig. 11.7):

CASE 1: The probability density $f(y)$ is centred in the middle of the tolerance T,
$\qquad m_y = 0$ and $k\sigma_y < T$
CASE 2: $m_y > 0$ and $k\sigma_y < T$
CASE 3: $m_y < 0$ and $k\sigma_y < T$
CASE 4: $m_y <> 0$ and $k\sigma_y^* > T$

Substituting the values for m_y and $k\sigma_y$ or $k\sigma_y^*$ into Equation (11.8) and calculating the IDA for these cases, one can establish that the proposed assessment of the dimensional accuracy by IDA provides objective and sensitive accuracy criteria. This is also a very basic parameter for controlling the dimensional accuracy in automated machining systems.

11.6 Dimensional Accuracy Transfer Function

A machining process responsible for generating a part with a desired shape and dimension is, as already explained, considered as a transformation process, in which a less accurate part (or a set of parts) is transformed into a more accurate one, with a new dimension.

This statement is valid both for the accuracy of the geometrical shape and for the surface roughnesses. Figure 11.2 illustrates this transfer process for various quality characteristics. It is related to the transformation of an input quality of parts defined by IDA_i into an output quality of parts described by IDA_o required by the designer.

The transformation of lower quality into higher quality by a machining process can be described by a Dimensional Accuracy Transfer Function (DATF). We will limit our explanation to the dimensional accuracy only.

The DATF is defined as the ratio between the Index of Dimensional Accuracy (IDA) at the output against that at the input [7, 8]:

$$\text{DATF} = \frac{IDA_o}{IDA_i} \tag{11.11}$$

It is known that the parameters m_y and σ_y of the distribution $f(y)$ vary as a function of time $|m_y(t), \sigma_y(t)|$. For this reason we can extend the concept of the IDA into the function of dimensional accuracy by introducing the time t in which disturbances, such as tool wear, thermal deflections, vibrations etc, that are changing with time affect the dimensional accuracy of parts.

The IDA_o for the output is

$$\text{IDA}_o(t) = \frac{2T_o}{2m_y(t) + k(\sigma_{yp}(t) + \sigma_y(t))} \cdot \frac{1}{\varphi^{n_o}} \tag{11.12}$$

The index of dimensional accuracy IDA_i at the input is a constant value, which is evident from Fig. 11.2. A set of parts with length $L_i > L_o$ varies dimensions from $L_{i\,min}$ to $L_{i\,max}$, so that $T_i = L_{i\,max} - L_{i\,min} = k\sigma_x$. Considering the length L_{max} and the size of T_i we determine from the ISO tolerance system (Table 11.1) the quality class IT to which the combination L and T belongs. The IDA_i is in this case

$$\text{IDA}_i = \frac{T_i}{k\sigma_x + c} \cdot \frac{1}{\varphi^{n_i}} \tag{11.13}$$

Because $c = 0$ and $k\sigma_x = T_i$ it follows that

$$\text{IDA}_i = \frac{1}{\varphi^{n_i}} = \text{constant} \tag{11.14}$$

The dimensional accuracy transfer function DATF is expressed by Equation (11.11). Considering this equation, we can write

$$\text{DATF} = \frac{IDA_o}{IDA_i} = \frac{\dfrac{2T_o}{2m_y(t) + k(\sigma_{yp}(t) + \sigma_y(t))} \cdot \dfrac{1}{\varphi^{n_o}}}{\dfrac{1}{\varphi^{n_i}}} \tag{11.15}$$

If $\varphi^{n_i}/\varphi^{n_o} = K_t$, Equation (11.15) may be written as

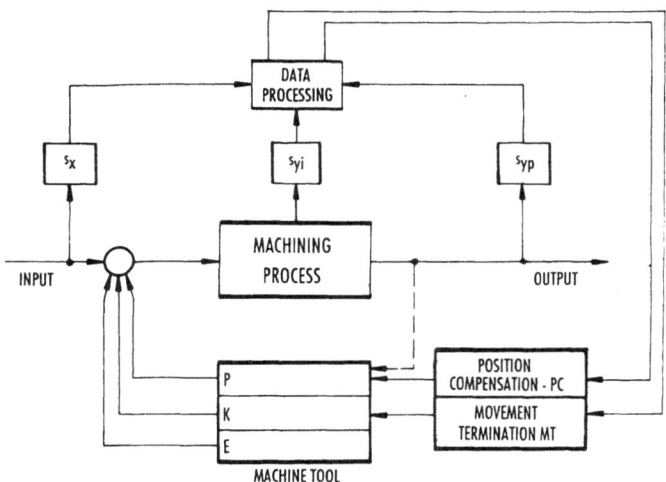

Fig. 11.8. Structure of the geometrical adaptive control system.

$$\text{DATF} = \frac{2K_t T_o}{2m_y(t) + k(\sigma_{yp}(t) + \sigma_y(t))}$$

(11.16)

The numerator represents the amplification of the transformation process from the lower to the higher quality. The denominator characterizes the dynamics of the geometrical quality transformation process and is used for adaptive quality control.

11.7   Geometrical Adaptive Control (GAC) System

In order to implement quality control in a Flexible Machining Cell (FMC) or a Flexible Machining System (FMS) it is necessary to develop the basic structure of the GAC system. The system approach in this field has been established by Peklenik [9] in analysing the existing systems of in-process control.

Figure 11.8 shows the cybernetic structure of a GAC system. The machining process transforming the input quality of a part (or parts) into a higher output quality and maintaining the output quality at the level required, is implemented by a machine tool. This device consists of three subsystems: the P (positioning), K (kinematics) and E (energy) subsystems. The parameters defining the relative position of the tool and the blank, the cutting speed and the feedrates are set in accordance with the specifications of the part at the input and the output. This is the configuration of the basic machining system.

The development of a GAC system requires the implementation of a closed-control system which incorporates three or four typical system components:

Measuring sensors for collecting the information;

Data processing unit;

Position compensation device; and/or

Movement stopping device.

Figure 11.8 shows the interconnections of these components in a GAC system. The classification of GAC systems proposed in [5, 6] is in accordance with the time taken for the information to be collected by a sensor. There are three possibilities of measurement:

Pre-process by a sensor S_x;

In-process by a sensor S_{yi}; and

Post-process by a sensor S_{yp}.

Figure 11.9 shows a schematic representation of these cases, with the corresponding accuracy diagrams.

The GAC pre-process system measures with S_x the input dimension of the part and the position of the tool, in accordance with the anticipated elastic deformation of the system due to the variation of cutting forces resulting from different depths of cuts.

The GAC in-process system measures with S_{yi} a dimension of the part during machining, and interrupts the movement of the tool by means of a movement-termination device (MT) when this dimension corresponds to the required specification.

The GAC post-process system measures with S_{yp} a dimension of the part. Should this dimension fall outside the tolerance limits, the position-compensating device (PC) introduces a correction Δy to compensate the systematic displacement of m_y caused by tool wear, thermal deflections etc. These random inputs affect the dimensional errors y, because the sensor S_{MT}, which terminates the movement of the tool, cannot estimate these systematic influences. When the spindle reaches a given position, the sensor S_{MT} terminates the spindle movement by means of a movement-termination device (MT). The compensation Δy is initiated by the sensor S_{yp} when the upper or lower tolerance limit (or the control limit) responds, so giving a signal for the position-compensation device (PC) to come into operation.

The information processing is provided by a microprocessor calculating on-line the statistical parameters as expressed in the DATF – equation (11.16).

Based on this generic model of GAC systems, various geometrical adaptive control systems have been developed for manufacturing. There are, of course, many devices for measuring, compensating and terminating movement, and these can all be applied in the design of GAC systems. Several reviews of these possibilities are available in the literature [10–14].

11.8 Conclusions

The automatic quality control of machining cells and systems is of paramount importance for successful operation of CIM. To control these systems, it is necessary to develop a dimensional accuracy transfer function (DATF) in order to obtain appropriate parameters for the implementation of geometrical adaptive control. Following on from this a generic structural model for geometrical adaptive control (GAC) emerged.

Future developments in manufacturing technology will ultimately require a much more intensive usage of the GAC system, particularly for round-the-clock operation.

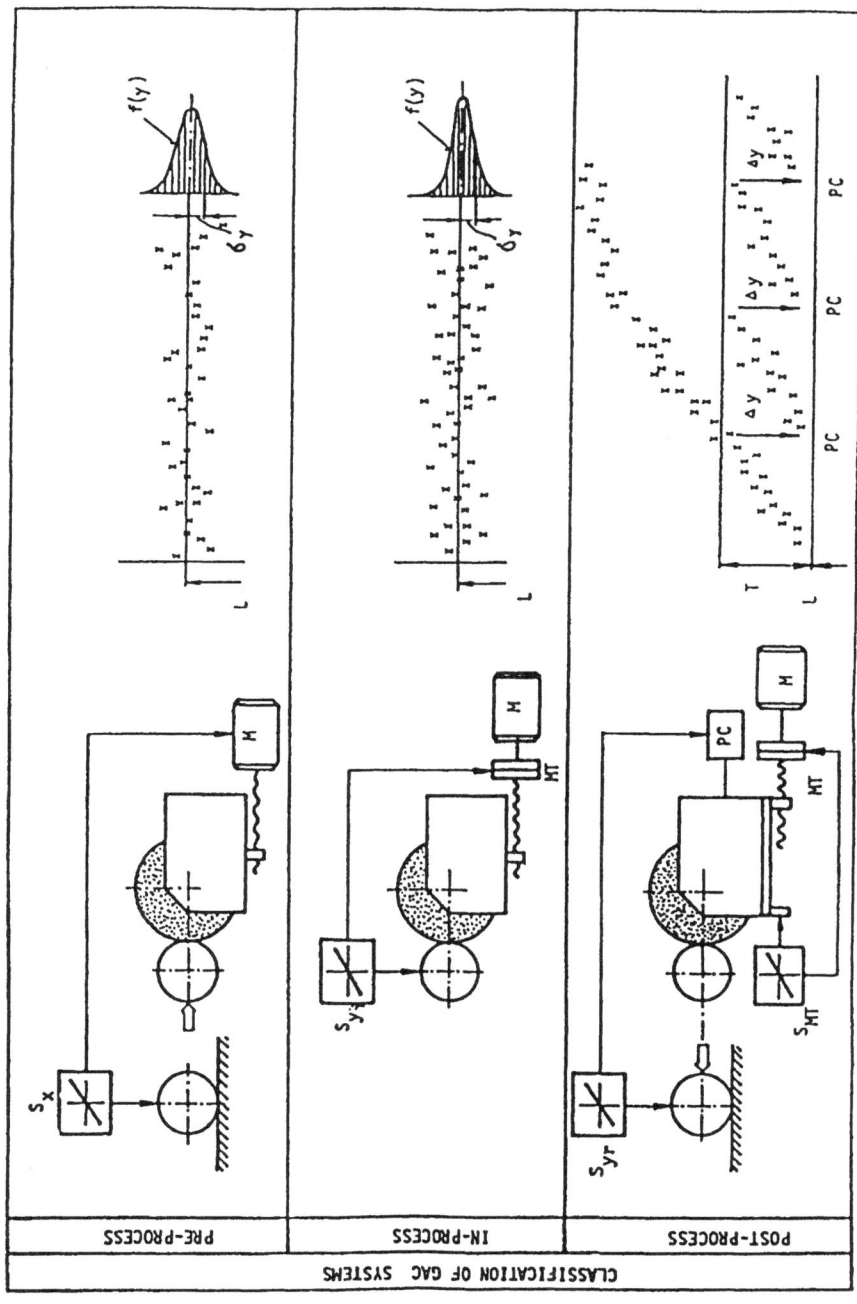

Fig. 11.9. Classification of GAC systems.

References

1. Peklenik J. Zur Fertigungsstabilität messgesteuerter Werkzeugmaschinen. Industrie Anzeiger 1959; 54: 839–846
2. Peklenik J. New developments in surface characterisation and measurement by means of random process analysis. In: Proc Intl Conf Properties and Metrology of Surfaces, Institute of Mechanical Engineers, Oxford, 1–4 April 1968. Proceedings 1967/68, 182(3K): 108–126
3. Peklenik J, Gartner J. Workpiece accuracy criterion for the dynamic machine tool acceptance test. Intl J Machine Tool – Design and Research 1966; 7(3): 189–203
4. GOST standard No 2789-73 of the former Soviet Union, Mašgis
5. Peklenik J. Untersuchung der Genauigkeitsfragen in der automatisierten Fertigung. Habilitationsschrift TH, Aachen, Westdeutscher-Verlag, Köln-Opladen, 1961
6. Peklenik J. Untersuchung der Genauigkeit in der automatisierten Fertigung. Industrie Anzeiger 1961; 83: 987–992
7. Peklenik J. Statistical concept of static and dynamic testing of manufacturing systems. SME paper MM71-221, Intl Engineering Conf, Philadelphia, Pennsylvania, 1971
8. Peklenik J. Fertigungskybernetik, Eine neue wissenschaftliche Disziplin für die Produktionstechnik. Festvortrag anlässlich der Verleihung des Georg-Schlesinger Preises 1988 des Landes Berlin, TU Berlin, 1988
9. Peklenik J. Geometrical adaptive control of manufacturing systems. Annals of the CIRP 1970; 18: 265–272
10. Presentation of data and control chart analysis. Special Technical Publication, American Society for Testing and Materials, Philadelphia, Pennsylvania 1976
11. Barkman WE. In-process quality control for manufacturing. Marcel Dekker Inc, New York, 1989, p 278
12. Corey HS. Surface finish from reflected laser light, Oak Ridge Y-12 Plant/Tennessee, Y/DAO7579, 1978
13. Domez MA et al. A general methodology for machine tool accuracy enhancement by error compensation. Precision Engineering 1986; 8(4): 187–196
14. Main RP. An interview with Alan Harmer on fiber optics, sensors and emerging markets. Optical Engineering Reports No 42, 1987

Glossary
Terms, Abbreviations and Definitions

M. Szafarczyk

AC (adaptive control) Adaptive control of manufacturing processes.

ACC (adaptive control constraints) Adaptive control constraints.

ACG Geometrical adaptive control.

ACO (adaptive control optimization) Adaptive control optimization.

ACT Technological adaptive control.

acoustic emission (AE) Stress waves released in a material undergoing deformation, fracture, friction, rubbing etc. It is claimed that AE in manufacturing processes has a frequency of over 100 kHz.

adaptability of manufacturing system Ability of a manufacturing system to eliminate or to diminish the influence of disturbances by appropriate changes in its operation.

adaptive control constraints (ACC) Adaptive control of a manufacturing process which attempts to keep the highest value of a quality index (to optimize the process) by increasing the chosen parameter (or parameters) of the manufacturing process, called the controlled variable (variables), but constrains the limit values of the chosen features of the process. For example, an ACC system may maximize the output of the turning process by increasing feed without exceeding the given value of the cutting force. An ACC system may optimize a manufacturing process when the relationship between the quality index and the controlled variable (the process parameter which is being increased) is a monotonic function.

adaptive control of manufacturing processes (AC) Control of a manufacturing process which attempts, in spite of disturbances, either to keep the chosen features of the process (or the features of the product) in the prescribed range or to obtain the highest quality index possible in the existing situation. The meaning of "adaptive control" in control theory is different.

adaptive control optimization (ACO) Adaptive control of a manufacturing process which attempts to find the highest possible value of a quality index (to optimize the process) somewhere inside the permissible range of change of the parameter (or parameters) influencing the manufacturing process, called the controlled variable (variables). An ACO system should be used when the relationship between the quality index and the controlled variable exhibits a maximum within the permissible range of the controlled variable.

AE (acoustic emission) Acoustic emission.

ASM (automatic supervision in manufacturing)   Automatic supervision in manufacturing.

automatic supervision in manufacturing (ASM)   Control which deals with the principal disturbances and allows an automatically supervised manufacturing process to be left alone without human supervision. An ASM system is usually composed of safety systems and adaptive control systems.

ASMT (automatic supervision of machine tools)   Control which deals with the principal disturbances influencing the proper functioning of a machine tool.

autonomic manufacturing system (cell)   A manufacturing system (cell) which may operate without human supervision.

break-down   A failure in operation which causes a stoppage of the process.

catastrophic failure of the tool   A sudden wear of the tool which makes further machining impossible.

clone tool   Sister tool.

compensating control   Adaptive control which is used to compensate for a specific, identified type of disturbance. Usually the measured value of the disturbance and a model of the process are used to decide how to adjust the input quantity (controlled variable) of the manufacturing system in such a way that the effect of the disturbance is reduced or eliminated. For example, the temperature of the NC machine tool body is measured, and the co-ordinate values are changed to compensate for the thermal deformations.

condition monitoring   Monitoring.

conditioning (of signal from a sensor)   Signal pre-processing.

controlled variable   A physical quantity being purposely changed by the control system of the manufacturing process with the aim of obtaining the chosen goal. The controlled variable is an output of the control system and an input to the manufacturing process.

corrective control   Adaptive control which is used to reduce or to eliminate the influence of disturbances on the specific output of the manufacturing system. It is a closed-loop system in which the output of the manufacturing system is measured and the value of a chosen input quantity (controlled variable) is adjusted in order to reduce the difference between the measured output value and the desired output value.

diagnose   Determine the situation from the point of view of the correct operation of the system (the machine) or the highest quality index of the process on the basis of observation of symptoms. **Diagnosis** may concern the present situation, the future situation (prognostic diagnosis) or the cause of the breakdown which has already occurred.

diagnosis   Diagnose.

disturbance   Anything which was not planned and influences the quality index of the system or the manufacturing process. Disturbances do not necessarily have an adverse influence on the quality index. Sometimes disturbances make it possible to obtain better results than planned, e.g. a smaller than planned diameter of the shaft before turning allows an increase of feed and thus an increase in the output of machining.

downtime   Time of interruption in machining caused by a failure (break-down).

emergency stop   Interruption of operation caused by the detection of a failure (break-down) or possibility of a break-down.

error   failure.

failure   State of manufacture described by the quality index equal to zero or negative. Error is a special kind of failure caused by the control system or the personnel.

feature extraction Signal processing in which the selected (important for monitoring purposes) features are exposed.

GAC (geometrical adaptive control) Geometrical adaptive control.

geometrical adaptive control (GAC) [ACG] Adaptive control dealing with the desired features of the product. In most cases a GAC system is aimed at achieving the proper dimensions of the workpiece after machining. It used to be designated by the abbreviation ACG.

identification Process of model verification. On-line identification may be a part of a sophisticated monitoring system.

intelligent monitoring system Self-improving monitoring system with some ability to learn on the basis of monitoring experience.

intelligent sensor Integrated sensor.

integrated sensor Sensor integrated with electronic circuits as one element (chip). Electronic circuits may be used for signal pre-processing, tare, self-diagnosing, self-calibration etc. The term intelligent sensor is justified only when artificial intelligence is used in signal processing.

monitoring On-line diagnosing of a system or of a process with the aim of recognizing malfunctioning or non-optimal functioning.

multi-sensor monitoring Sensor fusion.

quality index Number, the value of which expresses the quality of a product or a process according to a stated criterion of quality assessment. The bigger the number, the higher the quality level.

recovery procedure Method of operation after a break-down which allows manufacturing to recommence.

safety system System of automatic supervision which either prevents break-down or minimizes the damage caused by a break-down.

sensor Instrument which reacts to a certain physical condition by sending a signal (usually electrical in nature), transmitting information on this condition.

sensor fusion Collective use of signals from several sensors (from different sources) for monitoring purposes (multi-sensor monitoring).

signal conditioning Signal pre-processing.

signal pre-processing Set of operations performed on the signal in order to increase the ratio between information and noise. It is also called signal conditioning.

signal processing Set of operations performed on the signal in order to derive the desired information.

sister tool Identical tool placed in the tool magazine (of a machining centre) which may replace the tool should it become inoperative as a result of natural wear or catastrophic failure. It is also called a clone tool.

supervision Control dealing with disturbances, in order to safeguard against a catastrophic failure, assure correct operation or achieve the best results of the supervised process (highest quality index possible).

supervisory action Action (sometimes called reaction) taken by a supervisory system and influencing the manufacturing system with the aim of eliminating or diminishing the influence of disturbances.

supervisory system Control system used for supervision purposes.

symptom Phenomenon or circumstance accompanying malfunctioning (or non-optimal functioning) of a manufacturing system and regarded as evidence of malfunctioning.

syndrome Number of symptoms which collectively indicate malfunctioning of the system or non-optimal parameters of the process.

TAC (technological adaptive control) Technological adaptive control.

tare Value of the signal sent by a sensor when the measured quantity is equal to zero.

technological adaptive control (TAC) [ACT] Adaptive control which optimizes the manufacturing process. The value of the manufacturing process output, the inverted value of the manufacturing cost or the value of profit may be used as a quality index. It used to be designated by abbreviation ACT.

touch trigger probe (TTP) Device with a moving stylus sending a binary signal when the stylus is touched.

unmanned machining Machining in an autonomic manufacturing system (cell), without an operator. Also called untended machining.

untended machining Unmanned machining.

Index